Manufacturing Integrated Design

Peter Groche • Enrico Bruder
Sebastian Gramlich
Editors

Manufacturing Integrated Design

Sheet Metal Product and Process Innovation

 Springer

Editors
Peter Groche
Institute for Production Engineering
    and Forming Machines (PtU)
Technische Universität Darmstadt
Darmstadt, Germany

Enrico Bruder
Materials Science Department, Physical
    Metallurgy Division
Technische Universität Darmstadt
Darmstadt, Germany

Sebastian Gramlich
Institute for Product Development
    and Machine Elements (pmd)
Technische Universität Darmstadt
Darmstadt, Germany

ISBN 978-3-319-84890-7   ISBN 978-3-319-52377-4   (eBook)
DOI 10.1007/978-3-319-52377-4

Printed on acid-free paper

This Springer imprint is published by Springer Nature
The registered company is Springer International Publishing AG
The registered company address is: Gewerbestrasse 11, 6330 Cham, Switzerland

Preface

The commercial success of companies is closely related to their capability to develop innovative products. With regard to global markets and competition, product development needs to take a step forward in order to not just fulfill a set of product or customer requirements but to create products which are optimized with regard to their whole product life cycle. This task is becoming more and more demanding not only due to the fact that the complexity of products is growing but also due to the progress in technological know-how of the manufacturing processes that are involved. Regarding products for structural lightweight applications for instance, there is a variety of manufacturing processes available. Each of these processes has its individual potentials and restrictions, some of them being invariant whereas others keep changing in the course of research and development in manufacturing technology. Furthermore, most manufacturing processes for lightweight structures also affect the material properties, which might have a significant impact not only on the product performance but also on processes at later stages of the manufacturing chain, leading either to restrictions or to new potentials.

Keeping an overview of all these aspects and interactions in all their complexity requires expert level knowledge in a variety of fields, thus making product development a challenge with highly interdisciplinary character. To meet this challenge, a methodology and corresponding development tools are needed to help processing the continuously growing amount of technological knowledge. This becomes even more important when considering that new manufacturing technologies, which might unlock a new level of product performance, can also act as a trigger for product innovation.

A prime example for the complex interaction between manufacturing process innovation and product development is the technologies that were developed within a collaborative research center (CRC) on integral sheet metal design, which was established at Technische Universität Darmstadt in 2005. The aim of the CRC666 "Integral Sheet Metal Design with Higher Order Bifurcations" was not only the development of manufacturing processes for sheet metal products but also the beneficial use of process innovation for optimized products.

This book is based on the research activities within the CRC666 and aims at providing an integrated view on product and process development. While putting an emphasis on sheet metal products, the new design paradigm in which this book concludes is not limited to a specific product range but can be applied to a wide range of applications. The first two chapters of this book have an introductory character. They focus on the role of production technologies in current product development approaches and provide a motivation for a new approach that exploits the potentials provided by technological innovation to realize optimized solutions. In the third chapter, manufacturing technologies and process chains for branched sheet metal products are introduced which exhibit a high innovation potential and provide the technological basis for the new product development approach throughout the following chapters. The fourth chapter discusses the impact of manufacturing technologies on local material properties and their beneficial use to improve product performance. The aspect of optimization within the new development approach is addressed in the fifth chapter, from fundamental requirements and challenges for the formulation of optimization problems to application examples for the optimization of production sequences and process controls. The sixth chapter focuses on virtual product development methods and tools such as the information model, CAD modeling, and the numerical simulation of manufacturing processes and product properties. In the seventh and eighth chapter, two different scenarios for product innovation are discussed, being technological advances driven by marked demands and new product ideas that arise from manufacturing process innovation. Finally, the ninth chapter presents the new *Integrated Algorithm-Based Product and Process Development Approach* and illustrates its application and benefits based on case studies.

The CRC666 was funded by the German Research Foundation DFG, which is gratefully acknowledged by the research team. We would also like to recognize the additional financial support and encouragement that were given by the TU Darmstadt. We are very grateful for the advice and support from the panel of reviewers who monitored and evaluated the CRC. It is a privilege of the editors to thank all contributors for their excellent work and enthusiasm, without which this book would not have been possible. The tireless commitment of Vinzent Monnerjahn and all chapter coordinators has been invaluable. Among the long list of former colleagues within the research team, all of them having their share in making the joint project a success, there are some that stand out by giving direction to the research activities. Therefore, we would like to recognize Herbert Birkhofer, Andrea Bohn, Holger Hanselka, Alexander Martin, and Hermann Kloberdanz for their contributions within the CRC666.

Darmstadt, Germany   Peter Groche
November 2016   Enrico Bruder
                                                                              Sebastian Gramlich

Contents

Contributors

S. Abedini Institute of Construction Design and Building Construction (KGBauko), Technische Universität Darmstadt, Darmstadt, Germany

E. Abele Institute for Production Management, Technology and Machine Tools (PTW), Technische Universität Darmstadt, Darmstadt, Germany

L. Ahmels Physical Metallurgy (PhM), Technische Universität Darmstadt, Darmstadt, Germany

K. Albrecht Department of Computer Integrated Design (DiK), Technische Universität Darmstadt, Darmstadt, Germany

R. Anderl Department of Computer Integrated Design (DiK), Technische Universität Darmstadt, Darmstadt, Germany

A.-K. Bott Research Group Statistics (STAT), Technische Universität Darmstadt, Darmstadt, Germany

E. Bruder Physical Metallurgy (PhM), Technische Universität Darmstadt, Darmstadt, Germany

M. Gibbels Research group System Reliability, Adaptive Structures, and Machine Acoustics (SAM), Technische Universität Darmstadt, Darmstadt, Germany

S. Gramlich Institute for Product Development and Machine Elements (pmd), Technische Universität Darmstadt, Darmstadt, Germany

P. Groche Institute for Production Engineering and Forming Machines (PtU), Technische Universität Darmstadt, Darmstadt, Germany

M. Hansmann Research Group Statistics (STAT), Technische Universität Darmstadt, Darmstadt, Germany

B. Horn Research Group Nonlinear Optimization (NOpt), Technische Universität Darmstadt, Darmstadt, Germany

A. Hoßfeld Institute for Production Management, Technology and Machine Tools (PTW), Technische Universität Darmstadt, Darmstadt, Germany

I. Karin Research group System Reliability, Adaptive Structures, and Machine Acoustics (SAM), Technische Universität Darmstadt, Darmstadt, Germany

H. Kaufmann Fraunhofer Institute for Structural Durability and System Reliability LBF (LBF), Fraunhofer-Gesellschaft zur Förderung der angewandten Forschung e.V., Munich, Germany

M. Kohler Research Group Statistics (STAT), Technische Universität Darmstadt, Darmstadt, Germany

S. Köhler Institute for Production Engineering and Forming Machines (PtU), Technische Universität Darmstadt, Darmstadt, Germany

K. Lipp Fraunhofer Institute for Structural Durability and System Reliability LBF (LBF), Fraunhofer-Gesellschaft zur Förderung der angewandten Forschung e.V., Munich, Germany

H. Lüthen Research Group Discrete Optimization (DOpt), Technische Universität Darmstadt, Darmstadt, Germany

P. Mahajan Institute for Production Engineering and Forming Machines (PtU), Technische Universität Darmstadt, Darmstadt, Germany

I. Mattmann Institute for Product Development and Machine Elements (pmd), Technische Universität Darmstadt, Darmstadt, Germany

T. Melz Research group System Reliability, Adaptive Structures, and Machine Acoustics (SAM), Technische Universität Darmstadt, Darmstadt, Germany

Fraunhofer Institute for Structural Durability and System Reliability LBF (LBF), Fraunhofer-Gesellschaft zur Förderung der angewandten Forschung e.V., Munich, Germany

V. Monnerjahn Institute for Production Engineering and Forming Machines (PtU), Technische Universität Darmstadt, Darmstadt, Germany

C. Müller Physical Metallurgy (PhM), Technische Universität Darmstadt, Darmstadt, Germany

D. Neufeld Research group System Reliability, Adaptive Structures, and Machine Acoustics (SAM), Technische Universität Darmstadt, Darmstadt, Germany

M. Neuwirth Institute for Production Engineering and Forming Machines (PtU), Technische Universität Darmstadt, Darmstadt, Germany

J. Niehuesbernd Physical Metallurgy (PhM), Technische Universität Darmstadt, Darmstadt, Germany

M. Özel Institute for Production Engineering and Forming Machines (PtU), Technische Universität Darmstadt, Darmstadt, Germany

M. Pfetsch Research Group Discrete Optimization (DOpt), Technische Universität Darmstadt, Darmstadt, Germany

H. Pouriayevali Mechanics of Functional Materials (MfM), Technische Universität Darmstadt, Darmstadt, Germany

J. Reising Institute of Construction Design and Building Construction (KGBauko), Technische Universität Darmstadt, Darmstadt, Germany

M. Roos Institute for Product Development and Machine Elements (pmd), Technische Universität Darmstadt, Darmstadt, Germany

S. Schäfer Institute of Construction Design and Building Construction (KGBauko), Technische Universität Darmstadt, Darmstadt, Germany

S. Schmidt Institute for Production Management, Technology and Machine Tools (PTW), Technische Universität Darmstadt, Darmstadt, Germany

Y. Tijani Research group System Reliability, Adaptive Structures, and Machine Acoustics (SAM), Technische Universität Darmstadt, Darmstadt, Germany

A. Tomasella Research group System Reliability, Adaptive Structures, and Machine Acoustics (SAM), Technische Universität Darmstadt, Darmstadt, Germany

E. Turan Institute for Production Management, Technology and Machine Tools (PTW), Technische Universität Darmstadt, Darmstadt, Germany

S. Ulbrich Research Group Nonlinear Optimization (NOpt), Technische Universität Darmstadt, Darmstadt, Germany

R. Wagener Fraunhofer Institute for Structural Durability and System Reliability LBF (LBF), Fraunhofer-Gesellschaft zur Förderung der angewandten Forschung e.V., Munich, Germany

C. Wagner Institute for Product Development and Machine Elements (pmd), Technische Universität Darmstadt, Darmstadt, Germany

A. Walter Research Group Nonlinear Optimization (NOpt), Technische Universität Darmstadt, Darmstadt, Germany

T. Weber Martins Department of Computer Integrated Design (DiK), Technische Universität Darmstadt, Darmstadt, Germany

B.-X. Xu Mechanics of Functional Materials (MfM), Technische Universität Darmstadt, Darmstadt, Germany

A. Zimmermann Institute of Construction Design and Building Construction (KGBauko), Technische Universität Darmstadt, Darmstadt, Germany

Abbreviations

2D	Two dimensional
3D	Three dimensional
ANN	Artificial neural networks
API	Application programming interface
ARCex	Algorithm
B-Rep	Boundary representation
BTR	Bundle trust region algorithm
CAD	Computer-aided design
CAE	Computer-aided engineering
CAM	Computer-aided manufacturing
CAPP	Computer-aided process planning
CAx	Computer-aided
CEM	Cut-expand-method
CNC	Computerized numerical control
CSG	Constructive solid geometry
CSS	Cascading style sheets
DfM	Design for manufacture
DoF	Degree of freedom
DOM	Document object model
E	Young's modulus
ECAP	Equal channel angular pressing
EDO	Engineering design optimization
FE	Finite element
FEA	Finite element analysis
FEM	Finite element method
G	Shear modulus
GUI	Graphical user interface
HW	Supporting roll
HPT	High pressure torsion
HSC	High speed cutting

HSLA	High strength low alloy
HTML	Hypertext markup language
HW	Supporting roll
IPF	Inverse pole figure
IPPD	Integrated product and process development
KKT	Karush-Kuhn-Tucker
LBS	Linear bend split
LFS	Linear flow split
M	Mass
MBS	Multibody simulation
MINLP	Mixed integer nonlinear program
MIP	Mixed integer program
MLSS	Material law of steel sheet
MVS	Method of variable slopes
NC	Numerical control
ND	Normal direction
NLP	Nonlinear program
Nlopt	Nonlinear optimization
NURBS	Nonuniform rational B-splines
PCG	Preconditioned conjugate gradient
PDE	Partial differential equation
PEEQ	Equivalent plastic strain
PIDG	Process integrated design guidelines
PMI	Product and Manufacturing Information
R	Forming radius of splitting roll
RCF	Rolling contact fatigue
RD	Rolling direction
REST	Representational State Transfer
RFID	Radio-frequency identification
RP	Rotation point
SADT	Structured analysis and design technique
SE	Simultaneous engineering
SIMP	Solid isotropic material with penalization
SPD	Severe plastic deformation
SPR	Super convergent patch recovery
SQP	Sequential quadratic programming
SRCS	Smart Requirement Configuration System
STEP	Standard for the exchange of product model data
SW	Splitting roll
TD	Transverse direction
UDF	User-defined feature
UFG	Ultrafine-grained
UML	Unified Modeling Language
XML	Extensible Markup Language

Symbols

Chapter 3

a_e	Radial depth of cut
a_p	Axial depth of cut
a_z	Length of upper or lower flange
$a_{z,\text{tot}}$	Total flange Lenh
b_n	Strip tension
b_{web}	Broadness of the web
e	Elongation
f_z	Feed per tooth
F_{crit}	Critical load
$F_{y,\text{HW}}$	Component of supporting roll force in y direction
$F_{y,\text{SW}}$	Component of splitting roll force in y direction
h	Forming radius supporting roll
h_d	Dome height
j	Height of the leg
l	Length of unbuckled stringer
l_0	Length of bending line
L	Length of stringer edge after the forming process
n	Stand number
R_z	Average surface roughness
s	Thickness
s_0	Thickness of the web
s_f	Thickness of the flanges
S	Springback
$u_{y,\text{HW}}$	Displacement of reference point in y direction
v_{sheet}	Sheet metal velocity
v_c	Cutting speed
v_f	Feed rate

v_u	Circumferential speed of supporting rolls
w	Waviness
y_{inc}	Incremental infeed of the splitting roll
y_{tot}	Total splitting depth
α	Splitting angle
α_{crit}	Critical bending angle
β	Bending angle
γ	Rake angle
ΔX	Splitting roll offset
$\Delta\varepsilon$	Difference of strains
ε	Elastic strain
ε_x	Strain in direction x

Chapter 4

b	Length of Burgers vector
c	Fatigue strength exponent
d	Averaged grain diameter
$f^{(m)}$	Probability density function of $\delta^{(m)}$
$\hat{f}^{(m)}$	Estimated density of $f^{(m)}$
$g^{(m,\varepsilon_{a,t})}$	Probability density function of $N_f^{(m)}(\varepsilon_{a,t})$
$h_{1/2}$	Bandwidths
H	Bandwidth
K	Static strength coefficient
K_f	Fatigue notch factor
K_t	Theoretical stress-concentration factor
K'	Cyclic strength coefficient
$\widetilde{K}(x)$	Epanechnikov Kernel
M	Taylor factor
n	Static strain hardening exponent
n'	Cyclic strain hardening exponent
N	Cycles
N_k	Number of cycles at Knee point
N_i	Number of cycles to crack initiation
N_f	Number of cycles to fracture
$q_{N_f^{(m)},\alpha}(\varepsilon_{a,t})$	α-quantile of $N_f^{(m)}(\varepsilon_{a,t})$
$\hat{q}_{N_f^{(m)},\alpha}$	Estimated α-quantile of $\delta^{(m)}$
R	Load ratio
R_m	Tensile strength
$R_{p0,2}$	Yield limit for 0.2% residual elongation
$S - N$	Stress amplitude – Number of cycles to failure
T	Number of materials in the database
X	Random variable

α	Level of the α-quantile
$\Delta\sigma_{ys}$	Increase in yield strength
$\delta^{(m)}$	Random error term in the model of the number of cycles
$\hat{\delta}_i^{(m)}$	Estimated samples of $\hat{\delta}^{(m)}$
$\varepsilon_{a,t}$	Strain amplitude
ε_f'	Fatigue strength coefficient
$\mu^{(m)}(\varepsilon_{a,t})$	Expected value of the number of cycles
$\hat{\mu}(m, \varepsilon_{a,t})$	Estimated expected value of the number of cycles
ρ	Dislocation density
σ_0	Starting stress for dislocation movement
σ_a	Stress amplitude
$\sigma_{a.k}$	Stress amplitude at knee point
σ_f'	Fatigue ductility coefficient
σ_{ys}	Yield stress
$\sigma^{(m)}(\varepsilon_{a,t})$	Standard deviation of the number of cycles
$\hat{\sigma}(m, \varepsilon_{a,t})$	Estimated standard deviation of the number of cycles
φ_v	Degree of deformation
$\mathbb{R}$	Set of real numbers

Chapter 5

A_h	Stiffness matrix
$\bar{a}_j$	Relative blank holder force
$\hat{a}_j$	Relative internal pressure
B_h	Discrete normal trace operator
f	Volume load
F_b	Blank holder force
$\bar{g}$	Gap function in reference configuration
H_k	Hessian matrix
K	Fitness function
N_i^p	B-spline basis functions
u	Independent optimization variable
U_{ad}	Feasible set of u
y	Displacement
Γ_C	Potential contact boundary
$\Delta_k h$	Change of dome height in step k
Λ_h	Finite dimensional Lagrange multiplier subspace
λ_h	Discretized Lagrangian multiplier
ρ	Penalty parameter
Φ	Primal basis function
Ψ	Dual basis function
Ω	Domain
$\overline{\Omega}$	Closure of domain
$\partial\Omega$	Domain boundary

Chapter 6

a Loop
R_f Load ratio

Chapter 7

$dy1$ Deflection of the frame
$dy2$ Deflection of the tool cage
k Contact length radius R_1
k_s Springback ratio
l Contact length radius R_2
R_1 Radius of the convex section
R_2 Radius of the concave section
v_{sheet} Sheet metal velocity
X Point of rotation
y_{inc} Incremental infeed of the splitting roll

Chapter 8

g Initial gap of joining partners
p Pressure
$\sigma_{t,rem}$ Tangential remaining stress

Chapter 9

c_e Weight of edge
e Edge
G Weighted graph

Chapter 1
Introduction: Production Technologies and Product Development

V. Monnerjahn, S. Gramlich, P. Groche, M. Roos, C. Wagner, and T. Weber Martins

1.1 The Interaction Between Product Development and Production Technology

Many studies reveal that the development of products and manufacturing technologies are key factors for the success of industrial enterprises (Becheikh et al. 2006). Both development processes aim at the same goal: the creation of products that fulfill customer needs with a minimum of required resources. From the perspective of an enterprise the minimization of resources comes along with a maximization of productivity. According to (Tangen 2005) productivity can be increased by either higher efficiency or higher effectiveness. The first possibility is directed towards cost minimization, and the second towards higher quality, flexibility, and reduced lead time.

Due to the many options which are created and evaluated in product and manufacturing technology design processes and the multiple design criteria which have to be considered, a high complexity has to be handled. Complexity is even higher when the interdependences between product and process design are taken into account. Since the design processes are targeted at novel creations knowledge about the properties and limits of the newly designed products and processes is

V. Monnerjahn (✉) • P. Groche
Institute for Production Engineering and Forming Machines (PtU), Technische Universität Darmstadt, Darmstadt, Germany
e-mail: monnerjahn@ptu.tu-darmstadt.de

S. Gramlich • M. Roos • C. Wagner
Institute for Product Development and Machine Elements (pmd), Technische Universität Darmstadt, Darmstadt, Germany

T. Weber Martins
Department of Computer Integrated Design (DiK), Technische Universität Darmstadt, Darmstadt, Germany

© Springer International Publishing AG 2017
P. Groche et al. (eds.), *Manufacturing Integrated Design*,
DOI 10.1007/978-3-319-52377-4_1

fragmentary. This lack of knowledge leads to considerable uncertainties which have to be handled. Uncertainties in design processes are continuously increasing, because the numbers of available manufacturing technologies and relevant criteria are growing.

Complexity and uncertainties can be reduced by neglecting interdependencies. Therefore, products are often designed according to guidelines of existing and well-known manufacturing technologies. Also, manufacturing processes are often designed for products with (nearly) completely fixed properties. These limitations reduce the dimensions of available design spaces. Consequently, productivity can be potentially increased if both design processes are integrated.

In order to support the integration of product and manufacturing process design, powerful product life cycle management (PLM) tools have been created (Dekkers et al. 2013). PLM merges the various design activities by linking different software modules. Although the availability of relevant data is a useful tool relevant product and processes data still remain limited. Further developments of design methodologies are necessary. These should optimize the use of a selected manufacturing technology during product design. The book at hand describes a new approach which can meet this requirement.

1.2 Incorporation of Production in Current Product Development Approaches

Many established product development approaches provide procedures for dividing the development process into manageable subproblems. They often encourage stepwise concretization, from product idea to detailed product design. Within the product life cycle, production processes become relevant as soon as the technical product is physically realized in accordance with design specifications as a result of the development process. For this purpose, suitable manufacturing technologies are combined and arranged in manufacturing chains. Relevant process parameters that best meet the specifications of embodiment design are determined. To satisfy this causal relationship between production and product development, the majority of product development approaches consider production aspects already during the design process. However, the extent and type of incorporation vary considerably among the individual approaches.

Axiomatic design, a common and widely reported development approach, is based on stepwise concretization of the product and its realization. Core elements of axiomatic design theory and method are two axioms and horizontal decomposition into four domains: customer domain, functional domain, physical domain, and process domain. Beginning with the customer domain, each domain is characterized by a more concretized perspective of the product. The designers' task is modeled as a systematic mapping between these four domains, focusing on the key questions of "what we want to achieve" and "how we want to achieve it". The first three domains focus on meeting customer needs by providing a suitable functional

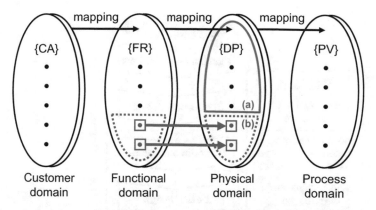

Customer   Functional   Physical   Process
domain   domain   domain   domain

Fig. 1.1 Four domains in axiomatic design (Tekkaya et al. 2015) according to Suh (2001)

and physical product design. Production issues are considered within the last domain once design parameters are determined (Suh 1998; Suh 2001). Tekkaya et al. (2015) expose the need for iterations in axiomatic design to perform design parameter modifications to ensure their feasibility in the chosen production technology. In addition, Tekkaya et al. (2015) highlight the need to focus on a special group of design parameters in the physical domain that are linked to material properties, especially to design parameters that can be varied by process control (Fig. 1.1 highlight b). The physical domain contains the group of design parameters that are achievable by conventional geometric product design (Fig. 1.1 highlight a). Only the equivalent and simultaneous incorporation of both groups of design parameters leads to a highly valued design and allows all functional requirements to be met in one-to-one mapping without exceptions (Tekkaya et al. 2015).

Engineering design (Pahl et al. 2007a) is a comprehensive and established approach to methodically support the entire product development process. The aim of this approach is a consistent procedure, based on structured concretization steps for developing technical products, supported by a defined system of methods and models. *VDI 2221* (1993) and *VDI 2206* (2004) also encourage such a stepwise concretization procedure. Pahl et al. (2007a), as well as VDI 2221 (1993), structure the development process in four phases: task clarification, conceptual design, embodiment design, and detail design. The early phases, task clarification and conceptual design, are dominated by objectives that focus on providing customer value and realizing the product function. Production processes are mainly deferred in these first stages. To ensure manufacturability, production technology restrictions are considered and supported in the later stages of the development process, e.g., with the help of technology-specific guidelines (Pahl et al. 2007a).

Rules and guidelines on issues like process limits, advantageous part geometries, machine capabilities, and sustainability are also methodical elements of Design for X. The general approaches Design for Manufacture (DfM), Design for Assembly (DfA), and Design for Manufacture and Assembly (DfMA) are complemented by technology-specific approaches, like Design for Forming (Altan and Miller 1990).

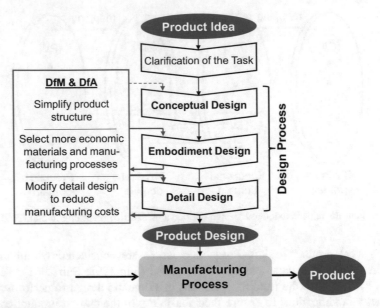

Fig. 1.2 Iterations of DfM in the process of VDI 2221 according to Groche et al. (2012)

The main intention is to feed back and integrate technological information into the design process to reduce manufacturing and assembly effort, time and cost, as well as time to market (Boothroyd et al. 2011; Bralla 1999). As shown in Fig. 1.2, the application of DfM approaches leads to iterations that mainly affect the later stages of the design process, particularly embodiment and detail design. These iterations are usually time consuming (Groche et al. 2012). Although designers are supported by a huge number of rules (Pahl et al. 2007a), guidelines (Boothroyd et al. 2011), design references (Roth 1996), and advices and tools (Meerkamm et al. 2012) to generate manufacturing and assembly-compliant design solutions, earlier stages, like the conceptual design phase, are only marginally supported (Groche et al. 2012).

Integrated product and process development (IPPD) is a more sophisticated approach to integrating technical processes of the product life cycle and thus of production processes (Fig. 1.3). In each step of the product development process the impacts on the processes of the whole life cycle have to be considered and integrated into product design. IPPD recommends continuous and consistent anticipation steps as well as conscious influencing of the product life cycle processes (Abele et al. 2007; Andreasen and Hein 1987; Birkhofer 2011). In this way, the designer is able to predict the possible environmental consequences of life cycle processes, which was the original focus of IPPD. IPPD is a fundamental approach that is linked to various methods and models (Heidemann 2001; Gramlich 2013) but is not operationalized completely.

Simultaneous engineering (SE) (Eversheim 1995) describes strategies for planning and executing processes during product development, with an emphasis on a reduction of time to market. Product development steps, like the steps suggested by

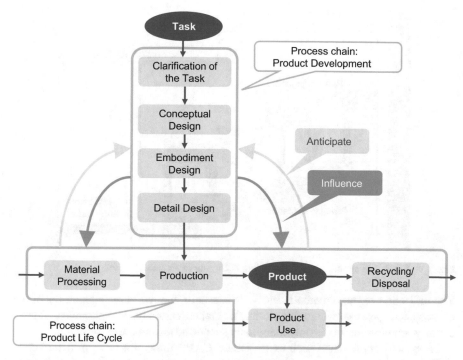

Fig. 1.3 Integrated product and process development (IPPD) (Birkhofer et al. 2012)

VDI 2221 (1993), are parallelized instead of sequential execution (Fig. 1.4). Product development and production process planning are considered at the same time. By parallelizing development steps, relevant information can be provided early to the following development and production steps. Synchronization between product design and production process planning is a challenge for operationalizing this approach.

Computer-aided process planning (CAPP) is an approach for supporting the operationalization of an integrated product and process development (Rollmann 2012) by consistently and efficiently utilizing CAx tools (CAD, CAM, etc.). CAPP is used for an automated generation of production process plans by providing the necessary interface between CAD and CAM (Lee 1999). By integrating the virtual product model, production tool models, and production parameters, the link between product design and production is established. While the integration of manufacturing technologies such as machining or roll forming is already realized, many existing tools are especially limited when considering complex product geometries, like integral bifurcated sheet metal profiles (Rollmann 2012).

Considering the aforementioned approaches, it is apparent how differently and heterogeneously production aspects are considered in product development. Many approaches take manufacturability into account only by addressing specific limitations and requirements of the determined production technologies. They provide various methods and tools, from generally applicable guidelines to systematic mapping.

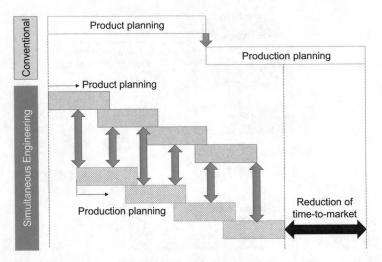

Fig. 1.4 Scheme of simultaneous engineering according to Eversheim (1995)

The utilization of these methods and tools is usually limited to the later stages of the development process, where the product's detailed geometry is determined. Consequently, solution finding is only marginally supported during the early stages.

Integrated approaches such as IPPD and CAPP focus not just on manufacturability. Information on specific production processes into design decisions is incorporated extensively and directly. Therefore, product design and process design are linked, which creates additional opportunities and benefits for the development process. Integrated approaches vary in their degree of operationalization, as well as their scope and primary focus. The comprehensive realization of manufacturing potential for product and process development on the basis of consistent models, methods, and tools has yet to be supplemented. Even though there are several other approaches, such as Ulrich and Eppinger (1995), Integrierte Produktentwicklung (IPE) (Ehrlenspiel and Meerkamm 2013), Integriertes Produktentstehungs-Modell (iPeM) (Meboldt 2008), characteristics-properties modelling (CPM)/property-driven development (PDD) (Weber 2005), and Münchener Produktkonkretisierungsmodell (Ponn and Lindemann 2011), they all address the two previously emphasized perspectives on incorporation of production in product development: ensuring manufacturability with low effort and costs and comprehensive integration of process aspects.

1.3   Market Pull vs. Technology Push

Product development largely contributes to a company's success by realizing innovative technical solutions that satisfy market demands and provide significant customer value. These innovative solutions in the form of products and processes are characterized by inventive and novel technical realizations of the product and process ideas and their successful market introduction (Pahl et al. 2007b). Market-driven strategies (*market pull*) and associated development approaches (Brehm and Voigt 2009) help structuring the development process with focus on realizing innovative solutions for a defined market. Market pull projects are usually characterized by incremental changes in products and processes, leading to minor necessary changes in customer behavior. The identified customer needs that have to be satisfied "pull" innovative technical solutions based on novel product or process ideas. Nickel-based superalloys are an example of innovative market-pulled solutions, having been developed to satisfy the identified need of higher efficiency gas turbines (Pollock and Tin 2006).

In addition, many successful innovations are the result of a systematic exploitation of technology-specific possibilities. Especially manufacturing technologies can be a possible initiator and driver for novel product and process ideas and the resulting innovative technical solutions that can be realized with the help of technology-driven development approaches (*technology push*). Knowledge of natural sciences is necessary to understand manufacturing technologies and to realize technology-specific solutions for technical problems (Trommsdorff and Steinhoff 2013). Technology push projects are characterized by an extensive search for possible applications and markets that suit the developed technology, making commercial use of this knowledge (Brehm and Voigt 2009). Technology push projects offer the potential for radical innovations (Brehm and Voigt 2009). The starting point can be for example new technologies in material science (e.g., shape memory alloys) or a new and innovative manufacturing technology that offers additional possibilities for product design that have not been provided by existing technologies yet, resulting in novel product and process ideas.

Market pull and technology push approaches are both represented in the Abell framework (Table 1.1). Although market pull approaches start from a given market, they still require the determination of a suitable technology to ensure that the product and process idea can be finally realized. Starting from new technologies and new combinations of existing technologies, technology push approaches allow the realization of completely new solutions; the potential market is yet to be determined. Since there is no concrete customer or market, radical technology-pushed innovations can fail when no suitable market is found (Abele 2013). Only by considering both the market and a promising manufacturing technology, product and process ideas can be successfully realized in the form of technical solutions. Consequently, most development projects apply a combination of market pull and technology push approaches (Hausschildt 2004) leading to innovative technical

Table 1.1 Distinction between technology push and market pull, based on the Abell framework (Abell 1980), according to Trommsdorff and Steinhof (2013)

	Technology	Customer needs	Customer markets
Technology push	Known	To be determined	To be determined
Market pull	To be determined	Known	Known

solutions that not only address customer demands from the market perspective but also comprehensively utilize the specific possibilities of new (manufacturing) technologies.

References

Abele E, Anderl R, Birkhofer H, Rüttinger B (2007) EcoDesign: Von der Theorie in die Praxis. Springer, Dordrecht

Abele T (2013) Suchfeldbestimmung und Ideenbewertung: Methoden und Prozesse in den frühen Phasen des Innovationsprozesses. Springer Gabler, Wiesbaden

Abell DF (1980) Defining the business: the starting point of strategic planning. Prentice-Hall, Englewood Cliffs

Altan T, Miller RA (1990) Design for forming and other near net shape manufacturing processes. CIRP Ann 39(2):609–620

Andreasen MM, Hein L (1987) Integrated product development. IFS (Publications), Bedford

Becheikh N, Landry R, Amara N (2006) Lessons from innovation empirical studies in the manufacturing sector: a systematic review of literature from 1993–2003. Technovation 26:644–664

Birkhofer H (2011) From design practice to design science: the evolution of a career in design methodology research. J Eng Des 22(5):333–359

Birkhofer H, Rath K, Thao S (2012) Umweltgerechtes Konstruieren. In: Rieg F, Steinhilper R (eds) Handbuch Konstruktion. Hanser, München

Boothroyd G, Dewhurst P, Knight WA (2011) Product design for manufacture and assembly, 3rd edn. CRC Press, Boca Raton

Bralla JG (1999) Design for manufacturability handbook, 2nd edn. McGraw-Hill, New York

Brehm A, Voigt KI (2009) Integration of market pull and technology push in the corporate front end and innovation management: insights from the German software industry. Technovation 29(5):351–367

Dekkers R, Chang CM, Kreutzfeldt J (2013) The interface between "product design and engineering" and manufacturing: a review of the literature and empirical evidence. Int J Prod Econ 144:316–333

Ehrlenspiel K, Meerkamm H (2013) Integrierte Produktentwicklung: Denkabläufe, Methodeneinsatz, Zusammenarbeit, 5. überarb. und erw. Auflage. Hanser, München

Eversheim W (1995) Simultaneous Engineering: Erfahrungen aus der Industrie für die Industrie. Springer, Berlin

Gramlich S (2013) Vom fertigungsgerechten Konstruieren zum produktionsintegrierenden Entwickeln: Durchgängige Modelle und Methoden im Produktlebenszyklus. Fortschritt-Berichte VDI, Konstruktionstechnik/Maschinenelemente, vol 423. VDI-Verlag, Düsseldorf

Groche P, Schmitt W, Bohn A, Gramlich S, Ulbrich S, Günther U (2012) Integration of manufacturing-induced properties in product design. CIRP Ann 61(1):163–166

Hausschildt J (2004) Innovationsmanagement, 3. völlig überarb. und erw. Aufl. Verlag Franz Vahlen, München

Heidemann B (2001) Trennende Verknüpfung: Ein Prozessmodell als Quelle für Produktideen. Fortschritt-Berichte VDI, Konstruktionstechnik/Maschinenelemente, vol 351. VDI-Verlag, Düsseldorf

Lee K (1999) Principles of CAD, CAM, CAE systems. Addison-Wesley Longman Publishing Co., Inc., Boston

Meboldt M (2008) Mentale und formale Modellbildung in der Produktentstehung: Als Beitrag zum integrierten Produktentstehungs-Modell (iPeM). Forschungsberichte. IPEK, vol 29. Karlsruhe

Meerkamm H, Wartzack S, Bauer S, Krehmer H, Stockinger A, Walter M (2012) Design for X (DFX). In: Rieg F, Steinhilper R (eds) Handbuch Konstruktion. Carl Hanser, München, Wien, pp 443–462

Pahl G, Beitz W, Feldhusen J, Grote KH (2007a) Engineering design: a systematic approach, 3rd edn. Springer, London

Pahl G, Beitz W, Feldhusen J, Grote KH (2007b) Konstruktionslehre: Grundlagen erfolgreicher Produktentwicklung, 7. Aufl. Springer, Berlin

Pollock TM, Tin S (2006) Nickel-based superalloys for advanced turbine engines: chemistry, microstructure and properties. J Propuls Power 22(2):361–374

Ponn J, Lindemann U (2011) Konzeptentwicklung und Gestaltung technischer Produkte: Systematisch von Anforderungen zu Konzepten und Gestaltlösungen. Springer, Berlin

Rollmann T (2012) Simultaneous Engineering von integralen Blechbauweisen höherer Verzweigungsordnung: Ein Beitrag zur Integration von Konstruktion und Produktionsprozessplanung. Forschungsberichte aus dem Fachgebiet Datenverarbeitung in der Konstruktion, vol 40. Shaker Verlag, Aachen

Roth KH (1996) Konstruieren mit Konstruktionskatalogen, 2. Aufl. Springer, Berlin

Suh NP (1998) Axiomatic design theory for systems. Res Eng Des 10(4):189–209

Suh NP (2001) Axiomatic design: advances and applications, MIT-Pappalardo series in mechanical engineering. Oxford University Press, New York

Tangen S (2005) Demystifying productivity and performance. Int J Product Perform Manag 54 (1):34–46

Tekkaya AE, Allwood JM, Bariani PF, Bruschi S, Cao J, Gramlich S, Groche P, Hirt G, Ishikawa T, Löbbe C, Lueg-Althoff J, Merklein M, Misiolek WZ, Pietrzyk M, Shivpuri R, Yanagimoto J (2015) Metal forming beyond shaping: predicting and setting product properties. CIRP Ann 64(2):629–653

Trommsdorff V, Steinhoff F (2013) Innovationsmarketing, 2. vollst. überarb. Aufl. Vahlen, München

Ulrich KT, Eppinger SD (1995) Product design and development. McGraw-Hill, New York

Verein Deutscher Ingenieure (1993) Methodik zum Entwickeln und Konstruieren technischer Systeme und Produkte, Richtlinie VDI 2221. VDI, Düsseldorf

Verein Deutscher Ingenieure (2004) Entwicklungsmethodik für mechatronische Systeme, Richtlinie VDI 2206. VDI, Düsseldorf

Weber C (2005) CPM/PDD: an extended theoretical approach to modelling products and product development processes. In: Bley H, Jansen H, Krause FL, Shpitalni M (eds) Proceedings of the 2nd German-Israeli symposium for design and manufacturing. Fraunhofer IRB Verlag, Stuttgart, pp 159–179

Chapter 2
The CRC666 Approach: Realizing Optimized Solutions Based on Production Technological Innovation

V. Monnerjahn, E. Bruder, S. Gramlich, P. Groche, S. Köhler, I. Mattmann, M. Roos, and C. Wagner

2.1 Motivation for and Goals of a New Development Approach

Finding technical solutions for given problems is one of a designer's key challenges. The task is especially demanding since the designer tries to find not only one possible solution but also the best possible solution, taking all existing conditions, limitations, and requirements into account (Pahl et al. 2007). There are many product development approaches that support the designer in this. The focus and drivers of the approaches differ:

- Reduction of complexity (Suh 1998)
- Integration of product development in company processes (Ehrlenspiel and Meerkamm 2013)
- Methodical approach based on analysis and synthesis steps (VDI 2221 1993)
- Cross-domain development of systems with a focus on mechatronic systems (VDI 2206 2004)
- Sustainable product design (Birkhofer et al. 2012)
- Effectiveness and efficiency (Lindemann 2009)
- Flexibility (Lindemann 2009)

V. Monnerjahn (✉) • P. Groche • S. Köhler
Institute for Production Engineering and Forming Machines (PtU), Technische Universität Darmstadt, Darmstadt, Germany
e-mail: monnerjahn@ptu.tu-darmstadt.de

E. Bruder
Physical Metallurgy (PhM), Technische Universität Darmstadt, Darmstadt, Germany

S. Gramlich • I. Mattmann • M. Roos • C. Wagner
Institute for Product Development and Machine Elements (pmd), Technische Universität Darmstadt, Darmstadt, Germany

© Springer International Publishing AG 2017
P. Groche et al. (eds.), *Manufacturing Integrated Design*,
DOI 10.1007/978-3-319-52377-4_2

- Cost and time reduction; quality improvement (Eder and Hosnedl 2010)
- Computer-aided automatization (Weber 2005)

In addition, there is one especially important driver of product development approaches: *manufacturing technologies* (Sect. 1.2). Manufacturing technological knowledge, along with further knowledge about engineering and natural sciences, is important for the designer to be able to find feasible technical solutions (Pahl et al. 2007). Especially in technology-pushed development projects, knowledge about manufacturing technologies is essential for finding innovative solutions with comprehensively realized manufacturing potentials (Sect. 1.3). This knowledge is not always available for the product designer.

Although a huge number of manufacturing technologies are available, many manufacturing possibilities have not yet been taken into account, resulting in a high potential for realizing additional benefits in product and process solutions (Sect. 1.1). Innovative manufacturing technologies or novel combinations of manufacturing technologies especially offer new design possibilities (Sect. 1.3). A comprehensive analysis of these manufacturing technologies is required to provide the designer with knowledge about how to systematically realize manufacturing technological potential.

Unfortunately, there is no development approach that combines the mentioned aspects to find the best possible technical product and process solution and to realize manufacturing technological potentials on the basis of manufacturing technological knowledge. Many approaches focus on manufacturability of product solutions. Often, manufacturing information is only considered in later stages of the development process, during embodiment and detail design, resulting in iterations. Actual solution finding in the early phases of product development is typically not affected by manufacturing information. There are also approaches that focus on integrating manufacturing information into the development process. These approaches address the product in the context of its life cycle processes (including manufacturing) and support concurrent development of products and processes. Many of these approaches lack comprehensive operationalization due to the lack of a consistent system of concepts, models, methods, and tools (Sect. 1.2).

The goal is to provide a consistently formalized approach for realizing *manufacturing technological potential*, leading to an *innovative* as well as an *optimal product and process solution*. Manufacturing is no longer just a "service provider" for product development. Instead, an *integrated view* of the product and its *life cycle processes* during the *early development phases* is the focus, which can result in the following benefits:

- Ensuring optimality of the product and process solution
- Realizing technological possibilities, leading to product and process benefits in time, cost, and quality
- Realizing product and process innovations
- Increased product and process maturity in early phases of the development process, especially when considering new manufacturing technologies

2.2   Options in Manufacturing Technologies

Today's industry offers many options in manufacturing technologies to produce a specific workpiece. According to Grote and Antonsson (2009) the term manufacturing is the production of workpieces of geometrically defined shapes, whereas manufacturing technologies allow for the production of products which are distinguished by material and geometric characteristics. According to DIN 8580 (2003) the numerous different manufacturing processes can be classified into six main groups: primary shaping, forming, cutting, joining, coating, and changing of material properties (Fig. 2.1).

For today's industry one of the main tasks according to Westkämper and Warnecke (2010) is to select the manufacturing process with the greatest possible economic efficiency, taking into account the numerous criteria and given boundary conditions. In selecting the process, it is particularly important to consider the entire manufacturing chain up to the finished workpiece. A manufacturing process, which is assessed unfavorable due to an isolated view of the process chain, can prove more economical in the case of high unit numbers. In Westkämper (1997) this fact is shown in an assessment of different manufacturing chains for a gear. For a holistic view of the economic efficiency of a manufacturing process, the assessment criteria in Westkämper (1997) have to be considered (Fig. 2.2). Partially some assessment criteria are difficult to quantify, so many decisions in today's industry are based on the employee's experience.

Fig. 2.1 Manufacturing processes after DIN 8580 (2003)

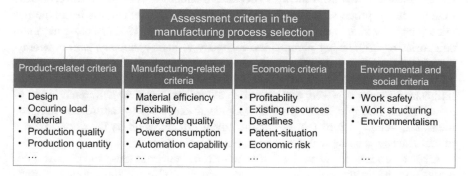

Fig. 2.2 Some assessment criteria in the selection of the suitable manufacturing process (Westkämper and Warnecke 2010)

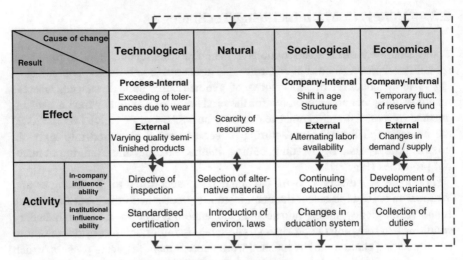

Fig. 2.3 Classification of uncertainties (Schmitt et al. 2012)

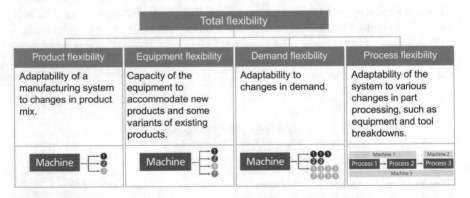

Fig. 2.4 Flexibility according to Son and Park (1987)

Uncertainties in Manufacturing Processes: Besides the decision of the economic manufacturing process, today's forming companies are influenced by numerous uncertainties which are characterized in Schmitt et al. (2012) into four main categories of influencing parameters: technology, nature, society, and economy. Considering all four categories with regard to the cause of change and the result, 15 different kinds of uncertainties are classified (Fig. 2.3). There are also interrelationships between effects and activities as well as between different causes of change. For example, ecological developments can lead to nature conservation measures, which in turn require technological effects, such as the demand for a manufacturing process with reduced pollutant emission.

One approach to encounter the numerous uncertainties in the field of manufacturing is the concept of an increased flexibility of the applied systems and processes (Groche et al. 2010). According to Son and Park (1987) total flexibility consists of four different kinds of flexibility, which is shown in Fig. 2.4.

Product flexibility is determined as the adaptability of a manufacturing system to variances in the product mix. Equipment flexibility is the capability of a system to integrate new products and variants of existing products. Demand flexibility describes the adaptability of a manufacturing system to changes in the market demand. Last, process flexibility is defined as the adaptability of the system to changes in part processing, for example caused by changes of technology.

The trend of today's manufacturing lies in the flexible design of the total process chain. This often requires a higher initial investment by the manufacturer, but the manufacturing system can encounter the numerous uncertainties in a better way and could lead finally to a cost reduction. Flexibility is also regarded as an important objective for the development of new manufacturing processes. Since the area of application is unknown in the beginning, possibilities to produce a variety of product geometries made out of different materials are essential. The book at hand will demonstrate the expansion of basic technologies into a highly flexible class of manufacturing opportunities.

Manufacturing Processes for Branched Profile Geometries: For the production of profiles with branched cross sections, various manufacturing processes of DIN 8580 (2003) can be considered. On the basis of a targeted double-T-profile geometry, some of the possible processes are shown in Fig. 2.5. Against the background of thin-walled profiles in mass production, the manufacturing technologies of primary shaping, cutting, or joining are not effective enough and can be neglected. On the other hand, forming technologies offer many advantages, due to the material utilization or the improvement of the material properties and represent the main technologies to manufacture thin-walled profile geometries.

In the roll-forming process, a flat sheet metal can be continuously formed through several roll-forming tools to an open or closed profile geometry with any length. At the same time, the thickness of the sheet metal shall not be reduced. The common speed for production is between 40 and 100 m/min. Sheet metals with a thickness of 0.3–12 mm can be manufactured (Lange 1990). However, branched profile geometries can only be manufactured by roll forming by doubling the material, which could be in conflict with a lightweight profile design. In the process

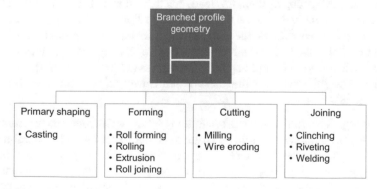

Fig. 2.5 Options in manufacturing of branched profile geometries

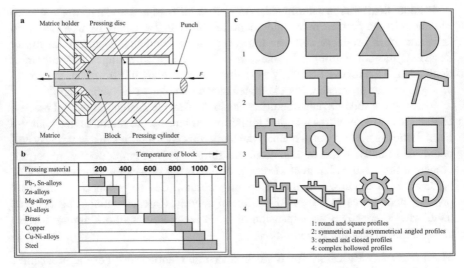

Fig. 2.6 Extrusion by Fritz and Schulze (2008): (**a**) process principles, (**b**) block temperature of pressing material, (**c**) possible profile geometries through extrusion

of rolling, slabs are formed by shaped rolls to the targeted profile geometry. The slab is usually heated and formed by several roll stands, whereas it is not possible to form a rectangular cross section directly to the final geometry. Rather a calibration sequence is necessary, which is often experience based (Lange 1988). However, rolling is usually used for thick-walled profile geometries. Other conventional forming technologies are represented by extrusion or roll joining, which are described in detail in the following section.

Extrusion: In the process of extrusion, a heated block is compressed by a punch in a pressing cylinder. The material starts to flow and exits the matrices as a continuous thread. The length of the manufactured threads is limited according to Fritz and Schulze (2008) to 20 m, whereas various profile geometries with undercuts or hollow spaces are possible. For extrusion well-formable materials are suitable, such as aluminum, copper, zinc, tin, lead, and their alloys. The temperature for the preheating of the block has to be adjusted to the material and is shown in Fig. 2.6.

Under certain conditions the extrusion of steel is also possible and can be realized by the *Ugine-Séjournet* process. Due to the high temperatures glass is used as lubricant. Very complex profile geometries are possible with the process of extrusion. The minimal wall thickness is limited to 3.5 mm for steel (Fritz and Schulze 2008).

Roll joining: The roll-joining process allows the continuous manufacturing of T-joint beams out of flat sheet metal strips and other materials and its combinations for lightweight constructions. New opportunities in the design of profiles are possible and are no longer limited to restrictions regarding sharpness or constant wall thickness such as in conventional roll-formed profiles (Lappe and

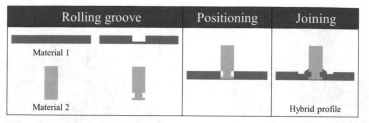

Fig. 2.7 Process principles of roll joining

Niemeier 2002). The process of roll joining consists of several stages, which are shown in Fig. 2.7.

In the first step, a continuous groove is rolled into the web of the profile and a discontinuous groove in longitudinal direction is rolled into the other joining part. In the second step, the two materials are positioned to each other. In the last step, the web of material 1 is locally rolled, whereby the material is plastified and flows into transversal direction. The gap between the two materials is closed and the result is a form- and force-locked hybrid connection.

Today's manufacturing processes are severely limited with regard to the production of branched profile structures made of sheet metal. Up to now, there is no manufacturing process of DIN 8580 (2003), which allows the production of thin-walled, branched profile geometries with any length out of high-strength steels in integral style. This gap is closed with the new manufacturing processes linear flow splitting and bend splitting described in the following Chap. 3.

2.3   Manufacturing-Induced Properties

The realization of a novel development approach that strongly integrates production and product design (Sect. 2.1) is accompanied by the following key question: What information is fundamental for the designer's perspective as well as for the manufacturer's perspective in order to realize an integrated view of the product and its manufacturing processes?

Answering this question is only possible by characterizing the conceptual differences between manufacturing technologies in a formalized way. Each manufacturing technology, along with its manufacturing processes, describes a distinct way of realizing products with defined properties, starting from the workpiece's initial state (Heidemann 2001). The product's final state is mainly dependent on the chosen manufacturing technology and its underlying procedural principle, as well as the material used. The procedural principle characterizes and describes the generally valid transformation procedure from workpiece to product for a specific manufacturing technology (Gramlich 2013).

Considering manufacturing technologies for realizing external threads, there are primarily two options, thread cutting and thread rolling. The procedural principle of

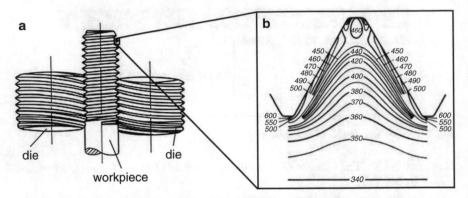

Fig. 2.8 (**a**) Die arrangement for thread rolling (through feed method), (**b**) grain direction and hardness pattern in rolled threads (alloy steel) according to Tschaetsch (2006)

thread rolling, which is used in the vast majority of applications, is based on plastic deformation of the blank using a set of rolling dies (Tschaetsch 2006) (Fig. 2.8a). The penetration of the rolling dies into the blank surface not only creates the thread geometry but also changes the local material properties (Fig. 2.8b). Due to work hardening, the yield strength of the material increases substantially, especially in the root and at the flanks. Furthermore, strain gradients during the manufacturing process result in the formation of subsurface compressive residual stresses in axial direction. In contrast to thread cutting, the grain flow remains intact, thus avoiding preferred crack initiation sites near the root (Tschaetsch 2006).

Based on the specific procedural principle, the manufacturing technology realizes characteristic and reproducible product properties. These characteristic properties are called manufacturing-induced properties (Groche et al. 2012). Manufacturing-induced properties are always assigned to manufacturing-induced design elements, such as the external thread. Consequently, manufacturing technologies are unambiguously characterized by a specific set of manufacturing-induced properties realized in specific products (Groche et al. 2012). The formal description of products by their properties gives the product designer the opportunity to formally model and document all technical products (Birkhofer and Wäldele 2008). Thus, manufacturing-induced properties are crucial for both the product designer and the manufacturer who focuses on manufacturing processes, especially their output. Within an integrated development approach, manufacturing-induced properties are a key element for establishing the link between designer and manufacturer perspectives.

The example of thread rolling clarifies that manufacturing technologies not only realize the geometric properties of products but also have a significant influence on material properties. Material properties involve mechanical properties (mechanical resistance, impact resistance, fatigue resistance, formability, residual stresses, etc.), surface properties (surface topography, wear resistance, corrosion resistance, etc.), and physical properties (electrical properties, magnetic properties, optical properties, etc.) (Tekkaya et al. 2015). Consequently, manufacturing-induced properties are comprised of geometric and material properties, which necessitates the investigation of both (Chaps. 3 and 4) when considering them within a manufacturing-integrated development approach. Utilizing the potential of manufacturing-induced properties can lead to products with higher performance and functionality (Sect. 4.3).

2.4   Mathematical Optimization of Product Geometries and Manufacturing Processes

Optimization is an important tool to improve the efficiency and effectiveness of mechanical devices. Depending on the scientific discipline, optimization processes are understood differently. From an engineering point of view, an optimized process often only means the improvement of one parameter like the production volume per hour (Pahl et al. 2006). Optimization in a mathematical sense implies finding the proven best solution under defined constraints using an optimization model (Jarre and Stoer 2013). This chapter provides an introduction to the mathematical optimization of product structures and manufacturing processes. Later on, we will introduce optimization procedures applied to the design of branched structures (Sect. 5.2) and the manufacturing processes to produce them (Sect. 5.3).

Optimal structures can be characterized by the following properties:

- Multifunctional
- Functional and stressable
- Maximized strength and operationally stable
- Weight minimized
- Optimized in terms of heat conduction and transfer
- Minimum number of individual parts

If technical branched structures with regard to these requirements are investigated, some deficits can be observed. This is due to the fact that in most structures only certain requirements were optimized. Furthermore, some requirements cannot be optimized at the same time since they are contradicting. Therefore, this leads to conflicts of objectives. For example, focusing on the minimum number of parts leads to geometrical limits caused by the used manufacturing process. These geometrical limitations can be in conflict with functional requirements. Here, multiobjective algorithms offer the possibility of finding an optimal solution for these complex problems not only for the product itself. Conflicting manufacturing process parameters can be optimized, too.

Optimality of a product often is not reached due to the fact that the development process of a new component typically strongly depends on the intuition and experience of the designer. For instance, it depends on intelligence and creativity of the developers. In a conventional development process, the subjectivity manifests especially in the evaluation of proposed solutions. The choice of evaluation

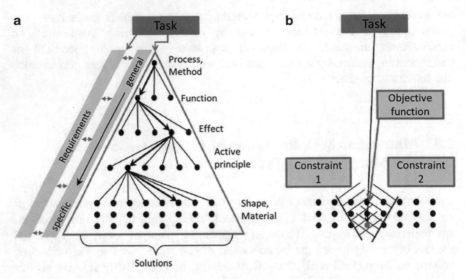

Fig. 2.9 (**a**) Conventional development, (**b**) algorithm-based development

criteria and subsequent evaluation of solutions can lead to people-related errors and is often the weak spot of a methodical development process (Pahl et al. 2006).

An algorithm-based product development combined with a mathematical optimization increases the degree of automation of this process and improves product properties. It also provides a decision assistance for the designer. Figure 2.9 shows the differences between a conventional (a) and an algorithm-based development process (b). Proceeding along the conventional development process, the number of possible solutions increases for the initial problem. With the variety of existing solutions, it is difficult to find the ideal solution. The rating of the solutions is, as mentioned before, strongly depending on the planner. In contrast, the algorithm-based approach to solve the design task leads directly to an optimal solution. The objective function rates the different solutions objectively and constrains the solution space. Therefore, the optimal solution for a design task can often be found quicker and more accurately. However, a mandatory requirement for a mathematical optimization-based development task is that we can model the problem.

Computational costs often limit the application of optimization methods. Simulation techniques, like finite element analysis (FEA) or computational fluid dynamics (CFD), are very expensive in terms of time consumption and system requirements (Roy et al. 2008). The calculation time for an optimization task depends on the problem size. The range of time to find the optimum varies from a few seconds up to years for complex algorithms. However, Bixby (2002) found that the combination of algorithm improvement and further development of computer chips continuously provides new opportunities to solve increasingly complex optimization tasks. In the time range of 10 years, both algorithmic speed and solving power increased by a magnitude of three orders and thus accelerate the solution of a complex linear problem by a factor of more than one million (Bixby 2002).

Finally, developments using mathematical optimization methods provide several additional benefits. Many experimental iteration steps in product developments are replaced by automatically conducted computational iterations. These procedures require fewer resources and man power. The formulation of the optimization model itself leads to a new perspective and a deeper understanding of the development task or the process. Even if the optimum cannot be found, the developer gets an overview on the influences and effects of design/process parameters. Although the algorithm-based development possesses a high degree of automation, the formulation of the optimization model and the choice of the algorithm still depend on personal know-how. An automatic setup of the optimization task is not possible in most cases. Algorithms are often derived from related problems.

Another advantage of mathematical optimization is the fact that optimality can be proven. A popular example for a requirement of an optimum are the Karush-Kuhn-Tucker conditions (Nocedal and Wright 2006). They generalize the necessary condition $F(x) = 0$ for a convex nonlinear first-order problem.

2.4.1   Classification of Optimization Tasks

This section classifies the optimization types in several categories. After a short introduction of the general optimization process, a classification of product optimization algorithms is presented, followed by a characterization of process optimization. Then, a brief mathematical view on the topic is given. Eventually, after the aspect of optimization under uncertainty is introduced, some examples for classical optimization tasks are given in the next section.

Adamy (2011) describes the general flow of an optimization process in three steps:

1. Mathematical formulation of the optimization problem
2. Selection of a suitable algorithm
3. Determination of a solution

Roy et al. give an extensive overview on the topic of engineering design optimization (EDO) (Roy et al. 2008). Table 2.1 shows several criteria for the classification of EDO algorithms.

The design variables are a possible classification scheme. The number, nature, or dependence of the variables are subcategories. The constraints lead to another way of categorization of an EDO. The number and type of constraints have a significant influence on the optimization process. Furthermore, the problem domain, characterized by the physics of the problem, or its environment with further subcategories provides another possible ranking. From a mathematical perspective, the objective functions are another possible classification category. Their number, nature, and separability lead to many categories of optimization problems (Roy et al. 2008).

Table 2.1 Classification of EDO algorithms (Roy et al. 2008)

Category	Subcategory
Design variables	Number of variables
	Nature of design variables
	Permissible values of design variables
	Dependence among design variables
Constraints	Existence of constraints
Objective functions	Number of objective functions
	Nature of objective functions
	Separability of objective functions
Problem domain	Physics of problem
Environment	Uncertainty
	Existing knowledge about the problem
	Designer confidence required
	Nature of the environment

Besides product optimization, process optimization is a category of its own. However, the former mentioned EDO classifications can be adopted by using the number of processes as a possible classification scheme. Besides single processes, an optimal order of a process chain can also be calculated with the aim of the smallest throughput time. This approach can also be added to be called process optimization. This optimization is also useful to shorten downtimes in production processes.

The mathematical representation of the task can be summarized in the minimization (or maximization) of an objective function $F(x)$ for the variables x, respecting the given constraints. From a mathematical perspective, optimization problems can be separated into three classes (Biegler 2010):

- Differentiable (derivative based)
- Nondifferentiable (derivative free)
- Discrete

Derivative-based strategies search for an optimal solution by using the derivative of the objective function $F(x)$ and try to find feasible solutions x that lead to $F(x) = 0$. If $F(x)$ is differentiable, this is the most common way to solve optimization problems, because of its speed and the fact that local optimality of the solution can be proven. Classical examples are the Newton method or the sequential quadratic programming (SQP) method (Nocedal and Wright 2006). However, practical problems are often of a complexity that the differentiation and evaluation of the objective function are too expensive. Derivative-free algorithms like the Nelder-Mead method are a possible alternative (Conn et al. 2009). Finding a local optimum with a derivative-free algorithm is often more time consuming because the solutions are found iteratively and built on top of each other, allowing to use less information about the problem. Discrete algorithms, dealing with integer problems, are the third class. The branch-and-bound algorithm, with its tree structure, is a typical algorithm for this class (Lawler and Wood 1966). They are often used to solve logical or combinatorial problems.

Furthermore, evolutional and genetic optimization strategies have to be mentioned as an important algorithm group that is often referred to as *heuristic*. They are derivative free but do not fit exactly in one of the three formerly mentioned classes. Evolutional and genetic algorithms are often used to find an approximation of a global optimum of a problem. There are no quality criteria for this class of algorithms. This type of optimization method is limited in its usefulness because it is impossible to determine the quality of the solution and whether or not it would be worth to continue with the optimization. However, their application in technical backgrounds has become more important as several examples show (Adamy 2011).

The definition of a global optimum and some other important mathematical definitions in optimization problems are depicted in Fig. 2.10a–c. In Fig. 2.10a, the difference between local, global, and robust optima is shown. Whereas a global optimum is a best solution for the entire function, a local optimum obtains the best objective value within a region. In practical problems it is often impossible to determine whether the found optimum is a global or just a local optimum, because computing a global evaluation is not feasible or it is too expensive to prove global optimality. For practical applications it is also very important that the found solution is robust. Robustness implies that relevant changes in design values do not lead to significant changes in the value of the objective function. A representation of multiple optimum solutions can be seen in Fig. 2.10b. Here, several points lead to the best solution for the optimization tasks. Therefore, one aim is to find as many optimal solutions as possible.

In multi-objective problems with more than one objective function, the amount of optimal solutions can be summarized in a Pareto front (see Fig. 2.10c). A change of the value that leads to a better solution for one objective function leads to a worse solution for the other objective function on the Pareto front.

Uncertainties in real-life problems, like unpredictable parameters or inaccuracies in manufacturing processes and products, lead to the necessity of robust optimization. Beyer and Sendhoff give an extensive review on the recent achievements in this field of research (Beyer and Sendhoff 2007). In some approaches, the uncertainty is represented in the objective function by statistical information like the standard deviation and the mean (Roy et al. 2008). Wiebenga et al. summarize that balancing the number of time-consuming FE simulations spent on the robustness evaluation and the accuracy of the robustness predictions themselves are remaining challenges in robust process optimization (Wiebenga et al. 2012). The determination and handling of uncertainties in the optimization of process chains and mechanical structures is one central research aim of the CRC805 at TU Darmstadt. Recent developments in this area can be found in the proceedings of ICUME 2015 (Pelz and Groche 2015).

Fig. 2.10 (a) Mathematical definitions of robust optimization, global/local optimum, (b) multiple optimum solutions, (c) Pareto front (Roy et al. 2008)

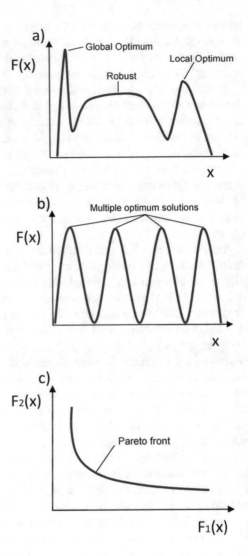

2.4.2   Examples for Optimization

This section introduces examples for applied optimization and the generated benefit for product and process development. While product optimization focuses on the properties of a product, we also often try to find ideal conditions for the production in a process optimization.

Product optimization: Marsden et al. present an example for a derivative-free product optimization. The surrogate management framework (SMF) technique is applied to optimize the trailing of an airfoil to reduce noise in unsteady laminar flow. The objective function that is minimized represents the acoustic density at a

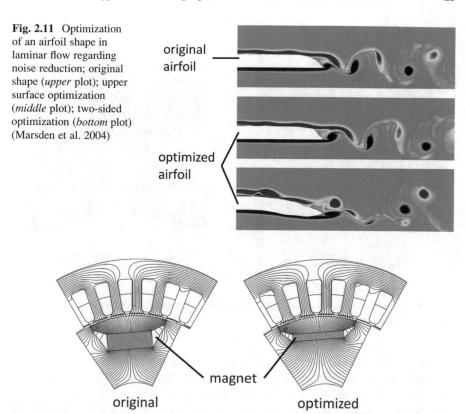

Fig. 2.11 Optimization of an airfoil shape in laminar flow regarding noise reduction; original shape (*upper* plot); upper surface optimization (*middle* plot); two-sided optimization (*bottom* plot) (Marsden et al. 2004)

original airfoil

optimized airfoil

magnet

original   optimized

Fig. 2.12 Original and optimized magnet design of an electromagnetic machine (Alla et al. 2015)

far-field position. The dimensionless lift and drag of the airfoil are used to set the constraints for the optimization problem. A reduction of 70% in vortex shedding noise could be achieved by an optimized airfoil shape. The resulting shapes compared to the original shape are plotted in Fig. 2.11 (Marsden et al. 2004).

Another example for a mathematical product optimization is given by Alla et al. In their case, the derivative-based SQP method is applied to optimize an electromagnetic machine with permanent magnets. In order to save expensive material, the optimization goal is to minimize the volume of the magnet for a given electromotive force. The problem is transformed from 3D into a 2D problem, and as we can see in Fig. 2.12, a volume reduction of more than 50% is achieved compared to the original design without a decrease in electromotive performance (Alla et al. 2015).

Mathematical optimization can also be utilized in products of much bigger dimensions as Koch et al. (2015) show. In their book "Evaluating gas network capacities" a nationwide gas network is examined by the use of discrete-nonlinear or mixed-integer nonlinear optimization algorithms. The primary goal of the optimization problem is to find a way to transport a certain amount of gas. A challenge in this task is the consideration of active elements, like valves, in the mathematical model (Koch et al. 2015).

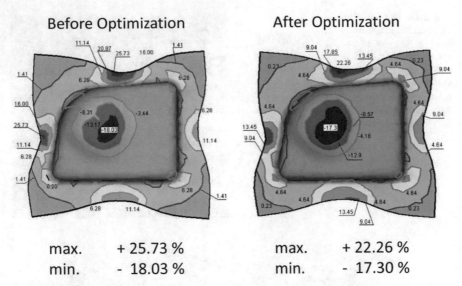

Fig. 2.13 Sheet metal thickness distribution of a deep-drawn dashpot cup before and after process optimization (Naceur et al. 2001)

Process optimization: Derivative-free optimization can also be used to find the optimal process conditions. Agapiou shows the application of the Nelder-Mead simplex method to find the best parameters for a machining process. He determines the optimum for the cutting speed and feed for a given cutting depth in a turning process. The aim is to minimize the production time and production costs which are affected by tool wear (Agapiou 1992).

Another mathematical process optimization is introduced by Naceur et al. They use the derivative-based SQP and Broyden-Fletcher-Goldfarb-Shanno (BFGS) method to improve a deep drawing process. The drawbead design and restraining forces are optimized to minimize the thinning distribution of the sheet metal part which is significant for the typical deep-drawing errors necking and wrinkling of the cup. The variations in thickness could be minimized by 97% in a square cup. The authors also present the application of the optimization on the production of a dashpot cup in a car (see Fig. 2.13). Hence, this work shows the usefulness of optimization for industrial processes (Naceur et al. 2001).

2.5 Integrated Algorithm-Based Product and Process Development

Product development and manufacturing are closely related not just because every product has to be produced. A symbiosis between both domains can be especially beneficial: An extensive exchange of information not only leads to better products but also to manufacturing processes which better fit to product design (Sect. 1.1). Existing approaches mainly focus on developing the product and ensuring its manufacturability. Rethinking is necessary to realize the full potential of both domains. Domain-specific methods and procedures have to give way to a multi-domain consideration of the entire technical system. Developing both domains requires a consideration of the product and its life cycle processes, with a focus on manufacturing (Ehrlenspiel and Meerkamm 2013). Integrated development approaches possess the highest potential for the development of innovative products and processes (Sect. 1.2), especially when considering technology-pushed development processes. Manufacturing technologies are the starting point of the development process. A development approach aiming at successful realization of a technology-pushed product or process solution requires comprehensive and equal development of the product and the processes that are based on the initial manufacturing technology (Sect. 1.3).

Innovative manufacturing technologies and novel combinations of manufacturing technologies create new possibilities for innovative product solutions (Sect. 2.2). A development approach has to be able to identify and process essential manufacturing technological knowledge to fully realize this potential. Manufacturing-induced properties that are realized with the help of the considered manufacturing technology play a major role in formalizing this knowledge (Sect. 2.3). The challenge of an integrated product and process development approach lies mainly in the high complexity that arises from the concurrent consideration of large amounts of product and process development information and restrictions. Implementing and utilizing mathematical optimization algorithms is an adequate way to handle this complexity (Sect. 2.4).

Based on this foundation, a new approach for an *integrated algorithm-based product and process development* that allows comprehensive consideration of product and process information with the help of mathematical algorithms is introduced. The approach has the following key elements:

- *Early determination of suitable manufacturing technologies*
  An early determination of suitable manufacturing technologies is mandatory for the combined consideration of product and process information. In particular, innovative technologies, like linear flow splitting, create new design possibilities that have to be determined by anticipating and analyzing manufacturing processes (Chap. 3). The characteristics of manufactured products, described with the help of manufacturing-induced properties, have to be processed specifically for product and process development purposes (Chap. 4).
- *Product and process solution finding by applying optimization algorithms*
  Finding optimal product and process solutions is a key element of the development process. Especially in the context of technical problems with complex product and process requirements, solution finding can be executed more efficiently with the help of mathematical optimization algorithms (Chap. 5).

- *Consistent integration*
  Computer-integrated methods, tools, and interfaces that are specifically tailored
  to the overall approach enable collective processing of product and process
  information during the whole development process (Chap. 6).
- *Product and process innovation*
  To develop innovative products in the context of their life cycle processes, the
  approach allows the development of market-pulled and technology-pushed
  solutions (Chaps. 7 and 8).

References

Adamy J (2011) Fuzzy Logik, Neuronale Netze und Evolutionäre Algorithmen. Shaker Verlag, Aachen

Agapiou J (1992) The optimization of machining operations based on a combined criterion, part 1: the use of combined objectives in single-pass operations. J Eng Ind 114(4):500–507

Alla A, Hinze M, Lass O, Ulbrich S (2015) Model order reduction approaches for the optimal design of permanent magnets in electro-magnetic machines. IFAC-PapersOnLine 48 (1):242–247

Beyer HG, Sendhoff B (2007) Robust optimization—a comprehensive survey. Comput Methods Appl Mech Eng 196(33–34):3190–3218

Biegler LT (2010) Nonlinear programming: concepts, algorithms, and applications to chemical processes. SIAM, Philadelphia

Birkhofer H, Wäldele M (2008) Properties and characteristics and attributes and...: an approach on structuring the description of technical systems. In: Vanek V, Hosnedl S, Bartak J (eds) Proceedings of AEDS 2008 Workshop. Pilsen, pp 19–34

Birkhofer H, Rath K, Thao S (2012) Umweltgerechtes Konstruieren. In: Rieg F, Steinhilper R (eds) Handbuch Konstruktion. Hanser, München

Bixby RE (2002) Solving real-world linear programs: a decade and more of progress. Oper Res 50 (1):3–15

Conn AR, Scheinberg K, Vicente LN (2009) Introduction to derivative-free optimization. SIAM, Philadelphia

Din 8580 (2003) Deutsches Institut für Normung e. V: Fertigungsverfahren—Begriffe, Einteilung

Eder WE, Hosnedl S (2010) Introduction to design engineering: systematic creativity and management. CRC Press/Balkema, Leiden

Ehrlenspiel K, Meerkamm H (2013) Integrierte Produktentwicklung: Denkabläufe, Methodeneinsatz, Zusammenarbeit, 5. überarb. und erw. Auflage. Hanser, München

Fritz AH, Schulze G (2008) Fertigungstechnik. Springer, Berlin

Gramlich S (2013) Vom fertigungsgerechten Konstruieren zum produktionsintegrierenden Entwickeln: Durchgängige Modelle und Methoden im Produktlebenszyklus. Fortschritt-Berichte VDI, Konstruktionstechnik/Maschinenelemente, vol 423. VDI-Verlag, Düsseldorf

Groche P, Scheitza M, Kraft M, Schmitt S (2010) Increased total flexibility by 3D Servo Presses. CIRP Ann Manuf Technol 59(1):267–270

Groche P, Schmitt W, Bohn A, Gramlich S, Ulbrich S, Günther U (2012) Integration of manufacturing-induced properties in product design. CIRP Ann 61(1):163–166

Grote K-H, Antonsson EK (eds) (2009) Springer handbook of mechanical engineering. Springer, Berlin

Heidemann B (2001) Trennende Verknüpfung: Ein Prozessmodell als Quelle für Produktideen. Fortschritt-Berichte VDI, Konstruktionstechnik/Maschinenelemente, vol 351. VDI-Verlag, Düsseldorf

Jarre F, Stoer J (2013) Optimierung. Springer-Verlag, Berlin

Koch T, Hiller B, Pfetsch ME, Schewe L (2015) Evaluating gas network capacities. SIAM, Philadelphia

Lappe W, Niemeier R (2002) Rollfügen—Der "coole" Weg zum Profil. Stahlbau 71(11):781–788

Lange K (1988) Umformtechnik: Handbuch für die Industrie und Wissenschaft. Band 2: Massivumformung. Springer, Berlin

Lange K (ed) (1990) Umformtechnik: Handbuch für die Industrie und Wissenschaft. Band 3: Blechbearbeitung. Springer, Berlin

Lawler EL, Wood DE (1966) Branch-and-bound methods: a survey. Oper Res 14(4):699–719

Lindemann U (2009) Methodische Entwicklung technischer Produkte: Methoden flexibel und situationsgerecht anwenden, 3. korr. Auflage. Springer, Berlin

Marsden AL, Wang M, Dennis JE Jr, Moin P (2004) Suppression of vortex-shedding noise via derivative-free shape optimization. Phys Fluids 16(10):L83–L86

Naceur H, Guo YQ, Batoz JL, Knopf-Lenoir C (2001) Optimization of drawbead restraining forces and drawbead design in sheet metal forming process. Int J Mech Sci 43(10):2407–2434

Nocedal J, Wright S (2006) Numerical optimization. Springer Science & Business Media, New York

Pahl G, Beitz W, Feldhusen J, Grote KH (2006) Pahl/Beitz Konstruktionslehre: Grundlagen erfolgreicher Produktentwicklung—Methoden und Anwendung. Springer, Berlin

Pahl G, Beitz W, Feldhusen J, Grote KH (2007) Engineering design: a systematic approach, 3rd edn. Springer, London

Pelz PF, Groche P (eds) (2015) Uncertainty in mechanical engineering II. Trans Tech Publications, Darmstadt

Roy R, Hinduja S, Teti R (2008) Recent advances in engineering design optimisation: challenges and future trends. CIRP Ann Manuf Technol 57(2):697–715

Schmitt SO, Avemann J, Groche P (2012) Development of manufacturing process chains considering uncertainty. In: Proceedings of the 4th International Conference on Changeable, Agile, Reconfigurable and Virtual production, Montreal, Canada, 2–5 Oct 2011, Springer, Berlin, , pp 111–116

Son YK, Park CS (1987) Economic measure of productivity, quality and flexibility in advanced manufacturing systems. J Manuf Syst 6(3):193–207

Suh NP (1998) Axiomatic design theory for systems. Res Eng Des 10(4):189–209

Tekkaya AE, Allwood JM, Bariani PF, Bruschi S, Cao J, Gramlich S, Groche P, Hirt G, Ishikawa T, Löbbe C, Lueg-Althoff J, Merklein M, Misiolek WZ, Pietrzyk M, Shivpuri R, Yanagimoto J (2015) Metal forming beyond shaping: predicting and setting product properties. CIRP Ann 64(2):629–653

Tschaetsch H (2006) Metal forming practise: processes—machines—tools. Springer, Berlin, Heidelberg

Verein Deutscher Ingenieure (1993) Methodik zum Entwickeln und Konstruieren technischer Systeme und Produkte, Richtlinie VDI 2221. VDI, Düsseldorf

Verein Deutscher Ingenieure (2004) Entwicklungsmethodik für mechatronische Systeme, Richtlinie VDI 2206. VDI, Düsseldorf

Weber C (2005) CPM/PDD: an extended theoretical approach to modelling products and product development processes. In: Bley H, Jansen H, Krause FL, Shpitalni M (eds) Proceedings of the 2nd German-Israeli symposium for design and manufacturing. Fraunhofer IRB Verlag, Stuttgart, pp 159–179

Wiebenga JH, van den Boogaard AH, Klaseboer G (2012) Sequential robust optimization of a V-bending process using numerical simulations. Struct Multidiscip Optim 46(1):137–153

Westkämper E (ed) (1997) Null-Fehler-Produktion in Prozeßketten: Maßnahmen zur Fehlervermeidung und -kompensation. Qualitätsmanagement. Springer, Berlin

Westkämper E, Warnecke H-J (eds) (2010) Einführung in die Fertigungstechnik. Vieweg+Teubner Verlag, Wiesbaden

Chapter 3
New Technologies: From Basic Ideas to Mature Technologies

M. Neuwirth, S. Abedini, E. Abele, P. Groche, S. Köhler, V. Monnerjahn, S. Schäfer, S. Schmidt, and E. Turan

Lightweight design aims at minimizing the weight of a structure while product requirements are completely fulfilled. Utilization of closed profiles and materials with higher strength but reduced material thickness often enables weight reduction. However, these approaches are limited. When the thickness becomes too low, instability phenomena like buckling and wrinkling become apparent (Groche et al. 2004).

It is well known that adding stiffeners like stringers to the walls is a very effective measure to reduce the unsupported length of a thin-walled construction (Groche et al. 2005).

In nature, bifurcations are used in many cases to increase structural strength. Additionally, the bifurcation frequently generates separate compartments for multiple purposes. Figure 3.1 presents leaf structures and horsetails as examples for the application of bifurcations in natural subjects (Groche et al. 2004).

The effectiveness of using bifurcations can be illustrated by the example of a mussel. Figure 3.2 presents the comparison between an arced shell structure of flat sheet metal and the same construction with additional stiffeners on the sheet. We can show that the quotient between the bearable critical load and the mass increases significantly by the application of stringers (Groche et al. 2010a).

M. Neuwirth (✉) • P. Groche • S. Köhler • V. Monnerjahn
Institute for Production Engineering and Forming Machines (PtU),
Technische Universität Darmstadt, Darmstadt, Germany
e-mail: neuwirth@ptu.tu-darmstadt.de

S. Abedini • S. Schäfer
Institute of Construction Design and Building Construction (KGBauko),
Technische Universität Darmstadt, Darmstadt, Germany

E. Abele • S. Schmidt • E. Turan
Institute for Production Management, Technology and Machine Tools (PTW),
Technische Universität Darmstadt, Darmstadt, Germany

© Springer International Publishing AG 2017
P. Groche et al. (eds.), *Manufacturing Integrated Design*,
DOI 10.1007/978-3-319-52377-4_3

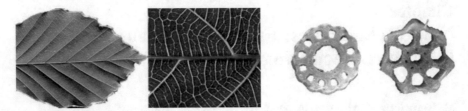

Fig. 3.1 Examples from nature: bifurcation for structural and other purposes (Groche et al. 2004)

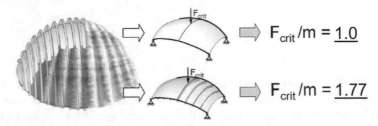

$$F_{crit}/m = \underline{1.0}$$

$$F_{crit}/m = \underline{1.77}$$

Fig. 3.2 Improved relation between load capability and mass due to bifurcations (Groche et al. 2010a)

Since the prevention of buckling and wrinkling is highly relevant for thin sheet metal or other shell structures, the capability to manufacture branched sheet metal products attracts attention. Bäcker et al. (2015) show (Fig. 3.3) possible production routes for stringer sheets with multiaxial curvature. For the manufacturing of specific parts multiple manufacturing methods and sequences come into consideration, covering processes of primary shaping, forming, and joining.

As an additional alternative to the yet established production processes, the forming processes of linear flow splitting and linear bend splitting enable the production of sheet metal parts with bifurcations in an integral style. These processes offer a huge potential for future lightweight structures. However, they are new and have to be developed and qualified.

3.1 Basics of Linear Flow Splitting and Bend Splitting

Within the continuous flow production line new forming methods have been implemented and established. Linear flow splitting and the process modification linear bend splitting are both capable of producing sheet metal products with bifurcations in an integral style. Section 3.1 contains the description of the behavior, the process principle, and the extension of the technologies derived from systematic process analyses and designs.

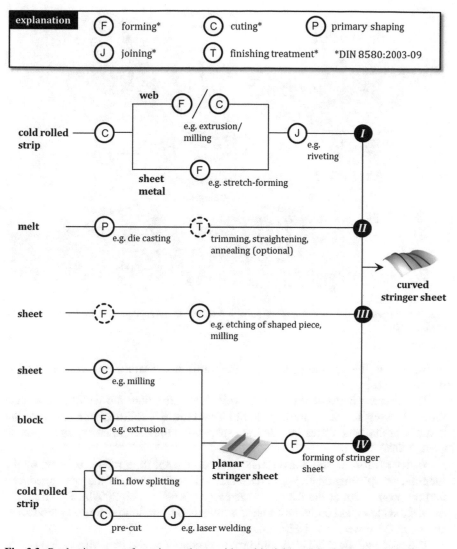

Fig. 3.3 Production routes for stringer sheets with multiaxial bending (Bäcker et al. 2015)

3.1.1   *Process Principles*

Linear flow splitting is a profile-rolling process. At room temperature, the flat sheet metal is fed through a tooling system consisting of an upper and lower supporting roll and lateral splitting rolls as shown in Fig. 3.4 (Groche et al. 2007b).

Feed motion of the sheet is provided by the supporting rolls, which also fix it in vertical direction. The obtuse-angled splitting rolls contain a forming radius R that

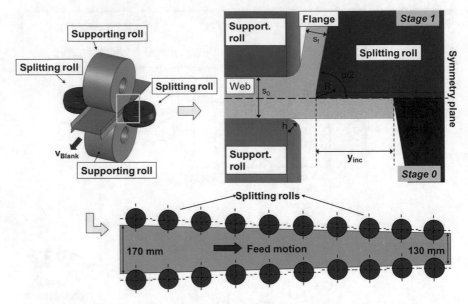

Fig. 3.4 Process principle and realization of a continuous multistage forming process (Groche et al. 2010b)

is located at the splitting angle α. The radii h determine the contour of the supporting rolls.

The distance between the splitting rolls is smaller than the width of the sheet metal. Driving the sheet through the tool system, the splitting rolls displace the material at the band edges into the free spaces and the flanges are created (Groche et al. 2007a).

With each further step the distance between the splitting rolls is reduced by the incremental splitting depth y_{inc} at each side. Thus, the width of the web is reduced and the span width of the flanges is increased (Groche et al. 2009b). By arranging multiple stands consecutively we have established a continuous forming process as shown in the lower part of Fig. 3.4.

The produced double-sided split profile is characterized by a bifurcation line at both band edges. The generated flanges exhibit a splitting angle that is determined by the splitting rolls (Groche et al. 2009a). The geometric characteristics of split profiles are presented in Fig. 3.5. The distance between the bifurcations is defined by the broadness of the web b_{web}. The total flange width $a_{z,tot}$ of the bifurcation is related to the total splitting depth y_{tot} and results in the length of the upper and lower flanges a_z. The total splitting depth y_{tot} is the sum of the incremental splitting depths y_{inc}. The initial thickness of the sheet is the thickness of the web s_0 and the reduced thickness of the flanges is defined as s_f.

The resulting geometry of flow split profiles is influenced by the setup of the roll gaps. A narrower gap between splitting and supporting rolls will result in thinner flanges. A shift of the splitting roll above the symmetry plane of the web will result

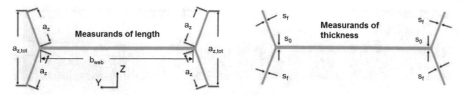

Fig. 3.5 Geometric characteristics of flow split profile (Vučić 2010)

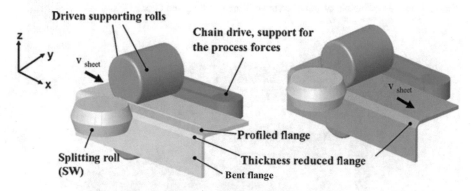

Fig. 3.6 Process principle of linear bend splitting (Groche et al. 2009a)

in thinner upper and thicker lower flanges and more material will flow into the lower flanges. A shift of a supporting roll in lateral direction, relatively to the other supporting roll, will result in asymmetric flanges.

While linear flow splitting generates the bifurcation at the edge of a sheet metal, linear bend splitting is capable to generate the bifurcation anywhere in the plane of the sheet. For that purpose the sheet is bent to a U- or an L-profile and the splitting operation is carried out at the bending line with the same tool system as applied for linear flow splitting (Ludwig et al. 2008b). In Fig. 3.6 we show that the process is capable of generating an additional flange at the upper side of the bending line or of reducing the thickness of the bent lower flank without generation of an additional flange.

Another differentiating factor of linear bend splitting compared to linear flow splitting is recognizable in the process principle. From the initial situation of a bent sheet three phases of the process are to discern in Fig. 3.7.

In phase 1, the gap between the splitting and the supporting rolls is reduced consecutively to thin the flank and fill the free space in the forming zone. The free space results from the contact relation between the forming radius of the splitting roll and the bending radius.

During phase 2, the gap between the rolls is kept constant and the width of the web is reduced to generate and form the additional flange. As of phase 3, the process principle is comparable with linear flow splitting. The gaps are kept constant and the width of the web is reduced consecutively to increase the length of the flanges.

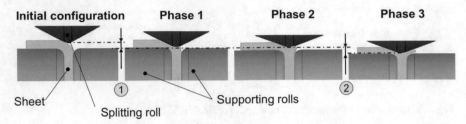

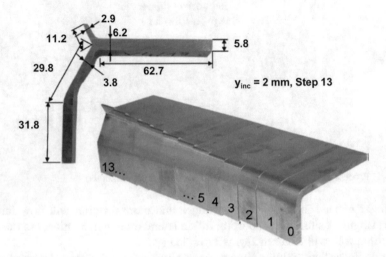

Fig. 3.7 Process principle of linear bend splitting (Ludwig and Groche 2009)

Fig. 3.8 Stage sequence and prototype of EN-AW 5754 (Groche et al. 2009a)

Figure 3.8 shows the final geometry of a linear bend split profile made of EN-AW 5754 after 13 stages as well as the stage sequence. The initial thickness s_0 was 6.0 mm. Stages 0–2 represent phase 1 of the process principle, stages 3 and 4 represent phase 2, and as of stage 5, phase 3 applies.

3.1.2 Technology

To design the tooling system of the process linear flow splitting, which is presented in Sect. 3.1.1, various system and process parameters have to be considered (Fig. 3.9). The geometry of the workpiece may vary, e.g., in the sheet metal thickness or in the sheet metal width. This influences directly the width of the supporting rolls or the distance between the supporting rolls and shows the demand of flexibility in the tooling system. High forming forces could also appear in the process which requires an accurate bearing and stiffness of the system. Additionally, the process should be adjustable, e.g., in the sheet metal velocity, which shows

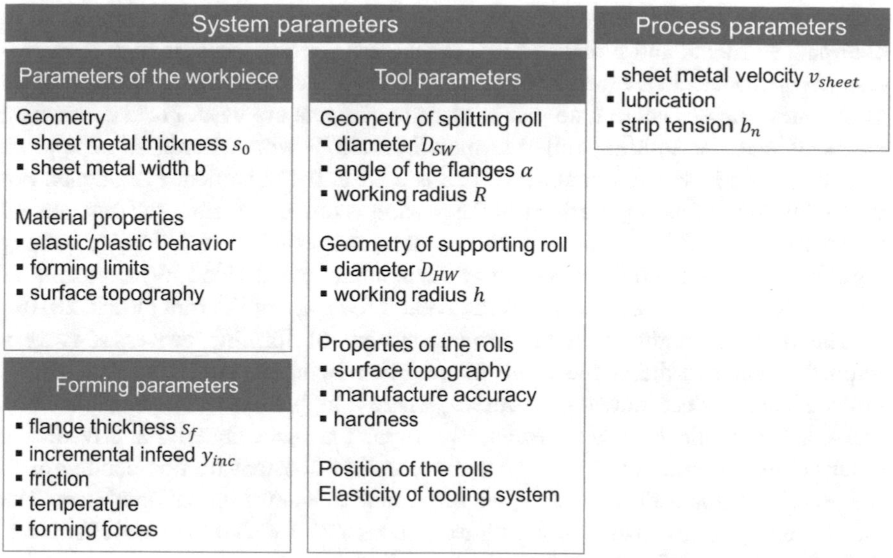

System parameters		Process parameters
Parameters of the workpiece	**Tool parameters**	• sheet metal velocity v_{sheet} • lubrication • strip tension b_n
Geometry • sheet metal thickness s_0 • sheet metal width b	Geometry of splitting roll • diameter D_{SW} • angle of the flanges α • working radius R	
Material properties • elastic/plastic behavior • forming limits • surface topography	Geometry of supporting roll • diameter D_{HW} • working radius h	
Forming parameters	Properties of the rolls • surface topography • manufacture accuracy • hardness	
• flange thickness s_f • incremental infeed y_{inc} • friction • temperature • forming forces	Position of the rolls Elasticity of tooling system	

Fig. 3.9 Parameters of linear flow splitting based on Groche et al. (2007b)

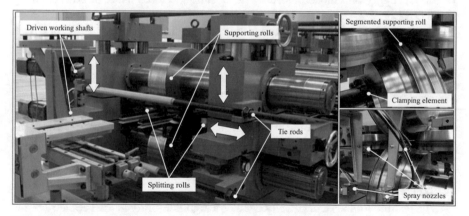

Fig. 3.10 Linear flow splitting stand

the need of a flexible process control. Against this background, we have designed a new tooling system for linear flow splitting, which meets its challenges.

To manufacture different flange lengths of the branched profile ten stands for the process of linear flow splitting were realized for scientific investigations. Each stand consists of a specific tool system of rolls (Fig. 3.10). The supporting rolls are mounted on an upper and a lower working shaft with a diameter of 150 mm, which is required due to the designed bearing. However, the splitting rolls are fixed in a cage between the upper and lower shaft, which affects the diameter of the supporting rolls. The supporting rolls have a diameter of 496 mm and are manufactured with hardened and unhardened segments, which are compressed by clamping elements and threaded rods. During the forming process high compressive stresses occur in the area of the supporting roll radius, due to the plastification of the sheet metal material, so that hardened segments are used. For the hardened segments and the splitting rolls the material 1.2379 with a hardness of approx. 60 HRC is used, whereas unalloyed steel is used in the unhardened segments. For the middle area of the supporting roll the loading is rather moderate and unhardened segments are used and thus result in a more cost-efficient and flexible tooling regarding different profile widths. The segmentation is divided in increments of 1 mm, which leads to a minimal incremental infeed y_{inc} of 0.5 mm (Vučić 2010).

The friction condition in the tooling system of forming processes is very important and can affect the result of the manufactured parts. This fact can be noticed, e.g., in deep-drawing processes, where a high friction between sheet and blank holder could lead to a decreased flow of the sheet material and results in cracks at the ground of the part. To achieve almost constant friction conditions in the process of linear flow splitting, we implemented a new lubrication system. The lubrication system consists of six block nozzles *CD-100 BSD (Raziol Zibulla & Sohn GmbH)* which lubricate the main contact areas of the splitting and supporting rolls in each splitting stand. The amount of lubricant can be adjusted by the pressure of the oil supply and fine adjusted by raster settings of the block nozzles (Ludwig et al. 2012).

To manufacture branched profiles with different sheet metal velocities each working shaft is driven with a power of 11.3 kW. This allows a process speed up to 100 m/min which sufficiently represents the process speed of conventional flow production lines of roll forming. The tooling system is controlled by the system *Siemens Simatic-S7*, which offers the opportunity to set the velocity of each shaft independently. Thereby, a strip tension b_n is possible which is normally used in roll-forming lines to straighten the profile. In the process of linear flow splitting, the strip tension b_n is used to compensate the inhomogeneous material flow of the material (Sect. 3.1.4). The forming forces of the splitting rolls can be measured by strain gauge bolts in the axis of the splitting rolls, which allow a measuring range of 0–300 kN (Vučić 2010).

The flexible tool system concept enables a high flexibility in the geometry of the branched profiles which can be manufactured with the same tool system (Fig. 3.11). The splitting rolls are adjustable in all three coordinate directions. The supporting rolls are flexible in the setting of the roll width up to 600 mm, due to the segmented roll design. The upper supporting roll is adjustable in vertical direction z, which leads to a flexibility in terms of the processed sheet metal thickness of up to 10 mm.

At the same time the high flexibility in the roll adjustment affects the tooling system rigidity, due to the fitting tolerances and large levers of the linear flow splitting stand assemblies. This leads to inaccuracies in the geometry of the branched profile when using these stands. To understand the elastic deformation behavior of the linear flow splitting stand, Vučić (2010) generated a numerical model of the splitting stand. In a sensitivity analysis numerous construction designs are compared by applying point loads to consistent structural points. The occurring

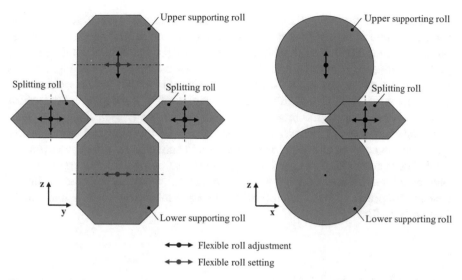

Fig. 3.11 Flexible roll adjustment and setting of a linear flow splitting stand based on Vučić (2010)

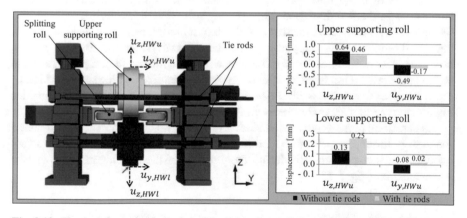

Fig. 3.12 Elastic deformation behavior of the linear flow splitting stand based on Vučić (2010)

elastic deformations in the splitting stand, caused by point loads of 107.2 kN on the supporting rolls and 150 kN on the splitting rolls, are shown in Fig. 3.12.

As a result of these point loads the displacements of the rolls are clearly visible. The upper supporting roll is displaced in horizontal direction $u_{y,\mathrm{HW}u}$ with −0.49 mm, whereas the position of the lower supporting roll is almost unchanged with −0.08 mm. In vertical direction $u_{z,\mathrm{HW}u}$ the same effect is visible. With the use of tie rods in the stand, the amount of displacements of the tool system is reduced significantly. The horizontal displacement $u_{y,\mathrm{HW}u}$ of the upper supporting roll is reduced to −0.17 mm and in vertical direction $u_{z,\mathrm{HW}u}$ to 0.46 mm.

The elastic deformation behavior of the splitting stand is essential for the profile tolerances, due to the fact that the form-shaping surfaces of the rolls are not planar parallel (Vučić 2010). The flexible tool concept of the linear flow splitting stands provides advantages regarding the system and process parameters of Fig. 3.9, but with limited tolerances of the branched profiles due to the elastic deformation behavior. Against this background a new tool system concept for linear flow splitting is designed in the project *319-12-19* of the *HA Hessen Agentur GmbH*. In the new tool concept the assemblies of supporting and splitting rolls are no longer connected to each other and ensure an easier adjustment. Additionally, a symmetrical deformation occurs under loading, due to the double-column stand design. Numerical results have shown that the deformation of the stands under loading is reduced to approx. 87%, which might lead to better tolerances of the manufactured branched profiles.

3.1.3   Process Characteristics

The process of linear flow splitting is characterized by a hydrostatic compressive stress state in the forming zone. The splitting rolls initiate compressive force into the sheet from transverse direction and displace material. Since the sheet is fixed in vertical direction by the supporting rolls and the material in the forming zone is bound in length direction by the sheet material, a hydrostatic compressive stress state is generated (Groche et al. 2009b). We could identify that the superposition of these compressive stresses increases the formability of the material and enables higher equivalent stains (Groche et al. 2007a).

As described in the fundamental studies of Jöckel (2005), the incremental flange increase is limited to process boundaries. An evaluation of the hydrostatic stress condition of the splitting rolls during the progress of the forming rolls shows two stress maxima (Fig. 3.13).

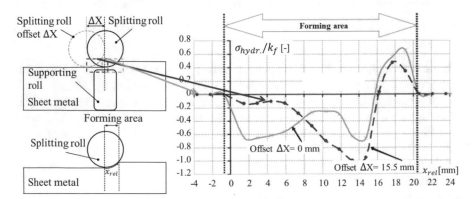

Fig. 3.13 Optimized stress condition by adjustment of the splitting roll offset ΔX (Vučić 2010)

The first maximum occurs near the contact area of the splitting roll with $x_{rel} = 15.5$ mm and the second with $x_{rel} = 0$ mm. In contrast, the highest rate of deformation is determined near the first touching point between the splitting roll and the sheet metal. Based on this context, the splitting roll is set to $\Delta X = 15.5$ mm in process direction relatively to the supporting rolls. With this offset, the first touching point of the splitting roll is aligned with the supporting roll axis and the hydrostatic stress in the forming zone is increased.

Figure 3.13 shows two minima for the offset $\Delta X = 0$ mm, which are optimized to one global minimum with the offset $\Delta X = 15.5$ mm. The stress at this point is higher than the previous minima, whereby further investigations show a decrease with higher offset values. Due to this fact, only one optimal offset is feasible, which depends on the incremental infeed and the diameter of the splitting roll. The adjustment of the splitting roll to this optimal offset value leads to higher possible flange lengths in the branched profile (Schmitt et al. 2011).

The flank angle of the tool system also has a significant influence on the stress distribution in the material and on the force level on the splitting (SW) and supporting rolls (HW) (Groche et al. 2007b). In Fig. 3.14 the influence of the flank angle on the force is shown from experimental and numerical results.

Hereby, a flank angle of 10° represents an angle of the bifurcation of 160°, while a splitting angle of 45° represents an angle of the bifurcation of 90°. A decreasing flank angle results in larger reaction forces on the rolls. The significantly largest reaction force $F_{y,sw}$ with a flank angle of 10° originates from the enlarged surface contact in z-direction (Groche et al. 2007b).

Regarding the diverse degrees of freedom in every stand of the multistage process and the parameters presented in Sect. 3.1.2, the process is conceivably sensitive to various parameters and subject to process errors.

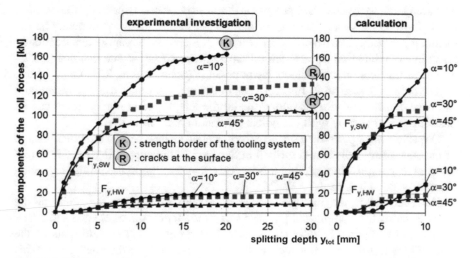

Fig. 3.14 Experimentally and numerically derived reaction forces (Groche et al. 2007b)

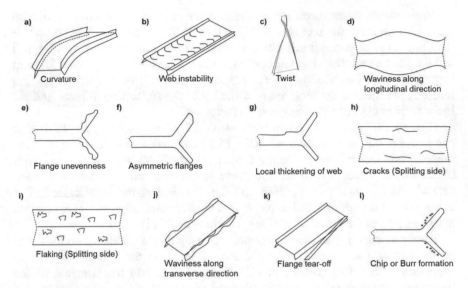

Fig. 3.15 Failure modes in linear flow splitting (Schmitt et al. 2011). (**a**) Curvature, (**b**) web instability, (**c**) Twist, (**d**) waviness along longitudinal direction, (**e**) flange unevenness, (**f**) asymmetric flanges, (**g**) local thickening of web, (**h**) cracks (splitting side), (**i**) flaking (splitting side), (**j**) waviness along transverse direction, (**k**) flange tear-off, (**l**) chip or burr information

Figure 3.15 presents failure modes of produced parts that can occur during the process of linear flow splitting. There are global effects like bending (a) or twisting (c) of the profile and those effects that affect either the web (b and g) of the profile or the generated flanges (d, e, f, h, i, j, k, and l) (Schmitt et al. 2011).

We have investigated the appearance of these defects within the research center, as well as effective methods to prevent them in production. Specific part defects can be traced back to mismatched process and tool parameters. Especially the position of the forming rolls and the setup of the roll gaps are sensitive parameters on the resulting geometry, like described in Sect. 3.1.1. Since varied parameters frequently result in a combination of different failure modes, a scientific and distinct correlation of parameters on part defects is not possible (Schmitt et al. 2011).

The defects described in Fig. 3.15 can occur in a stationary tooling system and can have constant occurrence along the profile length. Additionally, there can be unsteady process errors with periodical or random distribution.

One effect with periodical occurrence is related to the specific structural design of the supporting rolls. As described in Sect. 3.1.2, the supporting rolls feature a segmented composition of thin discs. This leads to high flexibility regarding changes of the width of the rolls (Vučić 2010).

However, due to the multiple contact zones between the tool segments and the bearing shafts, the uptight tool package suffers higher elastic resilience and a lack of accuracy compared to solid roller tools (Ludwig et al. 2013). Additionally, the perpendicularity of the roller tool package can deviate from the shaft plane. This results in a periodic transversal movement of the edge radii during rotation of the

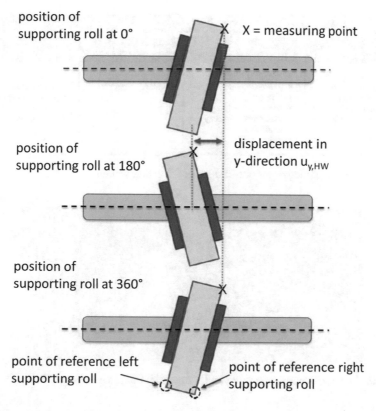

Fig. 3.16 Effect of tumbling motions of the supporting rolls due to setup of tools and bearing conditions (Vučić 2010)

rolls. Due to that effect, the upper and lower flange gap changes periodically along the profile length (Vučić 2010), which is shown in Fig. 3.16.

3.1.4 Process Qualification

With the fundamental studies of Jöckel (2005), Vučić (2010), and Ludwig (2016) the linear flow splitting process is analyzed in detail to understand the process behavior under certain conditions. The results show that the geometry and the material properties of branched profiles manufactured with the technology of linear flow splitting depend on numerous process and system parameters (Sect. 3.1.2). To transfer linear flow splitting from research to commercial manufacturing, a process qualification is necessary. In the following section we present the qualification of linear flow splitting concerning aspects of the profile and the process of Fig. 3.17.

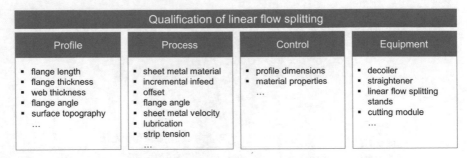

Fig. 3.17 Aspects of the process qualification of linear flow splitting

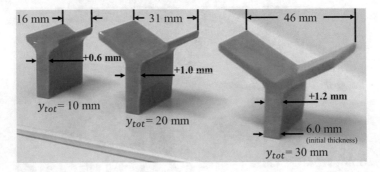

Fig. 3.18 Effect of thickening in split profiles from DD11 mild steel (Jöckel 2005)

Incremental Infeed: The geometry of the profile is mainly influenced by the geometry of the supporting and splitting rolls. On the contrary, the elasticity of the linear flow splitting stand has also a decisive influence. Near the supporting roll radius in the web of the branched profile a thickening of the sheet metal can be observed, which increases with a higher total incremental infeed y_{tot}. With a total incremental infeed of 10 mm and the use of the material *DD11* the thickness of the web is measured to 6.6 mm, which results in a thickening of 0.6 mm (Fig. 3.18).

With a total incremental infeed of 30 mm the thickening increases to 1.2 mm. Tests with the sheet metal material *AlMg3(w)* result in a lower thickening of the web. In comparison to the material *DD11*, the thickening of the web is determined to 0.25 mm with a total incremental infeed of 10 mm (Jöckel 2005).

Flange Angle: Besides the rigidity of the splitting stand (Sect. 3.1.2) and sheet metal material, the thickening effect of the web is also influenced by the roll geometry. With lower flange angles α of the splitting roll, the thickening of the web increases. Due to this thickening effect, the supporting rolls are positioned relatively to the sheet metal in an angle, which results in higher forces of the supporting rolls. This assumption is verified by Jöckel (2005) with the sheet metal material *DD11*:

Fig. 3.19 Effect of the offset ΔX to the incremental flange increase (Vučić 2010)

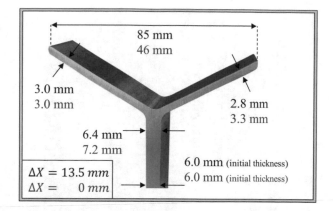

$\alpha = 10°$; $y_{\text{tot}} = 10$ mm; force $= 128.4$ kN; thickening $= 0.9$ mm
$\alpha = 30°$; $y_{\text{tot}} = 10$ mm; force $= 103.4$ kN; thickening $= 0.6$ mm
$\alpha = 45°$; $y_{\text{tot}} = 10$ mm; force $= 94.0$ kN; thickening $= 0.6$ mm

Offset: As described in the fundamental studies of Jöckel (2005), the incremental flange increase is limited to process boundaries. On the contrary Schmitt et al. (2011) have shown that the total flange length can be increased by the use of an optimal offset value ΔX (Sect. 3.1.3). The adjustment of the splitting roll to this optimal offset value leads to higher possible flange lengths in the branched profile (Fig. 3.19). This allows an increase of the flange length from 46 to 85 mm with the sheet metal material *DD11*.

Lubrication: Friction conditions could influence the result of forming processes in a decisive way. This fact is also seen in the linear flow splitting process. With the implemented lubrication system of Sect. 3.1.2 different lubricant conditions can be set, where the radii areas of supporting and splitting rolls are sprayed. For the process qualification, Ludwig et al. (2013) carried out long-term measurements of the profile geometry with different lubrication conditions. Figure 3.20 shows the achieved total flange length with ten splitting stands and a total incremental infeed of $y_{\text{tot}} = 20$ mm. The flange length is already changed by the use of low lubricant quantities. With the processed material *HC340LA* the total flange length a_z increases from approx. 20 to 25 mm. A reduction of the amount of lubrication from 0.034 to 0.162 g/min results only in minor changes of the flange lengths.

The change of the friction condition from solid friction to mixed friction leads to significant changes in the material flow and affects several profile tolerances at once. Considering the web of the profile this context is clearly visible. Figure 3.21 shows the thickness of the profile's web at seven measuring points, starting near the splitting area $s_{0,0}$ and ending at the middle of the profile $s_{0,\text{middle}}$. Without lubrication a major thickening of approx. 2.7 mm near the splitting area can be seen, while this thickening is reduced significantly to 2.5 mm by the use of lubricant. Through mathematical integration Ludwig et al. (2013) calculated and compared the decrease of the flanges to the percentage surface increase of the total

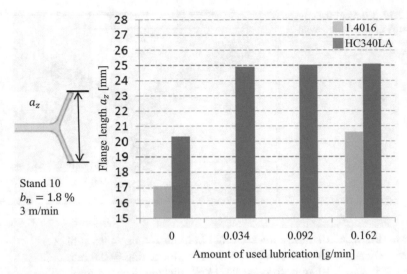

Fig. 3.20 Flange length a_z according to lubrication conditions (Ludwig et al. 2013)

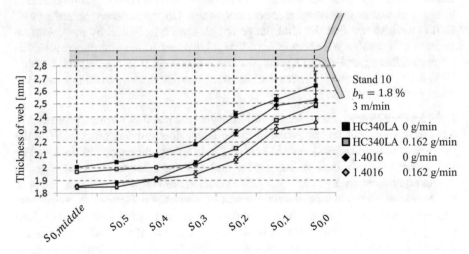

Fig. 3.21 Web thickness with different lubrication conditions (Ludwig et al. 2013)

cross section of the profile. The results show that approx. 90% of the flange increase results in the use of lubrication, where the remaining 10% of the material volume is linked to the change of longitudinal strains in the web.

Sheet Metal Velocity: Varying the sheet metal velocity also leads to changes in the geometry of the profile. Concerning this fact, Ludwig et al. (2013) analyzed the lubricant conditions 0.162 g/min and non-lubricant with different sheet metal velocities and sheet metal materials (Fig. 3.22). The measurement of the flange length a_z with the lubricant condition 0.162 g/min shows qualitatively constant

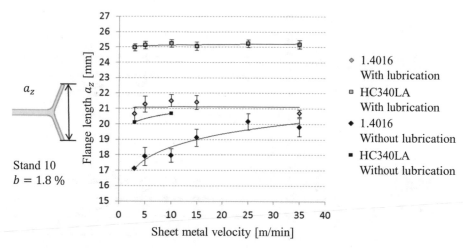

Fig. 3.22 Measured flange lengths a_z with different sheet metal velocities, lubrication conditions, and sheet metal materials (Ludwig et al. 2013)

values for the sheet metal material *HC340LA*. The influence of the sheet metal velocity with the lubricant condition 0.162 g/min to the flange length seems minimal. On the contrary, a significant influence of the sheet metal velocity on the flange length without lubrication can be determined. The flange length with the sheet metal velocity of 3 m/min and the material *1.4016* is measured to approx. 17 mm, which increases with the sheet metal velocity of 35 m/min significantly to approx. 20 mm.

Strip Tension: During the linear flow splitting process the plastified material flows not only within the cross section of the profile, but also in longitudinal direction. This fact underlines the requirement for the described flexible tool system concept of Sect. 3.1.2 clearly. To compensate this elongation in longitudinal direction a higher circumferential speed of the subsequent splitting stands is necessary. With the flexible tool concept of the linear flow splitting stand each shaft is driven and can be set independently. The percentage strip tension b of each stand n with the circumferential speed of the supporting rolls v_u is given by Eq. 3.1:

$$b_n = \frac{v_{u_n} - v_{u_{n-1}}}{v_{u_{n-1}}} * 100\%$$ (3.1)

Ludwig et al. (2013) varied the strip tension with the sheet metal material 1.4016 and showed that higher values of the strip tension lead to higher longitudinal strains in the profile's web (Fig. 3.23). With ten splitting stands and a strip tension of approx. 2% the strain ε_x of approx. 20% is measured. By the variation of the strip tension, the values of the flange length a_z and the thickness of the web s can be changed and offer the possibility for a quick change of the profile geometry.

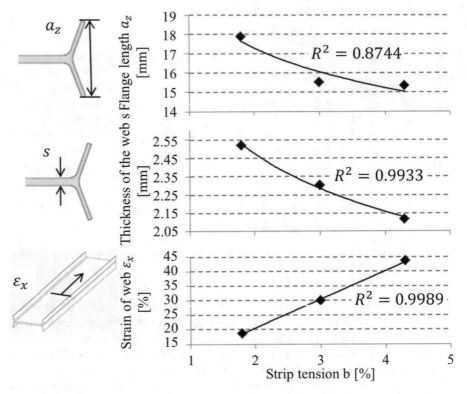

Fig. 3.23 Effect of strip tension to flange length, thickness of web, and strain of web with the material 1.4016 and the sheet metal velocity of 5 m/min (Ludwig et al. 2013)

Surface topography: The results in Ludwig et al. (2013) show that varying lubrication conditions and sheet metal velocities have no significant influence on the metallurgical properties in the flange of the profile. On the other hand, these process parameters affect the geometry of the profile and influence the appearance of cracks on the surface of the profile. This effect can be seen in Fig. 3.24, where the use of lubrication with minor sheet metal velocity of 35 m/min leads to significant cracks on the surface of the splitting area. Ludwig et al. (2013) suggest that higher values of the friction force lead to a higher compressive stress condition in the linear flow splitting process, which influences the forming capacity of the sheet metal material in a decisive way. The evaluation of a numerical model with different friction conditions confirms this hypothesis.

On the contrary, the experiments of Ludwig et al. (2013) with lubrication and a high sheet metal velocity of 35 m/min result in surface topographies without any cracks. Higher sheet metal velocities lead to lower friction coefficients, which is proven by Weber (1973) in experiments of a roll process. Besides this aspect, higher temperatures and strain rates have also to be considered when processing with high deformation rates. Both parameters influence the forming capacity of the material. Ludwig et al. (2013) conclude that higher sheet metal velocities lead to decreased

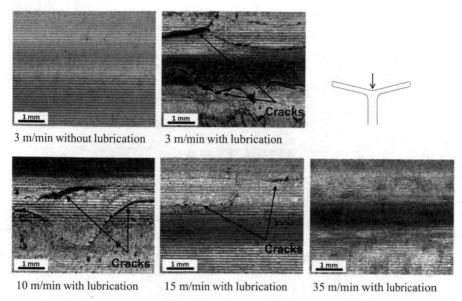

3 m/min without lubrication   3 m/min with lubrication

10 m/min with lubrication   15 m/min with lubrication   35 m/min with lubrication

Fig. 3.24 Microscopical pictures of the splitting center (Ludwig et al. 2013)

friction conditions and an increased temperature in the linear flow-splitting process, which result in surfaces without cracks and branched profiles with increased flange lengths.

Besides the process parameters of linear flow splitting, the surfaces of the used splitting rolls also have an influence on the surface topography of the manufactured profiles. This fact is shown in Fig. 3.25, where the surface texture of the branched profile is directly linked to the finishing operation of the manufactured rolls. The finishing process turning shows significant grooves on the surface of the roll, which depend on the feed rate and depth of the machining operation. Through the high compressive stresses in the forming area of the linear flow-splitting process, these grooves are coined into the surface of the split profile. Results in Monnerjahn and Fricke (2012) show that a turned surface of the splitting roll with an averaged roughness of $R_z = 8.22$ μm leads to an averaged surface roughness of $R_z = 8.38$ μm of the split profile (Fig. 3.25). A significant smoother surface of the roll achieved by a grinding or deep rolling operation leads to an equal surface roughness of $R_z = 2.91$ μm.

In the applications of branched profiles adjustable surface textures could be necessary to fit the requirements. Lubricant pockets in linear guides could be required due to the rolling contact between the rolls and the splitting center of the profile. Designed grooves in the splitting center of the branched profile, which are coined through the grooves of the splitting roll, could provide these requirements. This leads to a high flexibility in the surface topography of the manufactured profiles.

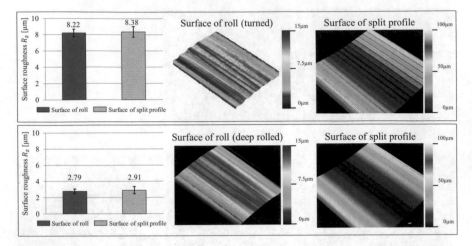

Fig. 3.25 Effect of the surface texture of the splitting roll on split profiles (Monnerjahn and Fricke 2012)

Fig. 3.26 Product example for the application of multiple production processes (Schmidt and Abele 2014)

3.2 Integrated Process Chains for Sheet Metal Structures

Figure 3.26 presents a multi-chamber profile that was manufactured by linear flow splitting, roll forming, milling, welding, and cutting in a continuous production line. For the realization of the production of a multifunctional finished part like the multi-chamber profile, a synchronized and fully integrated nonstop production line is required. The forming processes linear flow splitting and linear bend splitting are realized in such a flexible production line to produce diverse profile geometries with additional geometrical features (Fig. 3.27). Different manufacturing processes can be combined in arbitrary order in a continuous production line.

The novel production processes presented in Sect. 3.1.3 result in interdependency with the total production-orientated process chain. We have developed and investigated these interdependencies and the individual processes and process sequences within the research center. The processes and results are presented in Sect. 3.2.

Fig. 3.27 Flexible manufacturing facility of the CRC 666 (Groche et al. 2008)

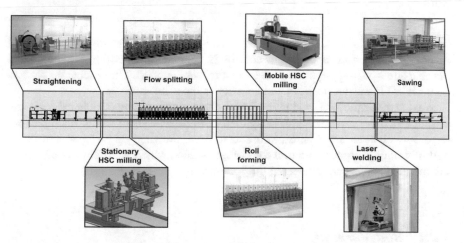

Fig. 3.28 Consecutive processing operations in the production line (Abele et al. 2012a)

The featured processes for integrated manufacturing are presented in exemplary order in Fig. 3.28. All the processes are designed for continuous and nonstop operation of the sheet metal and can be summarized as follows:

- Uncoiling and straightening of the sheet metal
- HSC milling for band-edge processing
- Linear flow/bend splitting
- Roll forming
- HSC milling for discrete geometric features
- Laser welding for closing the profile
- Cutting to length

For the combination and integration of multiple production processes, the processes have to be able to operate at nonstop sheet metal feed motion and to adapt to the sheet velocity. A common process control can be implemented to

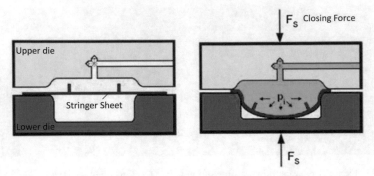

Fig. 3.29 Sheet metal hydroforming of a stringer sheet (section few of a tool) (Groche et al. 2010a)

comply with this task. For uncoiling, straightening, linear flow splitting, roll forming, and cutting a common control is applied. For milling, welding, and flexible flow splitting separate process controls are applied. These separate process controls use the sheet position and velocity as input information.

Next to the combined production line for the manufacturing of finished parts and assemblies, other processes can be applied on bifurcated sheet metal. Examples we have investigated are hydroforming and deep drawing of stringer sheets. A hydroforming process and a tool for stringer sheets, which were developed within the research center, are presented in Fig. 3.29.

3.2.1 Roll Forming

The process technology of roll forming is already used in the early of the twentieth century where sheet metal products from press brakes and other types of forming were replaced. In the last decades of the twentieth century, roll forming of automotive products became the fastest growing segment of the industry and is still growing strongly. In roll forming, a sheet metal strip is formed along bending lines with multiple pairs of contoured rolls without changing the thickness of the material at room temperature (Halmos 2006).

The geometries of the roll-formed profiles can be opened or closed, while the contours of the rolls change depending on the used bending stages and geometry of the profile (Lange 1990). Figure 3.30 shows the stands for a roll-formed frame structure.

Roll forming is a continuous production process and completes excellently the linear flow-splitting process to increase the variety of branched profiles. With the flexible tool concept of the continuous flow production line the process of roll forming can be integrated easily into the line. The branched profile of the linear flow-splitting process is then used as semifinished part, which can be roll formed to

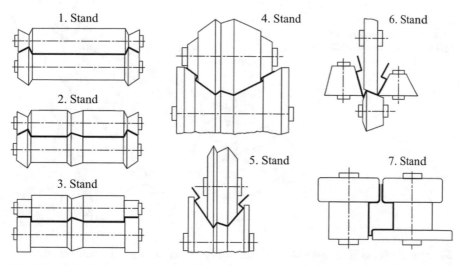

Fig. 3.30 Process sequence of roll-formed profile (Oehler 1963)

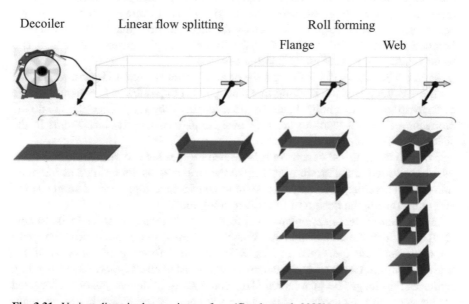

Fig. 3.31 Variant diversity by varying preform (Groche et al. 2009b)

different profile geometries (Fig. 3.31). The tool system of linear flow splitting is unchanged, whereas the configuration of the roll-forming process leads to different profile geometries. Groche et al. (2009b) showed that in contrast to the conventional process design, the variant diversity results from the preform instead of changing the complete roll-forming tool.

Fig. 3.32 Edge travel of a strip edge in the process of roll forming (Halmos 2006)

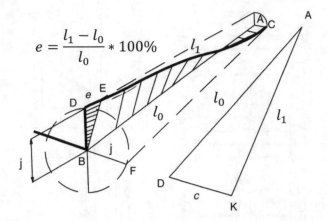

$$e = \frac{l_1 - l_0}{l_0} * 100\%$$

Flower Pattern: The sequence of the profile cross section of each roll-forming stand is called flower pattern and is very important for the manufacturing of profile geometries. In the process of roll forming the bending takes place between consecutive arranged rotating roll sets. The length of the forming path between two consecutive forming steps varies across the cross section, due to the continuous bending operation. This variation of the forming length causes longitudinal and shear strains in the sheet metal (Görtan et al. 2009).

Figure 3.32 shows the forming path of a strip formed into a U-geometry of the profile. The length of the bending line from point A to point B is l_0, where the edge of the strip travels in a helical pattern to the height of the leg j. This results into the elongation e, which leads to the previously explained longitudinal strains in the sheet metal (Halmos 2006).

Possible geometrical defects in roll-formed profiles are shown in Zettler (2007), where the longitudinal strain occurring at the strip edge is determined as the main influence. Therefore, it is very important to consider the longitudinal strains at the strip edge during the design of the roll-forming tools.

In contrast to flat sheet metals, roll forming of branched profiles leads to new challenges. For the roll design for branched profiles two main factors are very important and are presented in Fig. 3.33. The roll forming of the web of the branched profile can lead to an undercut and an adequate support of the bending radius can no longer be guaranteed. This could result in deviations of the targeted profile geometry. Nevertheless, to ensure a support of these bending areas, mandrels can be used. In Groche et al. (2009b) the critical bending angle α_{crit} is given in Eq. 3.2:

$$\tan(\alpha_{crit}) = \frac{\sin(\alpha_{crit})}{\cos(\alpha_{crit})} = \frac{f}{s} \tag{3.2}$$

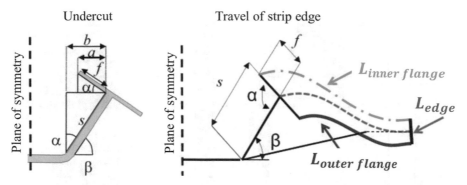

Fig. 3.33 New challenges in roll forming of branched profiles (Groche et al. 2009b)

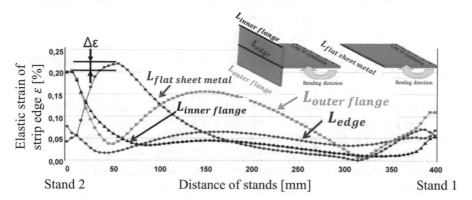

Fig. 3.34 Strains at the band edge (Groche et al. 2009b)

The second challenge in roll forming of branched profiles leads to the forming length of the band edge. In Fig. 3.34 the elastic strain ε of the band edge between two roll-forming stands is shown. In conventional roll forming of a flat sheet metal the forming length is located near the roll-forming stand (Bhattacharyya et al. 1984; Kiuchi et al. 1995), which can be seen in the elastic strain of $L_{\text{flat sheet metal}}$. In contrast, the elastic strain of the branched band edges $L_{\text{outer flange}}$ and $L_{\text{inner flange}}$ is distributed over the complete distance of the stands. This fact is explained by the higher stiffness of the profile (Groche et al. 2009b), due to the branched structure. Therefore, the forming length of the branched profile is higher and leads to lower elastic strains $\Delta\varepsilon$ at the band edge. For a targeted bending angle the number of roll-forming stands might be reduced, which could result in lower tool costs.

Flange forming: Roll forming of the strain-hardened flange material of split profiles also leads to new challenges, due to the limited formability. Bending tests in Niehuesbernd et al. (2016) show that higher bending angles β of the flange material lead to a significant higher surface roughness. Figure 3.35 shows this fact, where a bending angle of 90° results in a surface roughness R_z of approx. 60 μm. The cracks

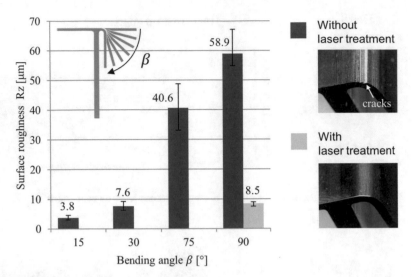

Fig. 3.35 Surface roughness of bent flanges with different bending angles β and heat treatments (Niehuesbernd et al. 2016)

on the flange surface are clearly visible and do not comply with the required surface quality of roll-formed profiles. To increase the formability of the flange material, several heat treatment approaches are available. The heat treatment by laser has a high potential, due to the unchanged material outside the laser-annealed zone. The positive effect of the strain-hardened material still remains and is presented in Sect. 4.1.2.

Roll-formed split profiles: The combination of the linear flow-splitting process with roll forming opens up new possibilities in the variety of branched profiles, which are exemplarily shown in Fig. 3.36. In the three-chambered profile (a) we continuously roll formed the web of the split profile with 16 stands to the closed profile geometry. Here, the chambers offer the possibility to guide different fluids such as gas or liquids for innovative solutions. On the contrary, we roll formed the design of the one-chambered profile (b) with only nine roll-forming stands. One of many possible geometries with roll-formed flanges is presented in (c), where the strain-hardened flanges are formed manually. The flanges of branched profiles do not only offer possibilities in the structure strength. In this example the flanges are used as connecting areas for other profile geometries, which is explained in detail in Sect. 9.2.2.

3.2.2 Cutting Technologies

Introduction: The manufacturing of complex sheet metal profiles within a continuous flow production line has high demands on the steel sheet band and the machining operations. Close tolerances of the semifinished product are necessary.

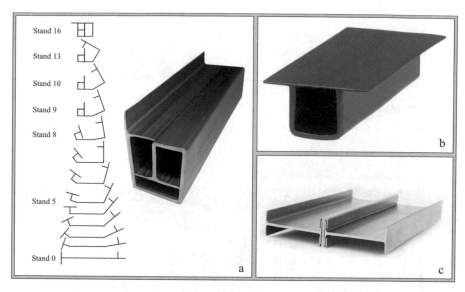

Fig. 3.36 Manufactured profiles with a combination of the linear flow splitting and roll-forming process: (**a**) Flower pattern of the three-chambered profile (Groche et al. 2009b), (**b**) one-chambered profile (Taplick et al. 2011), (**c**) profile connection of split profiles (Schäfer et al. 2013)

The sheet metal coil often shows deviations in band width and undefined edge geometries (Fig. 3.37). Both characteristics have major influence on the linear flow-splitting process. To ensure the demanded tolerances it is necessary to machine the band edges before the first bending stages.

Due to material inhomogeneity or variation in sheet metal thickness deviations in flange length of the split profiles occur. As the production of multichambered profiles needs a welding process, it is necessary to machine those flange edges to ensure a linear flange with close tolerances. The edge geometry has major influence on the welding process as well, so that the geometry should ideally be fitted to the desired welding situation.

For the discontinuous flow-splitting process it is necessary to prepare the sheet metal geometry (Fig. 3.42 and Sect. 7.1.2). There are several conceivable machining processes to machine the band or flange edge. Compared to laser cutting, water jet cutting, nibbling, or stamping the milling process is comparatively simple to integrate and provides a high flexibility. Additionally the required safety provisions are relatively low compared to processes like laser cutting. A study showed a negligible influence of the edge treatment on the result of the linear flow splitting (Neuwirth et al. 2015). Nevertheless, it is only possible to machine a defined band or flange edge like seen in Fig. 3.37 by using the milling process. The dimensional accuracy of the milling process compared to the other machining processes is comparatively high. Thus the edge preparation by linear milling process is already successfully used in many different applications, e.g., the band-edge preparation for welding of pipelines (Abele et al. 2011b).

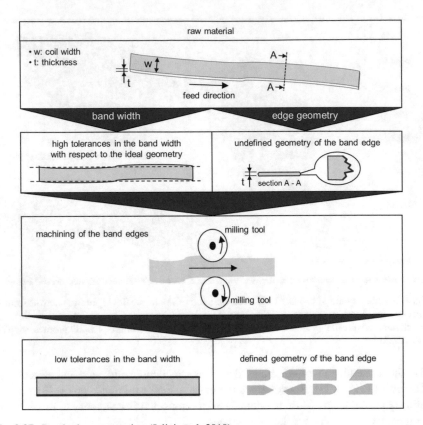

Fig. 3.37 Band-edge preparation (Jalizi et al. 2010)

Complex profiles often should contain multifunctional elements, e.g., for down-stream joining processes (Sect. 8.2.2), running surfaces, chamber connections, or profile merging. Similar to the machining of band and flange edge, the machining of those multifunctional elements could be done by various machining processes. Unlike, e.g., stamping the milling process can be used to machine continuous and discontinuous form elements (Fig. 3.38).

A huge advantage of the milling process is the ability to manufacture grooves, slots, and chamfers. Furthermore the machined milling features can be changed quickly and cost effectively without additional die costs. Discontinuous in-line milling operations in continuous flow production of metal profiles are not state of the art. In case of a designated milling operation, it is done after the cutting process by separate machine tools. The integration of discontinuous milling operations into the continuous flow production enables the retrenchment of those downstream operations. The necessary material handling, production time, and costs can be saved. As a consequence, the costs of the metal profiles can be reduced (Schmidt and Abele 2014).

Fig. 3.38 Features only possible by in-line discontinuous milling operations (Abele et al. 2008a)

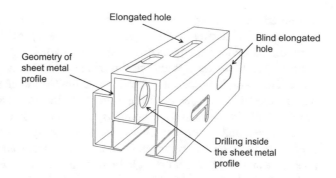

The profile velocity in typical continuous flow production lines is up to 100 m/min. Accordingly, the milling process has to be very fast. Typically the cutting speed in conventional cutting processes is about 100–150 m/min in normal strength steel. Depending on feed per tooth and amount of cutting edges milling feeds up to 4 m/min can be reached (tool diameter 12 mm, 3 cutting edges, feed per tooth 0.1 mm). That is sufficient for band or flange-edge milling processes, but for the intended discontinuous milling operation it is insufficient, because the machine tool has to perform tool paths relatively to the profile feed direction. Therefore the high-speed cutting (HSC) technology can be used. It provides cutting speeds about 5–10 times higher than conventional cutting speeds. With that, the needed HSC feed rates can be achieved. Furthermore, in contrast to conventional processing technologies, HSC offers a high potential for an increase in productivity and enhanced part quality.

The preliminary work of Schulz (1996) has shown the positive effect of the cutting speed increase on the material removal rate, the surface quality, and the cutting force reduction whereby cutting speed ranges for HSC processes are defined by the used workpiece material. In the case of the linear flow split profiles out of steel a minimum cutting speed of 750 m/min is necessary to realize an HSC process. Nevertheless, the cutting speed increase leads to a large amount of heat dissipation in the workpiece. As a result, the tool life is limited by a rapidly growing tool wear (Schulz 1996; Tschätsch and Dietrich 2011). Consequently, the challenges in HSC edge and sheet metal milling processes are on the one hand the increase of the material removal rate and the surface quality and on the other hand the reduction of cutting forces and heat dissipation.

Within our continuous flow production the aim of our research was to develop and build machine tools for continuous as well as discontinuous milling operations in continuous flow production lines and to equip them with specially designed HSC milling tools for highest reachable part quality and tool life.

Tool development for edge milling purposes: The use of the HSC technology in the edge milling process enables, due to the so-called HSC effect, the cost- and quality-optimized manufacturing of semifinished parts. In contrast to competing technologies, e.g., laser cutting or side trimming, the HSC edge milling process is characterized by a wide variety of band-edge geometries and a high-dimensional and geometrical accuracy. The edge milling process provides a high rationalization potential concerning the productivity increase by using the HSC technology due to the high flexibility via dynamic tool movements and almost any band-edge geometries. However, the economic efficiency of the HSC edge milling process depends on the tool life and the resulting burr after the machining process. Thus, the influence of the cutting edge geometry and the technological parameters, e.g., cutting speed and feed per tooth, on the productivity and cost-effectiveness of the milling process has to be investigated. A significant challenge is to define the process window which reflects a balanced compromise between tool life increase and burr reduction (Stein 2010).

According to this, cutting tests are accomplished to find the general influencing factors on cutting forces, surface quality, burr formation, and heat development in edge milling. For the experimental tests a conventional CNC machining center (type *Heller MCP-H250*) is used in order to exclude disturbing influences from the upstream and downstream machining units in the production line. As the gained knowledge in HSC edge milling has to be transferred to the milling process in the continuous flow production line, the high feed rates of the coil material, the vibration problem of thin sheet metals, as well as the workpiece material have to be considered within the basic investigations.

Therefore, the steel band material *HC 340 LA* is used. The workpiece has a length of 260 mm and a thickness of 2 mm (Stein and Abele 2007). By using a special fixture, connected with a three-component dynamometer from *Kistler* (type *9255A*), the sheet metal is mounted on the clamping device with four screws; see Fig. 3.39a. The recorded measuring signals are analyzed with a *National Instruments PCI 6133* measuring board and a PC unit using sampling rates between 3 and 5 kHz. Three different indexable inserts, named A, B, and C, are used in different face milling cutters to vary the rotational speed by using cutters with different diameters; see Fig. 3.39b. As a machining strategy down-milling is selected due to its positive effect on surface quality compared to up-milling (Stein and Abele 2007; Abele et al. 2007a). Due to the expected significant influence on cutting forces and surface quality, the cutting speed, the feed per tooth, and the radial depth of cut are varied systematically; see Table 3.1.

The cutting force measurement results confirmed that the decrease in cutting forces with increasing cutting speed up to 3.750 m/min can be reduced to the well-known HSC effect which is also valid for the sheet metal edge milling process (Stein and Abele 2007; Abele et al. 2007b). The decrease in cutting forces during the HSC can be traced to the small amount of forming due to the higher cutting temperatures (Schulz 1996). The surface roughness investigations have shown that the surface quality of the band edges can be enhanced by increasing the cutting speed from 1.000 to 3.750 m/min and decreasing the feed per tooth up to 0.3 mm. Burr height measurements revealed the correlation between the feed rate, the feed per tooth, the depth of cut, and the burr formation. The positive influencing factors on the burr formation are the feed rate and the feed per tooth. A burr width reduction can be achieved by increasing the feed rate and the feed per tooth. In contrast to this,

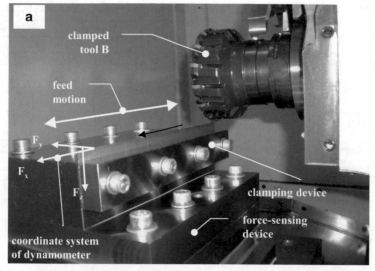

Fig. 3.39 (a) Experimental setup, (b) parameters of the used cutting tools (Abele et al. 2007b)

an increase in depth of cut leads to an increased burr formation converging to a maximum value (Stein and Abele 2007; Abele et al. 2007b). In conclusion, the increase in cutting speed results in lower cutting forces and higher surface quality whereas the burr formation can be reduced by higher feeds per tooth.

Developing a new tool concept for a high-productivity and surface-optimized HSC edge milling process requires a detailed analysis of the layout of the tool body and the inserts and its effect on the machining results. Based on the investigations in tool body design it can be concluded that the tool life and the burr formation can be

Table 3.1 Investigated technological parameters for band-edge milling (Abele et al. 2007b)

	Test series 1			Test series 2		Test series 3
Tool	A	B	C	A	B	C
Tool body diameter [mm]	200.08	132.97	125.10	200.07	132.97	132.97
Cutting speed [m/min]	1257–3750	1200–2500	1179–2350	2000		2500
Rotational speed [min^{-1}]	2000–5966	2873–5985	3000–5980	3182	4788	5985
Feed per tooth [mm]	0.3–0.10	0.17–0.08	0.3–0.15	0.18	0.10	0.08–0.40
Feed rate [mm/min]	7200			7200		7200–35,910
Radial depth of cut [mm]	1.0			0.5–5.0		1.0
Clamping height [mm]	10			10		10

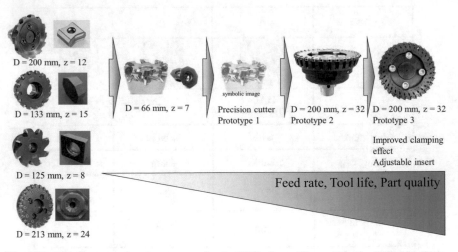

D = 200 mm, z = 12

D = 133 mm, z = 15

D = 125 mm, z = 8

D = 213 mm, z = 24

D = 66 mm, z = 7

symbolic image
Precision cutter
Prototype 1

D = 200 mm, z = 32
Prototype 2

D = 200 mm, z = 32
Prototype 3

Improved clamping effect
Adjustable insert

Feed rate, Tool life, Part quality

Fig. 3.40 Development of tool prototypes for the HSC edge milling within the CRC 666 (Stein and Abele 2008)

influenced by the tool body diameter, the design of the insert clamping system, the number of insert seats, and the residual unbalance. Considering these influencing factors we developed a new tool system for the HSC edge milling process in a continuous flow production. Characteristic features of this tool system are the greater tool body diameter, the improved insert clamping system, the higher number of insert seats, and the smaller residual unbalance of 4 gmm; see Fig. 3.40 (Stein and Abele 2008).

However, the development of a suitable tool body geometry is not enough to realize a burr- and tool wear-optimized edge milling process. Another important influencing factor on the surface quality, the burr height, and the tool life is the insert geometry and the coating. As the tool coating determines the tool life, five different carbide cutting materials are investigated under variation of the technological parameters; see Table 3.2.

The experiments are run on the CNC machining center of the type *DMG 75V linear*. The round inserts have a diameter of 12 mm and a thickness of $s = 3.97$ mm. The clearance angle measures 15°, the wedge angle 75°, and the rake angle $\gamma = 0°$. Due to the major influence of the cutting speed v_c on the tool wear, the cutting speed is varied in three stages 1.000 m/min, 2.000 m/min, and 3.000 m/min.

Furthermore, the feed per tooth f_z is also varied among 0.15 mm, 0.5 mm, and 1.0 mm as the feed per tooth affects the surface quality and the burr formation. In peripheral milling the sheet thickness restricts the axial depth of cut a_p which is 2 mm in this case. To hinder ploughing effects, which occur at very small radial depths of cut, a radial depth of cut a_e of 1 mm is selected (Abele et al. 2008b; Abele et al. 2008c).

The experimental results have shown that the cutting distance of all carbide cutting materials decreases with increasing cutting speed whereas the highest tool life is achieved at a cutting speed of $v_c = 1.000$ m/min and a feed per tooth of $f_z = 1$ mm. The increase in feed per tooth leads to a shift from an abrasive wear towards crater wear which causes a shorter tool life. In contrast to this, by increasing the cutting speed the abort criterion is reached by a tool breakage due to the large amount of heat dissipation. Nevertheless, the carbide cutting materials B and D are identified as the most suitable cutting materials for the HSC edge milling process due to its enhanced tool lives. The analysis of the burr height of the milled band edges exhibits the positive effect of an increase in feed per tooth on the burr formation. Consequently, higher feeds per tooth shall be chosen to reduce the burr height (Stein and Abele 2008; Abele et al. 2008b; Abele et al. 2008c). The results confirm the assumption that a further increase in cutting speed accelerates the formation of tool wear whereas a higher feed per tooth has a positive effect on the burr formation.

But there are still optimization potentials regarding the surface quality and tool life enhancement by modification of the cutting edge geometry of the inserts with the simultaneous variation of the technological parameters. By varying the cutting tool geometry elements such as protective chamfer angle and rake angle the influence of each cutting geometry element on the burr formation and the tool life can be determined. According to this, one geometry element is varied while the others are kept constant, as shown in Table 3.3.

The experimental results prove that the burr formation can be minimized up to 50% by choosing a constant feed per tooth and modification of the cutting geometry. The increase of the feed per tooth leads to a further reduction of the burr height of less than 10%. As the geometric elements such as protective chamfer angle, cutting edge radius, rake angle, and clearance angle affect the burr formation, the burr height is measured by varying the technological parameters.

Table 3.2 Investigated carbide cutting materials in HSC edge milling (Abele et al. 2008b)

Tool	A	B	C	D	E
Cutting material	Co, WC, TaC	Co, WC	Co, WC, TiC, TaNbC	Co, Carbide, WC	Not specified
Coating	TiAlN	TiAlN	TiAlN	TiN, Al2O3 Ti (C, N)	TiAlSiN
	PVD	PVD	PVD	CVD	CVD
Coating hardness [HV0.05]	3600	3300	3600	1420 HV30	3300

Table 3.3 Variation of the cutting tool geometry elements (Abele et al. 2012a)

Cutting geometry element	A	B	C	D
Protective chamfer angle [°]	10	15	20	15
Cutting edge radius [mm]	0.01	0.03	0.05	0.03
Clearance angle [°]	11	20	26	26
Rake angle [°]	0	5	17	0

The burr formation investigation of different protective chamfer angles reveals a proportional correlation between the burr height and the protective chamfer angle. An increase in protective chamfer angle causes a higher burr formation due to stronger squeeze and crushing effects. The minimum burr height is obtained at a protective chamfer angle of $10°$ and a higher feed per tooth of $f_z = 1.5$ mm. However, there is no significant influence of the cutting speed on the burr height.

Similar to the protective chamfer angle the burr formation becomes larger and irregular by increasing the cutting edge radius due to the displacement of the material in front of the cutting edge rounding. Thus, the risk of ploughing effect increases with increasing cutting edge radius. To hinder such effects the choice of a smaller cutting edge radius is to be preferred while the feed per tooth shall be increased. Also in this case the cutting speed variation does not have any significant effect on the burr formation.

Another factor influencing the chip and burr formation in edge milling is the rake angle. A flow chip formation is observed at a positive rake angle up to a certain feed per tooth whereas by using a neutral rake angle shear chips occur at smaller feeds per tooth. Due to the continuous flow of the material and the better material separation process during the flow chip formation a poor burr formation process results in a positive rake angle. Nevertheless, based on the results it is not possible to draw clear conclusions about the influence of the rake angle on the burr height.

The clearance angle has the greatest influence on the resulting burr height as a variation of the clearance angle leads to a burr height variation of more than 60% for the same technological parameter combination. Detailed investigations and simulations on the temperature, pressure, friction changes, and contact conditions in the quaternary shear zone are necessary to explain the great differences in the measurement results. However, the minimum burr height can be reached by using a clearance angle of $20°$ whereby a clearance angle of $26°$ is also suitable due to the smaller contact surface (Abele et al. 2012a; Turan and Abele 2012; Stein 2010). To sum it up, a significant burr reduction can be reached by a targeted combination of cutting geometry elements whereby these results can be transferred to any other steel band-machining process. The reason for this is that the technological correlations in the case of variation of geometry element values are the same.

To develop a tool life and burr-optimized cutting tool geometry based on the experimental results of the burr height measurements, the investigation of burr-optimized geometry combinations concerning the tool life is essential. For that reason, different geometry combinations are tested with regard to the tool life and the results are compared with the results of a reference geometry.

The tool with the reference geometry has a protective chamfer angle of 10°, a protective chamfer length of 0.20 mm, a cutting edge radius of 0.03 mm, a clearance angle of 26°, and a rake angle of 0°. Geometry combinations 1–4 differ from the reference geometry with regard to the rake angle which measures 5°. Geometry combination 1 and 3 has a sharper cutting edge due to the reduced cutting edge radius of 0.01 mm. As a clearance angle of 20° leads to a smaller burr height, a clearance angle of 20° was chosen for geometry combination 3 and 4 while geometry combination 1 and 2 has the same clearance angle as the reference geometry. The other geometric elements are kept constant.

The comparison of geometry combination 1 and 2 with the reference geometry shows that burr minimization can only be realized by modification of the cutting tool geometry but there is no significant enhancement in tool life. The lowest tool life is observed for geometry combination 3 because the abort criterion is reached earlier due to tool breakage. For the verification tests geometry combination 4 with a tool life twice as long as the reference geometry is used although the burr formation cannot be minimized compared to geometry combinations 1–3; see Fig. 3.41a.

The verification tests with geometry combination 4 show that the burr formation is reduced to 61% and the cutting distance is enhanced by a factor of 2.5 from 365 to 976 m; see Fig. 3.41b. As a result, a burr formation reduction and tool life enhancement can be achieved by using a positive instead of a neutral rake angle because the mechanical loads on the cutting edge are much smaller (Abele et al. 2012a; Turan and Abele 2012; Stein 2010).

Conclusion: The research results in HSC edge milling processes serve as the basis for designing new tool concepts for the use in continuous flow production lines. The gained findings concerning the effect of cutting tool geometry and technological parameter variation on the cutting forces, surface quality, and tool life correspond to the literature. However, the developed tool system is designed for the edge milling of thin bands out of steel. This means that the tool system and the technological parameters have possibly to be fitted if, for example, thicker sheet metal bands with different material properties are machined.

Machine tool development for band-edge milling purposes: During the machining in continuous flow production the profile feed must not be stopped. Therefore a conventional clamping of the sheet metal profile is not possible. So part of the main requirements on a machine tool for band- and flange-edge machining are to guide the workpiece in the machining area with sufficient retention forces because of the cutting forces and position preservation.

Figure 3.42 shows examples of possible edge-machining operations. To perform a linear band- or flange-edge milling process, the milling tools remain in position, and the profile guidance does not need to support varying structures.

To perform nonlinear band- or flange-edge machining operations in continuous flow production, it is necessary to know the actual workpiece velocity. It can be captured by contacting or contact measuring methods. With the knowledge of the actual sheet metal profile velocity, the axes infeed of the machine tool can be

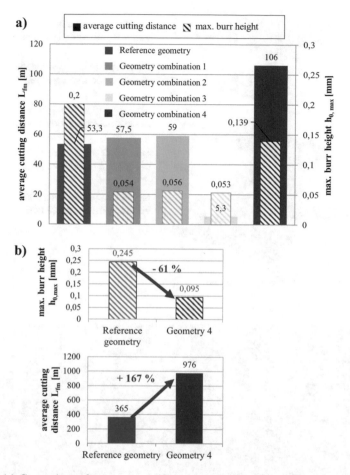

Fig. 3.41 (**a**) Comparison of average tool life and burr height of the reference geometry and cutting geometry combinations ($z = 2$, $v_c = 1.000$ m/min, $f_z = 1$ mm), (**b**) burr reduction and tool life increase by modification of the tool geometry ($z = 32$, $v_c = 1.000$ m/min, $f_z = 1$ mm) (Stein 2010)

calculated. Because of the high velocity of the profile, the machine tool axes have to provide high dynamics.

To ensure a high-bandwidth tolerance independently from tool calibration or tool wear an additional process control can be set up (Sect. 3.2.5).

Within our continuous flow production line we developed and built a special machine tool (Fig. 3.43). It consists of a stationary machine bed and two spindle stocks and can be placed in any section of the production line to provide its intended flexibility.

Depending on the designated machining operation, the motor spindles can be mounted either in a horizontal or in a vertical way. Figure 3.43 shows the machine tool with vertically mounted spindles for machining of horizontal sheet metal edges

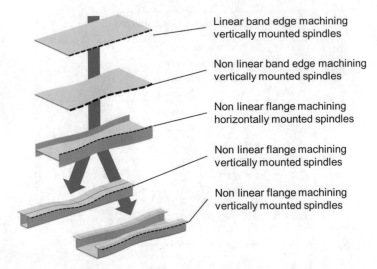

Fig. 3.42 Possible machining operations

Fig. 3.43 (**a**) Machine tool for band-edge and flange preparation, (**b**) first-generation profile guidance (Ludwig et al. 2008a)

or flanges. For the machining of vertically aligned edges or flanges the spindles are mounted horizontally. Additionally, it's possible to swivel both spindles around the Y-machine axis in the range of ±30°.

Both headstocks provide a traveling range of 300 mm in the Y-machine axis and 50 mm in the Z-machine axis. All axes are independently NC controlled. The reachable feed in all axes is 60 m/min with an acceleration of 5 m/s^2 to provide the needed dynamics to perform milling operations within a continuous flow production. The high-frequency spindles have 26 kW output at a rotation speed of 30.000 min^{-1}, so HSC cutting speed could easily be reached.

The profile velocity is captured by a friction wheel with an attached rotary encoder and processed as an additional machine tool axis within the machine control.

Tool development for sheet milling purposes: The sheet metal milling is an indispensable part of the manufacturing of multifunctional modules from steel profiles due to the possibility of integration of different notches and holes into the parts. The realization of a tool life-optimized and high-productivity sheet metal milling process requires the development of a suitable coated end-milling cutter which enables the milling at high feed rates of minimum $v_f = 10$ m/min. Thus, experimental investigations are carried out to determine the influence of the tool geometry on the tool stability, the surface quality, and the tool wear.

The combination of a cutting edge rounding and a neutral hollow grinding ensures a high surface quality and an enhanced tool stability. A smaller helix angle leads to less abrasive wear due to smaller friction between the rake face of the tool and the workpiece. The evaluation of the different tool geometries is made by analyzing its effect on the draw-in, the surface quality, the straightness, and the parallelism of slot holes and grooves. Based on these investigations, an end-milling cutter with three cutting edges and a diameter of 12 mm is developed for the HSC sheet metal milling (Stein et al. 2010).

As the tool coating has a significant influence on the tool life different coatings A (*AlTiN*), B (*AlCrN/Si₃N₄*), C (*TiAlCrN*), D (*AlCrN+TiAlN*), and E (*TiAlN*) are investigated in further tests with the aim of enhancement of the tool life. Figure 3.44 shows the milled workpiece and also the experimental setup (Stein et al. 2010).

Although all the investigated tools have the same cutting geometry the tool wear results differ from each other significantly. Nevertheless, the tool life decreases with increasing the cutting speed for every tested coating. Coating C yields the best tool life results at a cutting speed of 700 m/min and 1.000 m/min. Coating D achieves the longest cutting distance at a cutting speed of 1.300 m/min; see Fig. 3.45a. The analysis of the tool wear mechanisms has shown that with the increase in cutting speed the flank wear is shifted towards the cutting edge corner. As a consequence, a higher corner wear occurs which leads to a premature achievement of the abort criterion or in extreme cases to cutting tool failure. In this context, the coatings A and E exhibit the highest corner wear regardless of the cutting speed due to their lower hot hardness. An increased flank and rake-face wear is also

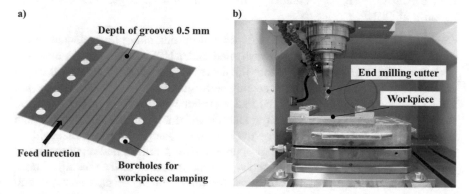

Fig. 3.44 (**a**) End geometry of the milled workpiece, (**b**) experimental setup (Stein et al. 2010)

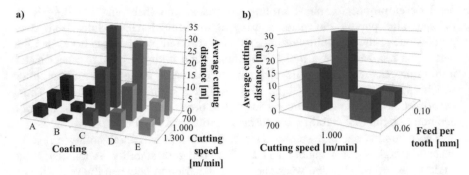

Fig. 3.45 (**a**) Cutting distances of different coatings under variation of the cutting speed, (**b**) the influence of the increase in feed per tooth on the cutting distances for coating C (Stein et al. 2010)

observed already at a cutting speed of $v_c = 700$ m/min. In contrast to this, the tool wear investigation of coating D reveals adhesive wear and built-up edges which increases with growing cutting speed (Stein et al. 2010).

Due to the longest tool life only coating C is tested at higher feed rates concerning the tool life and productivity enhancement. The test results demonstrate different effects of an increased feed per tooth on the cutting distance under variation of the cutting speed. At a cutting speed of 700 m/min the reached cutting distance rises about more than 70% up to 59.6 m whereas the increase of the cutting speed to 1.000 m/min results in a cutting distance decrease of about 40% to 12.2 m. This effect can be reduced to the rapidly increasing cutting temperatures during incremental enhancement of cutting speed. The technological parameter combination $v_c = 1.000$ m/min and $f_z = 0.10$ mm induces, compared to the smaller feed per tooth of $f_z = 0.06$ mm, a significant decrease in tool life, caused by enhanced thermal and dynamic loads; see Fig. 3.45b. However, the tool wear behavior under variation of the feed per tooth shows no significant differences with regard to flank, rake-face, and cutting edge corner wear. The dominant wear mechanisms are material adhesion and the formation of built-up edges. The increment of feed per tooth leads to a smaller formation of built-up edges and notch wear (Stein et al. 2010).

Conclusion: With the use of the developed end-milling cutter the influence of the tool coating on the tool life was investigated under HSC conditions. As expected with increasing cutting speed the tool life decreases significantly regardless of the coating. It could be shown that the coating performance, the toughness, and the hot hardness are decisive factors for the tool wear behavior. A high coating hardness does not show a positive effect on the tool life while a tougher coating with a high hot hardness results in higher tool lives. In conclusion, the tool life could be increased from 6 m at a feed rate of approximately 3 m/min and a cutting speed of 700 m/min to 34 m. It has been shown that the influence of the coating is much bigger than the cutting geometry. Thus, the development of new coatings will increase the productivity in HSC sheet metal milling.

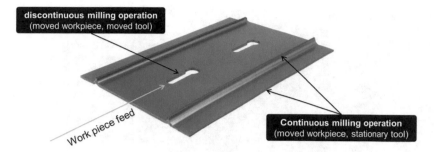

Fig. 3.46 Examples for continuous and discontinuous milling operations

Machine tool development for discontinuous in-line HSC milling operations:
Similar to the band- and flange-edge machining, the profile feed must not be
stopped during the machining of discontinuous milling features. In contrast relative
movements of the tool relative to the profile feed have to be performed in two
directions. Therefore a respective machine tool has to provide very high dynamics.

Figure 3.46 shows an example for continuous and discontinuous milling operations on a moved workpiece. There are three main challenges for this machine
tool. At first, machining of a continuously moving workpiece is only possible by the
synchronization of the machine tool movements with the workpiece feed. The NC
program operations have to be superposed by the workpiece feed. Otherwise the
dimensional accuracy of the milling features cannot be ensured. The profile velocity can be captured by contacting or contact measuring methods. Second, the profile
has to be guided in the machining area. That guidance has to offer a high stiffness
because of the milling forces. Additionally the relative movement of the machine
tool to the profile is a particular challenge for the guidance. Because of the high
profile velocity the machine tool will need a large traveling range in profile feed
direction. So the third challenge is to support the profile across the machine bed to
avoid gravity-based deflections.

To ensure the dimensional accuracy of the milling features, its absolute and
relative position on the sheet metal surface, and the surface quality of the machined
area a process control can be set up (Sect. 3.2.5). To meet the mentioned demands,
we developed and built a special gantry milling machine collaboration with *Wissner
Gesellschaft für Maschinenbau mbH* (Fig. 3.47). This machine tool offers a tra-
versing range of 7 m in directory of the workpiece feed which is up to 100 m/min.
To perform milling operations on a moved workpiece, the machine tool needs to
provide high-dynamic machine axes. The gantry reaches an acceleration of 17 m/s^2
and a velocity of 120 m/min in workpiece feed direction, which enables the gantry
to run ahead the workpiece (Abele et al. 2012a).

To guide the workpiece in the machining area a specially built material guiding
system (Fig. 3.48b) was developed (Abele et al. 2012b). It is connected to the
machine tool gantry and locks all degrees of freedom of the workpiece except the
one in its feed direction. As a further result a separate feed drive is not needed.

gantry milling machine tool	• **Travelling range X / Y / Z:** 7000 / 1500 / 400 mm
	• **Drive units:** X/Y: linear direct drives, Z: ball screw
	• **Machine control:** IndraDrive (Bosch Rexroth)
	• **Feed rate:** v_{max}: 120 m/min
work piece feed	• **Acceleration:** a_{max}: 17 m/s²
	• **Jerk:** r_{max}: 150 m/s³
	• **Spindle speed:** n_{max}: 42.000 min^{-1}

Fig. 3.47 Machine tool for HSC sheet milling processes

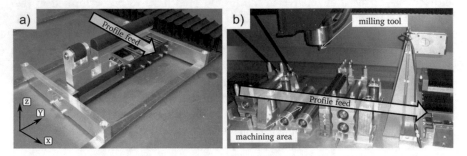

Fig. 3.48 (**a**) Workpiece rest, (**b**) workpiece guiding system (Schmidt and Abele 2014)

To avoid the gravity-caused deflection of the profile in the area of the machine base, various additional workpiece rests were developed; refer to Fig. 3.48a. They are mounted on the machine bed located below the moving workpiece. Because of the restricted amount of space on the machine bed, the workpiece rests have to turn down and up again every time they are ran over by the guidance system.

The machine tool runs on a *"Bosch Rexroth IndraMotion MTX"* CNC control unit, which provides the special feature *"system axes coupling."* This function uses a master-slave concept to synchronize the movement of two independent machine tool axes. The profile velocity is captured by a rotary encoder and processed as the actual position of an additional axis in the CNC control unit. The generated workpiece axis is selected as the master axis, and the machine tool *x*-axis as the slave axis. The system axis coupling can be executed in velocity compensation coupling mode or position correction coupling mode. In order to generate equidistant milling features, it is necessary to choose the position correction mode. Otherwise the velocity compensation mode is sufficient. Both coupling modes have successfully been implemented (Schmidt and Abele 2014; Abele et al. 2011a; Abele et al. 2014).

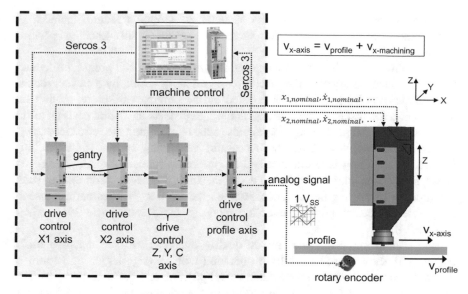

Fig. 3.49 Synchronization concept (Abele et al. 2012a)

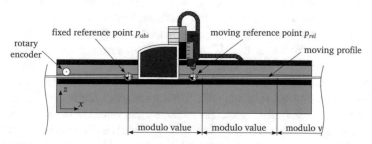

Fig. 3.50 Reference points and modulo values in coupled machining mode (Schmidt and Abele 2014)

Figure 3.49 shows the synchronization concept. Between the individual machining operations, the coupling has to be deactivated in order to not exceed the machine tool axis ranges. Because of the digital data storage, the actual position value of the workpiece axis is restricted. Hence the axis is selected as a "modulo axis." Its modulo value represents the maximum positioning area of this axis (Fig. 3.50). Exceeding that maximum axis position causes an automatic zeroing of the axis actual position (Schmidt and Abele 2014; Abele et al. 2014).

Figure 3.50 shows the side view of the gantry machine tool and the metal profile workpiece on it. The workpiece feed direction is equivalent to the machine tools x-axis. The rotary encoder is located on the left. Due to the selection as a modulo axis the workpiece axis is divided into equidistant segments. The machine gantry (slave) and the workpiece axis (master) are coupled in position correction coupling mode. Therefore the command value of the machine tool x-axis is superposed by the actual position of the workpiece axis. The system axis coupling feature includes an interpolator, which increases the slave axis velocity till reaching position synchronicity.

The activation of the system axis coupling replaces the NC program reference point p_{abs}, which is equal to the machine tool reference point, by the synchronization reference point p_{rel}, which is located on the moving workpiece. Within the modulo area the distance between p_{abs} and p_{rel} represents the actual position of the workpiece axis. All positioning commands of the NC program are referenced to p_{rel} during active coupling. After the successful machining of the desired milling feature, the coupling is terminated and p_{abs} becomes reference again. The gantry is set back to its initial position and the machining process starts again. Thus, the modulo value sets the distance between the milling features and represents the reference for the position coupling mode (Schmidt and Abele 2014; Abele et al. 2014).

The contouring error is a major quality characteristic for machine tools. Figure 3.51 shows a square milling operation with an edge length of 30 mm in the machine tool x–y layer, the related actual position and the contouring error of the x-axis plus the actual position value of the workpiece. After reaching the modulo value of 200 mm its actual position is zeroed, which initializes the coupling of the machine gantry and workpiece axes. With the successful synchronization the NC program starts a movement in the machine tool y-axis. Simultaneously the machine tool x-axis is moved synchronously to the workpiece axis. After reaching the destination in the y-axis the NC program starts a movement of 30 mm in the machine tool positive x-axis. This movement and the workpiece movement are superposed. The machine gantry runs at a higher speed than the workpiece. During the following movement in the machine tool y-axis, the gantry and the workpiece are moving synchronously again. After further movements in the machine tool negative x-axis and positive y-axis, the milling process ends by reaching the initial point p_{rel}. The axis coupling is terminated and the machine gantry is set back to its initial point p_{abs}. A comparison of the contouring error during coupled and uncoupled milling operation shows no negative influence of the axis coupling on the contouring error and with that on the machining accuracy. The strong rising after canceling the axis coupling results out of the rapid feed's acceleration. The high contouring error has no influence on the machining quality because there is no contact between milling tool and workpiece anymore at this time (Schmidt and Abele 2014).

Figure 3.52 shows a workpiece machined in coupled mode. The profile velocity was 5 m/min. With that the discontinuous milling operation during continuous workpiece feed was proved successfully. Nevertheless the milling features exhibit deviations between actual and reference geometry.

The research work shows no influence of the axis coupling on the machine tool contouring error (Schmidt and Abele 2014). The influence of the milling process forces on the contouring error in coupled and uncoupled machining operation has to be examined. In addition to the machining process forces, there are further perturbations affecting the milling process (Fig. 3.53).

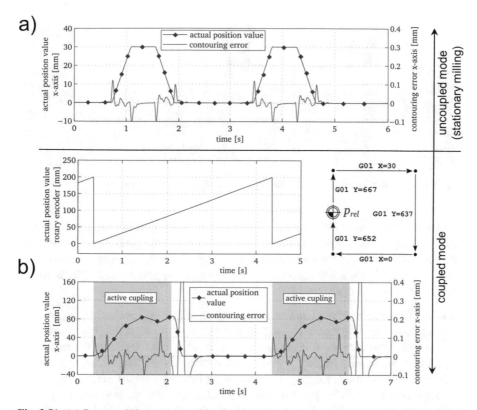

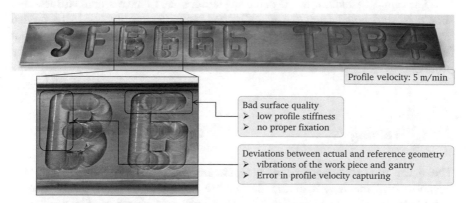

Fig. 3.51 (**a**) Square milling process with related contouring error in uncoupled and (**b**) coupled machining mode (Schmidt and Abele 2014)

Fig. 3.52 In coupled mode machined workpiece

At times during the test machining operations, vibrations of the machine tool gantry occur in its feed direction. Investigations showed vibrations which arise out of upstream and downstream processes together with the vibration excitation out of the milling process. These vibrations are captured by the rotary encoder, so not only

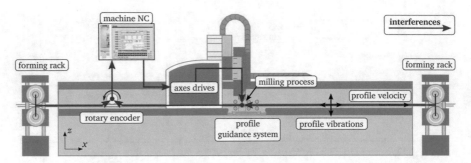

Fig. 3.53 Perturbations affecting the milling process (Schmidt and Abele 2014)

the workpiece motion but also the vibrations are superposed to the gantry motion and with that to the material guiding system. As a result the guiding system induces the vibrations back to the workpiece again. The whole system gets unstable. The vibration magnitude is only limited by the gantry drive performance (Schmidt and Abele 2014; Abele et al. 2014).

Advices for future work: The results lead to the following conclusion. In general a discontinuous milling operation within a continuous flow production is possible. Without deviations of the milling process and a sufficient fixation of the profile the achievable machining quality is equal to conventional milling operations. As the profile fixation in the machining area has major influence on the machining quality it should be a moved clamping system. A guiding system consisting out of guiding coils is not able to provide the necessary fixation support. To avoid the capturing of disturbances the profile velocity capturing system should be as imprecise as possible. The support structure of the capturing system has to have a high stiffness and should be decoupled from the machine bed. Still the profile has to be guided properly in the measuring area. The use of self-adjusting workpiece rests is not reasonable. Every time they get in upright position there is a burst excitation on the profile.

3.2.3 Welding

The linear flow-splitting process manufactures branched profiles in which the thickness and the length of the flanges are adjustable (Jöckel 2005). Further processing by roll forming (Sect. 3.2.1) allows new closed profile geometries which need an optional welding operation in some applications. Here, the number of welding areas depends on the number of chambers and flanges which are presented in Fig. 3.54.

Closing these complex profile geometries brings new challenges such as limited accessibility of the welding point or minimal influence to the material properties of

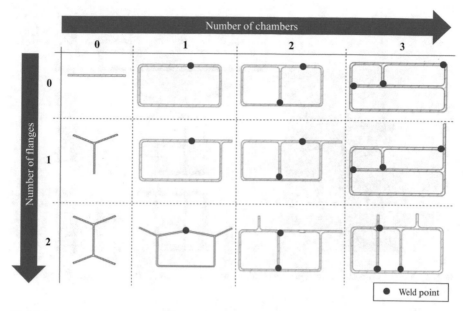

Fig. 3.54 Possible cross sections of profiles with different number of flanges and chambers (Groche and Vucic 2006)

the profile. The laser technology offers many advantages compared to conventional joining techniques and is practically used in all sectors of economic relevance, e.g., cutting, drilling, welding, ablation, hardening, brazing, or polishing (Poprawe 2005). The use of laser radiation offers advantages in terms of producing any form of geometry, manufacturing of hard materials, or minimal tool wear, but it is also very complex regarding process parameters (Eichler and Eichler 2003). Additionally, the material processing is also influenced by dynamic processes, which are described in Poprawe (2005) in detail.

In the flexible concept of the continuous flow production line we implemented a laser machining center for welding and cutting operations. The system utilizes a *YLS-300-S2T* laser as a beam source and has a power of 3000 W. An open cabin integrates the machining center into the manufacturing line and offers in-line welding or heat treatments of the branched profiles. The welding optic has a focal distance of 300 mm, so that even parts with low accessibility of the welding area can be processed. As presented in Sect. 3.2.1, different variants of profile geometries are manufactured within the flexible process chain. Figure 3.55 shows a manufactured three-chambered profile geometry with different seam types. Groche et al. (2009b) analyzed the fundamental requirements for good welding results and determined the weld preparation and welding parameters as key factors. For laser welding, zero gap is necessary which requires a perpendicular and straight band edge. Through the previous milling operation of the band edge (Sect. 3.2.2), these requirements are met. A specific roll-clamping system prevents invalid part

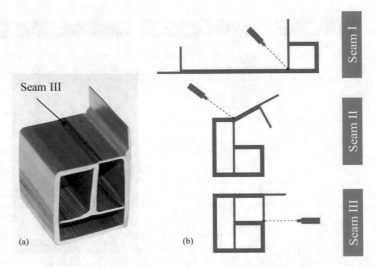

Fig. 3.55 (a) Three-chambered profile (Ludwig and Groche 2009), (b) seam types (Groche et al. 2009b)

positions during the welding process. To generate good welding results, the selection of the process parameters plays an important role. A sensitivity analysis is realized regarding welding speed and intensity on the parts' surface. With the use of nitrogen as inert gas and the determined process parameters, seam III (b) is successfully welded in the continuous manufacturing line (Groche et al. 2009b).

3.2.4 Deep Drawing

Introduction: In order to generate spatially curved bifurcated sheet metal parts, the development and implementation of a new process chain are necessary. Thus, this chapter shows the design and properties of the new process chain. The technical functions are explained showing the tool setup and the numerical simulations. Manufacturing induced properties, process characteristics, and technical risks as well as possible technical upgrades and process enhancements are demonstrated.

Process idea and motivation: The common way of stiffening spatial sheet metal parts is beads and braces (Wiedemann 2007). Stringers provide an alternative to this classical way. Stringers are elements arranged perpendicular to the ground sheet. They lead to a higher geometrical moment of inertia of the cross section. Possible applications for stringers can be found in air planes, buildings, or car body parts. Compared to beads, the stiffening effect of stringers grows with increasing stringer height. The reason for this behavior is the fact that the geometrical moment of inertia increases cubically with rising the stringer's height (Bäcker 2015). At the moment, the available methods to manufacture spatially curved stringer sheets, like milling (Denkena and Schmidt 2007), casting (Luo 2013), or etching (Brimm 1988), are often inefficient or have big losses of material. So there is a need for a new process chain to produce these parts efficiently.

The new approach for a manufacturing route to produce spatially curved stringer sheets generates bifurcated flat sheet metal parts before forming them into the required 3D form. The two investigated alternatives of bifurcating flat sheet metal are, on the one hand, linear bend splitting (integral way), which is explained in Sect. 3.1. On the other hand, joining stringer and flat sheet metal by laser welding provides a differential way of creating bifurcations. In both cases a complex 3D joining process and the material losses of a milling process can be avoided. Another advantage is a higher output compared to a pressure die-casting process and less energy consumption compared to an extrusion process.

The best chance to form flat stringer sheets in a 3D contour is hydroforming which provides high geometrical flexibility and homogeneous stress as well as strain distributions in the whole drawing part. We will demonstrate the feasibility of forming stringer sheets with solid tools and thus higher output, too.

Technology and tools: In the following section, the tools and the process characteristics are explained. First, the differential technology for producing bifurcated sheet metals is presented by describing the forming tools and the technological properties.

The production of the semifinished parts for the stringer sheet-forming process by laser welding is more flexible than by split bending. Hence this differential way is more suitable for experimental investigations. The process principle of this joining process is illustrated in Fig. 3.56. The stringers and the sheet metal are clamped together and we join them by a keyhole weld seam produced by a laser beam. The result of this process is a flat stringer sheet with visible welding seams on the opposite side of the sheet metal.

The following 3D forming process is shown in Fig. 3.57. Here, the die contains the negative of the aimed final part shape. The blank holder closes the die cavity, while it is filled with a pressure liquid (oil or water) that forms the stringer sheet into the die. The geometry of the stringer and its height are limited by the contour and the depth of the die cavity.

Figure 3.58 shows examples for geometries which we formed by hydroforming. The sheet thicknesses of the demonstrators vary from 1 to 2 mm.

In addition to the classical deep-drawing process failures, several new process errors can occur in a hydroforming process of stringer sheets (Fig. 3.60) (Ertugrul 2010). Due to the notch effect of the stringers, bursting of the sheet metal next to the stringer ends is one failure mode (Fig. 3.60a). Different blank holder forces in the deep-drawing process lead to alternating strain peaks at the end of the stringer. As can be seen in Fig. 3.59, higher blank holder forces lead to bigger strain peaks and thus to a high risk of rupture (Ertugrul 2010). Tensile and burst tests with different shapes of the stringer end show that the influence of the geometry of the stringer end on the notch effect is negligible (Bäcker 2015).

Fig. 3.56 Laser welding process for generating stinger sheets (Ertugrul 2010)

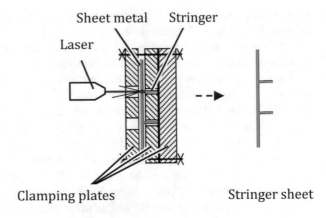

Fig. 3.57 Stringer sheet hydroforming (Groche and Bäcker 2013)

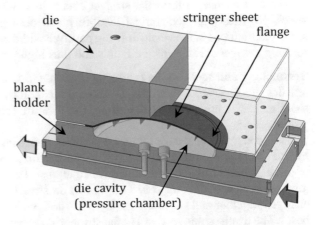

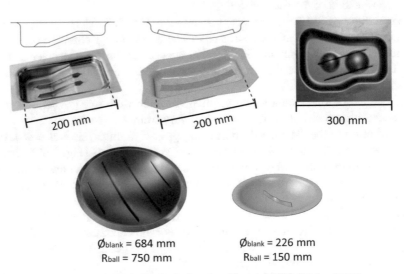

Fig. 3.58 Stringer sheets produced by hydroforming (Ertugrul 2010; Bäcker 2015)

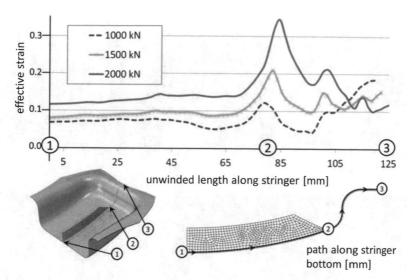

Fig. 3.59 Strain along stringer for different blank holder forces (Ertugrul 2010)

Other occurring process errors are rupture (Fig. 3.60i) or turn-down of the stringer (Figs. 3.60d, e), geometrical inaccuracies (Fig. 3.60b), and failure of the welding seam in case of a laser-welded stringer (Figs. 3.60g, h). This welding seam is also visible on the backside of the stringer sheet. In order to expand the application potential for stringer sheets, it can be polished off before the forming process. A useful manufacturing-induced property of split-bended semifinished parts is the fact that no welding seam is visible on the backside.

In addition, we can conduct numerical simulations to estimate process limits or to understand and predict failure. Depending on the investigation aim, the simulations need to be carried out with a dynamic implicit or explicit solver. For the 3D stress investigations a hexagonal element type is recommended while the buckling problem of the stringer should be observed in a submodel with shell elements. Python files can be used as input files to guarantee a consistent model setup for parameter variations and the communication with the mathematical optimization algorithms. The use of symmetry boundary conditions has to be well considered so that they do not influence the buckling behavior of the stringers.

High negative strains in the stringer can lead to buckles (Figs. 3.60c, f). The waviness w of the stringer edge is calculated as a criterion to detect buckles. With the length of the unbuckled stringer l and the length of the stringer edge after the forming process L, the waviness is defined as

$$w = \frac{L - l}{l} \tag{3.3}$$

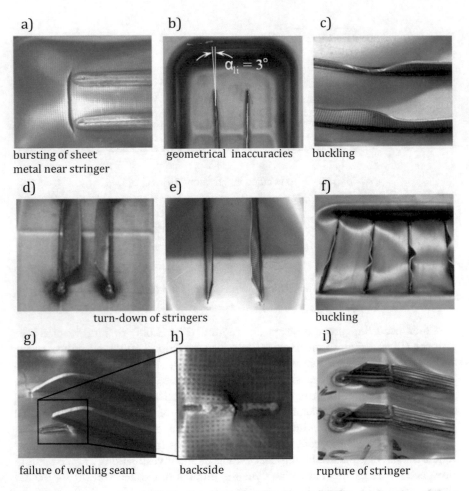

a) bursting of sheet metal near stringer

b) geometrical inaccuracies

c) buckling

d, e) turn-down of stringers

f) buckling

g) failure of welding seam

h) backside

i) rupture of stringer

Fig. 3.60 Process errors of stringer sheet hydroforming (Ertugrul 2010). (**a**) Bursting of sheet metal near stringer, (**b**) geometrical inaccuracies, (**c**) buckling, (**d, e**) turn-down of stringer, (**f**) buckling, (**g**) failure of welding seam, (**h**) backside, (**i**) rupture of stringer

A stringer can be assumed as buckled if the waviness w exceeds the value of 0.002 (Bäcker 2015). We can investigate the failure mode under several different process conditions. A process window for stretch drawing of stringer sheets regarding buckling is shown in Fig. 3.61.

The experimental results demonstrate that the buckling risk grows with rising stringer height and falling curvature, whereas growing stringer length and stringer thickness can avoid buckling (Bäcker et al. 2012).

The occurring tensions and strains can be described both analytically and numerically. The results of the calculations demonstrate a change from negative to positive strains with growing dome height h_d during the process, as shown in Fig. 3.62. When this changing point in strain is overcome, no buckling error occurs

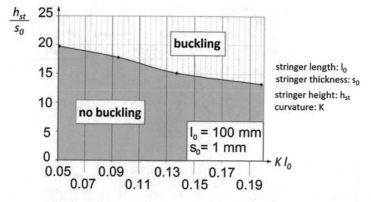

Fig. 3.61 Process window for stretch drawing of stringer sheets (Bäcker et al. 2012)

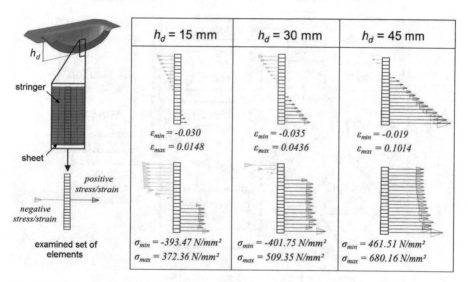

Fig. 3.62 Numerical results of the strain and stress investigation (Bäcker et al. 2012)

in the free-forming stage of the process. Later in the process, so-called secondary buckles can arise due to the contact of the sheet metal with the die. The strains in the stringer edge can change back into the negative area which causes the buckling failure (Ertugrul 2010).

In order to avoid buckling, nonlinear stringer heights can be used. Section 5.2.2 describes the mathematical optimization of the stringer height regarding buckling. The optimized nonlinear stringer heights reach higher average stringer heights which lead to a higher part stiffness.

The blank holder force has a significant influence on the buckling failure mode in the deep-drawing process. If the force is too small, buckling of the stringer occurs at lower stringer heights. However, if we set the force too high, rupture occurs in the

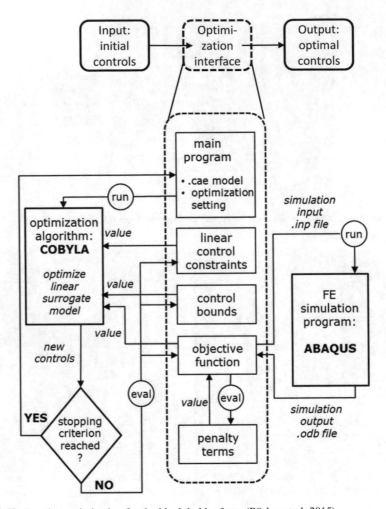

Fig. 3.63 Iterative optimization for the blank holder force (Bäcker et al. 2015)

sheet metal. Thus, a conflict of interests can be derived between the avoidance of buckling and sheet metal rupture. A mathematical optimization is set up to solve this problem. The iterative optimization of the blank holder force is depicted in Fig. 3.63 and described in detail in Sect. 5.2.2. The algorithm that is programmed in MATLAB starts with the input of the initial controls for the blank holder force. Aiming at an error-free form filling the optimization algorithm in COBYLA generates new input parameters for iterative simulations that are conducted in ABAQUS. These simulations have defined boundary conditions that must not be violated. The algorithm stops when the complete form filling is reached for a set of controls. The generated output includes optimal controls for the blank holder force that lead to a buckling and bursting-free stringer sheet deep-drawing process with the highest possible stringer height.

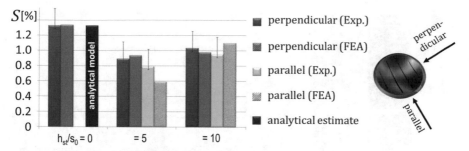

Fig. 3.64 Springback of stretch-formed stringer sheets (Groche and Bäcker 2013)

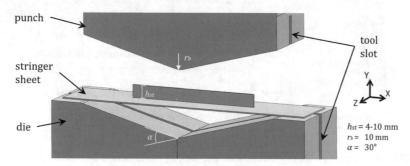

Fig. 3.65 Tool setup for stringer sheet forming with solid tools (Köhler et al. 2016)

The use of stringers influences the springback behavior. In experimental and numerical investigations, we can see an anisotropic behavior regarding springback. Figure 3.64 shows the value of the springback S parallel and perpendicular to the stringers. For stringers with lower height, the amount of springback perpendicular to the stringer is bigger.

An iterative numerical positioning algorithm for the stringer on the semifinished part can be utilized to improve geometrical accuracy in the final shape of the drawn stringer sheet. Therefore, the desired course of the stringer on the drawn part is defined as starting position for the algorithm. The drawing process is simulated and the contour of the stringer in the simulation result is compared to the desired course afterwards. The deviations lead to a new starting geometry of the stringer on the semifinished part, as the amount of deviation of each stringer node is offset. The algorithm stops when the deviations to the desired geometry are smaller than the defined amount (Bäcker and Groche 2014).

Solid tools can be used to fasten up the forming process of stringer sheets. For the first feasibility investigation of forming stringer sheets with solid tools, a die-bending process is set up. The process principle is shown in Fig. 3.65.

A slot in the punch and in the die guarantees that the tools do not damage the stringer during the forming process. This tool concept enables bending of stringers with linear layout.

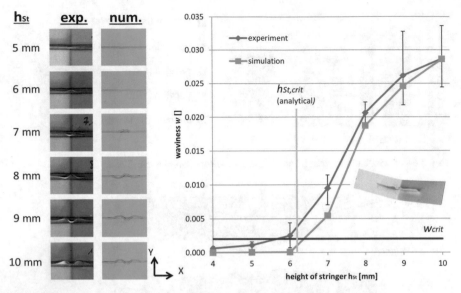

Fig. 3.66 Buckling investigation in stringer sheet die bending (Köhler et al. 2016)

Stringer sheets with different stringer heights are bended by 30°. At first, an analytical estimation for this process that utilizes a Euler buckling load problem leads to a critical stringer height of 6.164 mm regarding buckling. If the stringer height exceeds this value, it is estimated that buckling occurs. This value correlates with the results of the numerical and experimental investigations very well, as can be seen in Fig. 3.66. Here, the maximum stringer height with a waviness value $w < 0.002$ is also about 6 mm. The small deviations between the numerical and the experimental results are explainable by inaccuracies such as imprecise inlet positions during the experiments. A higher waviness leads to an increasing spread of the waviness values in the experiments (Köhler et al. 2016).

Stiffness tests of the die-bended stringer sheets show that the stiffening advantage of the bifurcated sheet metals does not increase with growing stringer height when the stringers are buckled. The stringer sheets are loaded by a body that moves a defined distance, while the force needed for this movement is measured in the test. Another fact is the increasing spread of the stiffness value when the stringer is buckled. Figure 3.67 shows the stiffening factor of stringer sheets with different stringer heights compared to a bended sheet metal without a stringer. The stiffening factor is calculated by the quotient of absolute stiffness of the stringer stiffened part (force/displacement) with the absolute stiffness of a part without stringer.

Thus, we can conclude that stringer sheet forming with solid tools is a possible alternative to hydroforming. Future investigations will transfer this knowledge into a deep-drawing process with solid tools.

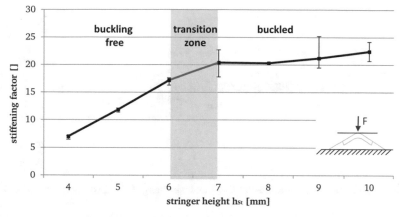

Fig. 3.67 Stiffness test of die-bended stringer sheets (Köhler et al. 2016)

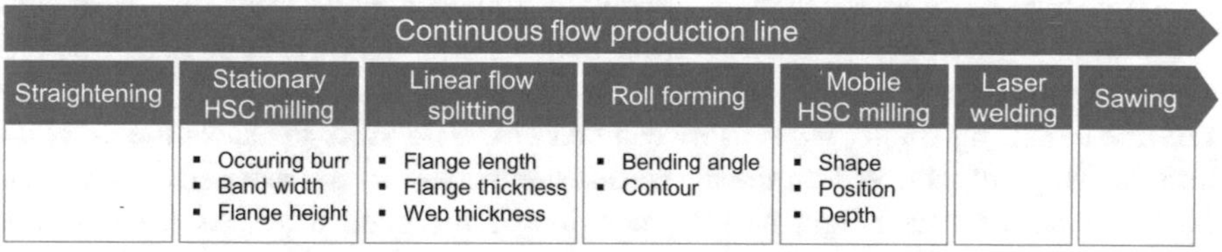

Continuous flow production line						
Straightening	Stationary HSC milling	Linear flow splitting	Roll forming	Mobile HSC milling	Laser welding	Sawing
·	• Occuring burr • Band width • Flange height	• Flange length • Flange thickness • Web thickness	• Bending angle • Contour	• Shape • Position • Depth		

Fig. 3.68 Controlled properties in the continuous flow production line

3.2.5   Process Control

Continuous flow production lines have high demands on manufacturing and machining quality of its single stations. One way to meet those demands is to implement process control installations. For example, within the continuous flow production line of our research project, we implemented process control installations in the HSC edge milling machine tool, the discontinuous in-line milling machine tool, as well as the linear flow splitting and roll-forming units (Fig. 3.68).

Band-Edge and Flange-Edge Milling Operations: Within linear flow production lines, edge milling operations are used to ensure a predefined sheet bands width and flange height as well as to achieve certain edge geometries (Sect. 3.2.2). The linear flow-splitting process demands high width accuracy of the coil material. The inhomogeneity of the material leads to nonlinear flange heights during the linear flow-splitting process. For downstream processes such as laser welding, a certain constant height is required. The quality criteria for band-edge and flange-edge milling operations within continuous flow production lines are the occurring burr as well as the bandwidth or flange height, respectively. The occurring burr has a major influence on downstream forming processes as well as on the laser welding process. Detached burr filaments can cause serious damage of the forming rolls.

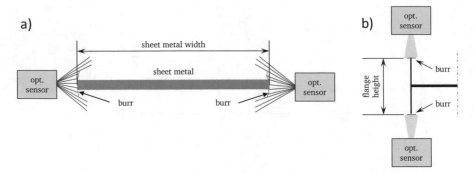

Fig. 3.69 (**a**) Measuring of sheet metal width and burr flange height, (**b**) measuring of burr

Additionally, a high burr on the machined edge causes welding errors. The burr height could be used as an indicator for tool wear as well.

To ensure the demanded width accuracy independently from tool wear and machine tool referencing, the occurring burr and the bandwidth or flange height can be measured by using two 2D displacement sensors (Fig. 3.69). Due to the flexible mounting frame, the sensors can be used to measure the sheet metal width and the flange height. Within the measuring controller, certain reference points can be set to detect burr height and the position of the band edge. As the absolute distance between the sensors is known, the criteria can be calculated easily.

By sending those information to the machine control, the operator or the distributed control system can be notified when predefined values are not met. To ensure a constant sheet metal width and flange height, these information can also be used to implement a closed-loop controller for the regarded criteria (Fig. 3.70). It is superposed to the machine axis upright to the machined edge. The reference value is the designated sheet metal width or flange height, respectively. The delay depends on the workpiece feed. Furthermore, the measured burr height can be used to start an autonomous tool change (Abele et al. 2012a).

In-line Milling Operations: The quality criteria for discontinuous in-line milling operations are the dimensional accuracy of the milling features, its absolute and relative position on the sheet metal surface, and the surface quality of the machined area. To monitor those characteristics, a 2D displacement sensor can be attached to the machine tool workpiece guidance system (Sect. 3.2.2). Fig. 3.71 shows a sensor mounted above the sheet metal profile.

The sensor captures a two-dimensional section of the scanned milling feature. The 2D information has to be composed to 3D data using the actual workpiece feed (Fig. 3.72). By mounting the sensor at the workpiece guiding system, it is moved relatively to the sheet metal profile depending on the machine gantry movements.

Therefore, the actual positions of the machine gantry and the sheet metal profile have to be considered in the processing. By using that information, a comparison of the actual machined feature and the reference geometry can be realized. With that,

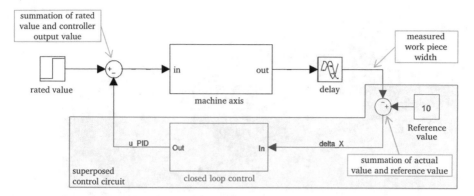

Fig. 3.70 Closed-loop controller to ensure a constant sheet metal width and flange height (Abele et al. 2012c)

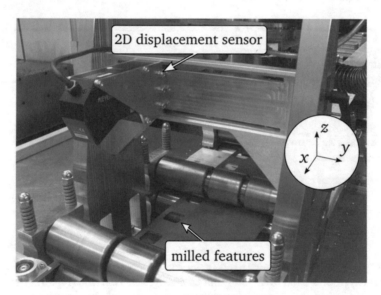

Fig. 3.71 2D displacement sensor to capture the workpiece surface

systematical failures in shape, position, and depth of the machined features can be compensated by the machine control to ensure high machining accuracy.

Quality Control in the Process of Linear Flow Splitting: In conventional roll-forming lines, a profile cross section is cut out of the profiled strip to examine the accuracy of the manufactured profile geometry in a profile projector. The springback of the profile's cross section and the undefined cutting lead to significant deviations in the measuring results. This procedure causes time-consuming adjustment of the tool system due to the misinterpretation of the measured profile geometry (Henkelmann 2009).

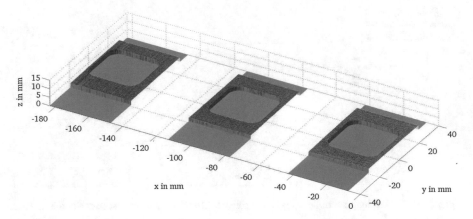

Fig. 3.72 Captured workpiece surface

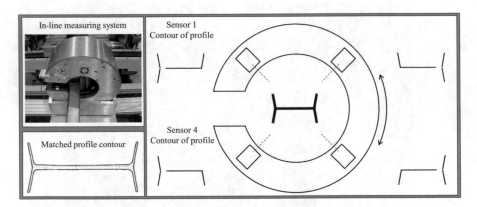

Fig. 3.73 In-line measurement system

For quality control of linear flow split or roll-formed profiles, an in-line measurement system is used in the continuous flow production line (Fig. 3.73), which is based on the principle of laser triangulation. In an opened chassis, four sensors with a laser and a detector are mounted and the contour of the measured profile is determined. To optimize the angle at which the sensors detect the part, the system is able to rotate. The measured profile contours of the four sensors are matched together to a closed profile shape. The measurement system has a sampling rate of 5 Hz with a resolution of 1 μm and offers a continuous monitoring of the manufactured profile.

The main advantage of the system is the direct measuring of the profile contour without any cutting operation. With the integrated laser measurement system, we carried out an in-line quality control of profiles and adjustment instructions for the tool system. In the software of the measurement system, a template is implemented with the most important profile tolerances. While the measurement is running, a comparison between the target and the actual geometry can be done. Thereby, for

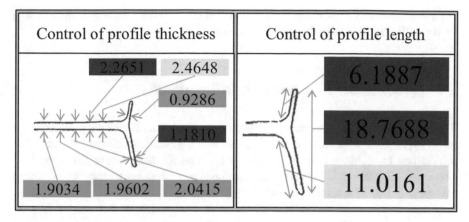

Fig. 3.74 Control of profile tolerances

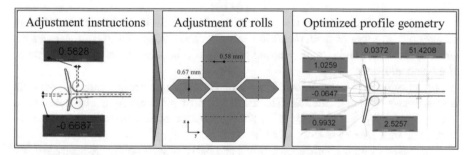

Fig. 3.75 Measured profile geometry for adjustment instructions

all measured values, an upper and lower warning limit (yellow) with control limits (red) can be implemented (Fig. 3.74). If the measured profile geometry is out of given limits, a signal can be sent to the process control system of the manufacturing line to prevent the production of large quantities of metal waste.

For an industrial use of linear flow splitting, besides the process time the setup and adjustment times are very important for a cost-efficient production. A secure and easy setup of the tool system contributes to the cost reduction. The object for the setup and adjustment of any roll tool system is the precorrection of the roll displacements under loading. Here, the challenge is the correction of the position of a roll in one degree of freedom, which leads to a change of several geometrical parameters of the profile.

With the presented in-line measurement system, adjustment instructions can be determined and given to the machine operators as well (Fig. 3.75).

As reference points for determining the respective position of the rolls, prints of the supporting and splitting rolls in the measured profile geometry are used. With the coordinates of the center of the radii, the distances or rather the position of the rolls to each other can be calculated. This value corresponds to the adjustment value of the rolls and equalizes the roll displacements.

3.3   Product Benefits Through Bifurcations

The processes and process chains presented in Chap. 3 facilitate the production of sheet metal parts with integral bifurcations. In some product examples the benefits, which can be achieved for products by the use of these bifurcations, will be presented. Section 3.3.1 regards bifurcations generated by linear flow splitting and Sect. 3.3.2 bifurcations generated by linear bend splitting.

3.3.1   Bifurcation at the Edge of Sheet Metal

Linear flow splitting enables the application of integral bifurcations and additional flanges on the edges of sheet metal. With further processing of the semifinished split profile, products can be manufactured that exploit the advantages of these bifurcations and additional flanges.

As introduced in Chap. 3, bifurcations can be used to increase structural strength and stiffness of a construction. The application of bifurcations is also suitable to prevent buckling in the plane of a sheet metal or of a thin wall or to prevent wrinkling of a free edge from a profile.

For closed or multichambered profiles and parts, branch points are necessary. By using integral bifurcations, additional joints and processing steps can be omitted.

For example, the attachment of pipes usually requires pipe clamps, as shown in Fig. 3.76a. This necessitates additional connecting components (Groche et al. 2009b) and the connecting mechanism is only based on friction. A safe connection has to be ensured even under varying conditions like temperature or lubrication conditions.

Compared to that, a duct with integral external flange can be securely fastened with a regular bolted connection without any additional connecting components or loose parts (Groche et al. 2009b), as shown in Fig. 3.76b, c.

Next to the geometric features of linear flow split and linear bend split profiles, the processes introduce a grain refinement. The severe plastic deformation leads to a refinement of the microstructure as well as an increase of durability and fatigue strength as introduced in Sects. 4.1.1 and 4.1.3.

The special features of improved surface quality (Sect. 3.1.4), freedom of geometric design, and increased fatigue properties make linear flow split flanges especially suitable as rolling contact areas (Groche and Vucic 2006), due to the special microstructure (Sect. 4.1.4). Figure 3.77 presents a telescopic linear guiding made of linear flow split profile, with which we exploit geometry and mechanical properties for rolling element contact.

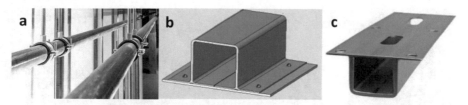

Fig. 3.76 (**a**) Attachment of a pipe using additional pipe clamps (MÜPRO GmbH 2016) and (**b, c**) using integral bifurcations (Groche et al. 2009b; Abele et al. 2014)

Fig. 3.77 Telescopic linear guiding made of split profile

3.3.2 Bifurcation in the Plane of Sheet Metal

Surface structures are sensitive against buckling. Groche et al. (2007a) investigated the buckling stress of the ratio to the mass of a surface structure. Based on a value of 1 for a simple plate an optimized web plate with two bifurcations received a value of 21. This finding can also be transferred to curved surface structures.

Conventionally manufacturing methods for surface structures with bifurcations usually are realized by differential design with very tight positioning tolerances (Zaeh et al. 2008) or by integral design with an extremely high metal removal volume (Fiedler 2002).

The continuous cold-forming process of linear bend splitting described in Sect. 3.1 in combination with the deep-drawing process described in Sect. 3.2.4 enables the manufacturing of curved surface structures with bifurcations. But the restriction from the manufacturing limits the maximum theoretical stiffness. Section 3.2.4 for example describes the limitation of the height of the bifurcations.

However, the potential of the technology can be shown in a real demonstrator made in the shape of a sphere whose bearing segments are the hexagons of a truncated icosahedron. The so-called Football-Demonstrator was built with a diameter of 1500 mm. The missing pentagons enable an insight into the inner part of the sphere. Each hexagonal element is stiffened by four parallel bifurcations of 20 mm height.

Bäcker et al. (2012) worked out the load capacity compared to the mass ratio. In the finite element software *ABAQUS* two balls were modeled as shell structures:

Fig. 3.78 Load capacity-to-mass ratio of the "Football-Demonstrator"

one ball with stringers and the other one without. The balls were positioned between two parallel plane analytical rigid surfaces. During the simulation these surfaces were moved towards each other by squeezing the ball 20 mm (Fig. 3.78). As a result the reaction forces at the analytical rigid surfaces were compared. While the stringer ball is 13% heavier the force-to-mass ratio increases by 260% (Fig. 3.78).

Besides the increasing load capacity of the surface structure, the continuous production process of the bifurcations and the material hardening in the web (Sect. 4.1) are very important benefits. From the perspective of the construction industry another main advantage of these products is the missing machining traces on the visible surface of the component. For more product ideas of bifurcated sheet metal panels refer Sect. 7.2.

References

Abele E, Stein S, Munirathnam M (2007a) Ultra-HSC-Bearbeitung von Bandkanten—Grundlagenuntersuchungen zur Gratbildung bei der spanenden Kantenbearbeitung von Blech. wt Werkstattstechnik online 97(1/2)

Abele E, Stein S, Schramm B, Fröhlich B (2007b) High speed edge milling of sheet metal. In: International Conference of High Speed Machining, San Sebastian, 2007

Abele E, Stein S, Quirmbach K (2008a) Einfluss der Werkzeugaufnahme bei der Blechbearbeitung mittels Hochgeschwindigkeitsfräsen. Zeitschrift für wirtschaftlichen Fabrikbetrieb (ZWF) 103 (7–8):470–474

Abele E, Stein S, Sieber M (2008b) Hartmetallschneidstoffe für die fräsende Bandkantenbearbeitung. wt. Werkstattstechnik online 98(1/2):24–30

Abele E, Stein S, Fröhlich B, Roth M (2008c) Joint cutting and forming processes for multi-chambered sheet metal profiles. In: Proceedings of the 9th International Conference on Technology of Plasticity Gyeongju, Korea, 2008

Abele E, Jalizi B, Baklouti F, Schiffler A (2011b) Herstellung multifunktionaler Strukturen mittels Integration des Zerspanprozesses in Walzprofilieranlagen. Zeitschrift für wirtschaftlichen Fabrikbetrieb (ZWF) 106(7–8):522–526

Abele E, Baklouti F, Jalizi B, Fischer J (2011a) HSC-Bandkantenbearbeitung von Grobblech in der Schwerindustrie. Zeitschrift für wirtschaftlichen Fabrikbetrieb (ZWF) 106(10):747–751

Abele E, Schmidt S, Turan E, Jalizi B, Haehn F (2012a) Milling and quality monitoring of sheet metal edges in continuous flow production. In: Proceedings of the 9th International Conference on Machine Tools, Automation, Technology and Robotics, Czech Republic, 2012

Abele E, Bohn A, Jalizi B, Haydn M, Lommatzsch N, Coutandin P (2012b) Steigerung der Prozessqualität einer fliegenden HSC-Fräsmaschine. wt. Werkstattstechnik online 102 (1/2):27–33

Abele E, Groche P, Ludwig C, Monnerjahn V, Schmidt S, Jalizi B (2012c) Konzepte zur Qualitätssicherung von spanenden und umformenden Prozessen in der Fließfertigung. In: Tagungsband 4. Zwischenkolloquium SFB 666, TU Darmstadt, pp 65–76

Abele E, Schmidt S, Hoßfeld A (2014) HSC-Fräsbearbeitung in der Fließfertigung—Integration diskontinuierlicher Fräsoperationen in die Fließfertigung von Blechprofilen. wt Werkstattstechnik online 104(1/2)

Bäcker F (2015) Formgebung mehrachsig stark gekrümmter Stahlbleche mit lastangepassten Versteifungsrippen. Dissertation, TU Darmstadt

Bäcker F, Groche P (2014) Ein Algorithmus zur Positionierung von Funktions- und Strukturelementen auf Halbzeugen für die Umformtechnik. In: Tagungsband 5. Zwischenkolloquium SFB 666, Darmstadt, 2014

Bäcker F, Groche P, Abedini S (2012) Stringer sheet forming. In: Proceedings of NAMRI/SME 40, Notre Dame, Indiana, 2012

Bäcker F, Bratzke D, Groche P et al (2015) Time-varying process control for stringer sheet forming by a deterministic derivative-free optimization approach. Int J Adv Manuf Technol 80(5):817–828

Bhattacharyya D, Smith PD, Yee CH, Collins IF (1984) The prediction of deformation length in cold roll-forming. J Mech Work Technol 9(2):181–191

Brimm DJ (1988) Method of forming integrally stiffened structures. US Patent, US 4725334 A

Denkena B, Schmidt C (2007) Experimental investigation and simulation of machining thin-walled workpieces. Prod Eng 1(4):343–350

Eichler J, Eichler H-J (2003) Laser: Bauformen, Strahlführung, Anwendungen; mit 57 Tabellen. Springer, Berlin

Ertugrul M (2010) Wirkmedienbasiertes Tiefziehen von verrippten Schalen. Dissertation, TU Darmstadt

Fiedler U (2002) Prozesssicherheit beim HSC-Fräsen von Aluminium-Knetlegierungen. Dissertation, TU Darmstadt

Görtan O, Vucic D, Groche P, Livatyali H (2009) Roll forming of branched profiles. J Mater Process Technol 209(17):5837–5844

Groche P, Vucic D (2006) Multi-chambered profiles made from high-strength sheets. Prod Eng Ann WGP 3(1):67–70

Groche P, Bäcker F (2013) Springback in stringer sheet stretch forming. CIRP Annals—Manufacturing Technology 62(1):275–278

Groche P, Vucic D, Fritsche D, Jöckel M (2004) Steifigkeitsoptimierter Leichtbau durch neue Umformverfahren für Blech. In: Neuere Entwicklungen in der Blechumformung. Paper presented at the international Symposium "Neuere Entwicklungen in der Blechumformung", IFU, Institut für Umformtechnik der Universität Stuttgart, Fellbach, 11–12 May 2004

Groche P, Vucic D, Jöckel M (2005) Herstellung einteilig verzweigter Blechstrukturen. Wt Werkstatttechnik online 95(10):753–758

Groche P, Ringler J, Vucic D (2007a) New forming processes for sheet metal with large plastic deformation. Key Eng Mater 344:251–258

Groche P, Vucic D, Jöckel M (2007b) Basics of linear flow splitting. J Mater Process Technol 183 (2–3):249–255

Groche P, Vucic D, Ludwig C (2008) Herstellen und Weiterverarbeiten verzweigter Profile. Wt Werkstatttechnik online 98(10):843–848

Groche P, Ringler J, Abu Shreehah T (2009a) Bending–rolling combinations for strips with optimized cross-section geometries. CIRP Annals—Manufacturing Technology 58 (1):263–266

Groche P, Vucic D, Jöckel M (2009b) Herstellung multifunktionaler Blechprofile. wt. Werkstattstechnik online 99(10):712–720

Groche P, Bäcker F, Ertugrul M (2010a) Möglichkeiten und Grenzen der Stegblechumformung. Wt Werkstatttechnik online 100(10):760–766

Groche P, Mueller C, Beiter P et al (2010b) Future trends in cold rolled profile process technology. Confederation of British Metalforming: cbm 19:16–19

Halmos GT (2006) Roll Forming Handbook. Manufacturing engineering and materials processing, vol 67. Taylor & Francis, Boca Raton

Henkelmann M (2009) Entwicklung einer innovativen Kalbrierstrecke zur Erhöhung der Profilgenauigkeit bei der Verarbeitung von höher- und höchstfesten Stählen. Dissertation, TU Darmstadt

Jalizi B, Bauer J, Fischer J, Abele E (2010) HSC—Bandkantenbearbeitung in der Schwerindustrie. In: Tagungsband 3. Zwischenkolloquium SFB 666, TU Darmstadt, 2010

Jöckel M (2005) Grundlagen des Spaltprofilierens von Blechplatinen. Dissertation, TU Darmstadt

Kiuchi M, Abe K, Onodera R (1995) Computerized numerical simulation of roll-forming process. CIRP Ann Manuf Technol 44(1):239–242

Köhler S, Groche P, Baron A, Schuchard M (2016) Forming of stringer sheets with solid tools. Adv Mater Res 1140:3–10

Lange K (ed) (1990) Umformtechnik: Handbuch für die Industrie und Wissenschaft, Band 3: Blechbearbeitung. Springer, Berlin

Ludwig C, Groche P (2009) Herstellung neuartiger Profile aus spaltprofilierten Halbzeugen. Paper presented at the 3. Dortmunder Kolloquium Rohr- und Profilbiegen, Dortmund, 8 Oct 2009

Ludwig C, Schmitt W, Groche P, Stein S, Abele E (2008a) Herstellung verzweigter Profile. In: Tagungsband 2. Zwischenkolloquium SFB 666, TU Darmstadt, 2008

Ludwig C, Vucic D, Ringler J, Groche P (2008b) Branched semi-finished products for a new class of multi-chambered profiles. In: ICTP 2008, Korea.

Ludwig C, Monnerjahn V, Jalizi B, Schmidt S, Groche P, Abele E (2012) Konzepte zur Qualitätssicherung von spanenden und umformenden Prozessen in der Fließfertigung. In: 4. Zwischenkolloquium des Sonderforschungsbereichs 666: Integrale Blechbauweisen höherer Verzweigungsordnung – Entwicklung, Fertigung, Bewertung. Meisenbach, Bamberg, pp 65–75

Ludwig C, Hammen V, Groche P et al (2013) Manufacturing of quality optimized linear flow split parts by variation of fast modifiable process parameters and their influence on material characteristics. Mater Werkst 44(7):601–611

Ludwig C (2016). Erhöhen der Technologiereife des Spaltprofilierverfahrens. Dissertation, TU Darmstadt

Luo AA (2013) Magnesium casting technology for structural applications. J Magnesium Alloys 1 (1):2–22

Monnerjahn V, Fricke S (2012) Potenziale des Festwalzens zur Optimierung von Oberflächen beim Walzprofilieren. In: 8. Fachtagung Walzprofilieren und 4. Zwischenkolloquium SFB 666, Darmstadt, 14–15 Nov 2012

MÜPRO GmbH (2016) Befestigungslösung mit Schiebeschlitten. http://www.muepro.de/unternehmen/presse/downloads.html. Accessed 19 May 2016

Neuwirth M, Ahmels L, Schmidt S, Hegemann M, Groche P, Müller C (2015) Research on the influence of the band edge processing on the process of linear flow splitting. Mater Sci Eng Technol 47(1)

Niehuesbernd J, Monnerjahn V, Bruder E, Groche P, Müller C (2016) Improving the formability of linear flow split profiles by laser annealing. Materials Science & Engineering Technology (submitted)

Oehler G (1963) Biegen unter Pressen: Abkantpressen, Abkantmaschinen, Walzenrundbiegemaschinen, Profilwalzmaschinen. Hanser, München

Poprawe R (2005) Lasertechnik für die Fertigung. Springer, Berlin

Schäfer S, Abedini S, Groche P, Bäcker F, Ludwig C, Abele E, Jalizi B, Müller C, Kaune V (2013) Verbindungstechniken durch die Technologie des SFB 666. Band 88, Bauingenieur

Schmidt S, Abele E (2014) Integration of discontinuous milling operations into the flow production of sheet metal profiles. In: MM Science Journal, Special Issue I HSM 2014, 11th International Conference on High Speed Machining, Prague

Schmitt W, Rullmann F, Ludwig C, Groche P (2011) Entwicklungsstufen des Spaltprofilierens. In: Tagungsband zum 1. Erlangener Workshop Blechmassivumformung 2011, DFG Transregio, vol 73, pp 249–255

Schulz H (1996) Hochgeschwindigkeitsfräsen metallischer und nichtmetallischer Werkstoffe. Carl Hanser Verlag, München

Stein S (2010) HSC-Kantenbearbeitung von Blech. Dissertation, TU Darmstadt

Stein S, Abele E (2007) Spanende Kantenbearbeitung von Blech. In: Tagungsband 1. Zwischenkolloquium SFB 666, TU Darmstadt, 2007, pp 79–84

Stein S, Abele E (2008) Werkzeugentwicklung in der HSC-Bandkantenbearbeitung. In: Tagungsband 2. Zwischenkolloquium SFB 666, TU Darmstadt, 2008, pp 73–78

Stein S, Thole J, Abele E (2010) Verschleißverhalten von Beschichtungsvarianten bei der HSC-Blechbearbeitung. In: Tagungsband 3. Zwischenkolloquium SFB 666, TU Darmstadt, 2010, pp 53–60

Taplick C, Schmitt W, Rullmann F, Ludwig C, Groche P (2011) High strength hydroformed steel profiles with free flanges. Weinheim, Wiley

Tschätsch H, Dietrich J (2011) Praxis der Zerspantechnik. Verfahren, Werkzeuge, Berechnung, 10. Auflage. Vieweg + Teubner Verlag, Wiesbaden

Turan E, Abele E (2012) Modifikation der Schneidkantengeometrie hinsichtlich Gratminimierung und Standwegerhöhung bei der HSC-Bandkantenbearbeitung. In: Tagungsband 4. Zwischenkolloquium SFB 666, TU Darmstadt, 2012, pp 59–64

Vučić D (2010) Methoden zum Herstellen und Weiterverarbeiten von Spaltprofilen. Dissertation, TU Darmstadt

Weber K (1973) Grundlagen des Bandwalzens. VEB Deutscher Verlag für Grundstoffindustrie, Leipzig

Wiedemann J (2007) Leichtbau: Elemente und Konstruktion. Springer, Berlin

Zaeh MF, Huber S, Gerhard P, Ruhstorfer M (2008) Bi-focal hybrid laser beam welding and friction stir welding of aluminum extrusion components. Adv Mater Res 43:69–80

Zettler A-O (2007) Grundlagen und Auslegungsmethoden für flexible Profilierprozesse. Dissertation, TU Darmstadt

Chapter 4
Manufacturing Induced Properties: Determination, Understanding, and Beneficial Use

L. Ahmels, A.-K. Bott, E. Bruder, M. Gibbels, S. Gramlich, M. Hansmann, I. Karin, M. Kohler, K. Lipp, T. Melz, C. Müller, D. Neufeld, J. Niehuesbernd, M. Roos, A. Tomasella, S. Ulbrich, R. Wagener, and A. Walter

Based on its procedural principle, every manufacturing technology affects a variety of properties of the workpiece or product in a characteristic way (Sect. 2.3). The sum of all those properties which comprise geometrical as well as material-related ones is considered as manufacturing-induced properties. While the geometric manufacturing-induced properties are often the reason why a specific technology is chosen by the designer for the manufacturing of a certain product, the material-

L. Ahmels (✉) • E. Bruder • C. Müller • J. Niehuesbernd
Physical Metallurgy (PhM), Technische Universität Darmstadt, Darmstadt, Germany
e-mail: L.ahmels@phm.tu-darmstadt.de

A.-K. Bott • M. Hansmann • M. Kohler
Research Group Statistics (STAT), Technische Universität Darmstadt, Darmstadt, Germany

M. Gibbels • I. Karin • D. Neufeld • A. Tomasella
Research group System Reliability, Adaptive Structures, and Machine Acoustics (SAM), Technische Universität Darmstadt, Darmstadt, Germany

S. Gramlich • M. Roos
Institute for Product Development and Machine Elements (pmd), Technische Universität Darmstadt, Darmstadt, Germany

K. Lipp • R. Wagener
Fraunhofer Institute for Structural Durability and System Reliability LBF (LBF), Fraunhofer-Gesellschaft zur Förderung der angewandten Forschung e.V., Munich, Germany

T. Melz
Research group System Reliability, Adaptive Structures, and Machine Acoustics (SAM), Technische Universität Darmstadt, Darmstadt, Germany

Fraunhofer Institute for Structural Durability and System Reliability LBF (LBF), Fraunhofer-Gesellschaft zur Förderung der angewandten Forschung e.V., Munich, Germany

S. Ulbrich • A. Walter
Research Group Nonlinear Optimization (NOpt), Technische Universität Darmstadt, Darmstadt, Germany

© Springer International Publishing AG 2017
P. Groche et al. (eds.), *Manufacturing Integrated Design*,
DOI 10.1007/978-3-319-52377-4_4

related manufacturing-induced properties are often seen as by-products of the process. With regard to metal forming, all manufacturing processes inherently influence the mechanical properties of the manufactured material. In many cases, these mechanical manufacturing-induced properties are merely regarded in terms of restrictions in product development. However, with respect to a manufacturing-integrated product development approach, the mechanical properties are of special interest, since we aim at utilizing their full potential to maximize the product performance.

In general, the change in mechanical properties of a material increases with the strain it is subjected to in a cold-forming process. The processes linear flow splitting and linear bend splitting, which we focus on in terms of manufacturing technology, introduce severe plastic strains into the material. This implies significant changes in the mechanical properties of the formed material such as the strength. By the deduction of general rules for the alterations of mechanical properties caused by the manufacturing process, it becomes possible to predict manufacturing-induced properties. If processed appropriately, the knowledge of these manufacturing-induced properties can be taken into account in early stages of product development allowing the designer to exploit the full potential of a manufacturing process, thus improving the product performance. Therefore, this chapter focuses on the determination of manufacturing-induced properties, their processing for product development, as well as examples that demonstrate benefits of this procedure.

The manufacturing-induced properties of linear flow splitting and linear bend splitting have been investigated on different predominantly ferritic steels: high-strength low-alloy steels with ferritic matrix containing Fe_3C precipitates (*HC 480 LA* and *HC 340 LA*), stainless steel with ferritic matrix containing carbide precipitations (*1.4016*), ferritic-perlitic structural steel (*DD11*), and microalloyed fine-grain steels (*RAWAEL 80s*). The changes in properties have been found qualitatively comparable for all investigated materials. In this chapter, the characteristic properties of linear flow split or linear bend split profiles are discussed on varying steel grades. However, the depicted trends in manufacturing-induced properties are of general validity and can be transferred to other materials of the above-mentioned categories.

4.1 Effects of Severe Straining in Integral Sheet Metal Design

Ultrafine-grained materials by severe plastic deformation: It is widely known that cold-forming processes change the yield strength of the formed material by increasing its dislocation density. It is standard to consider this effect in simulations of forming processes by implementing empirical flow curves. From a physical point of view, the correlation between dislocation density and yield strength is described by the Taylor equation (Hornbogen 2006; Rösler et al. 2008):

$$\Delta\sigma_{ys} = \alpha M G b \sqrt{\rho} \tag{4.1}$$

where $\Delta\sigma_{ys}$ is the increase in yield strength due to strain hardening, M is the Taylor factor, G is the shear modulus, b is the Burgers vector, α is a material constant, and ρ is the dislocation density. From this equation it is obvious that the increase in yield strength during deformation caused by the increase in dislocation density saturates at high dislocation densities, i.e., high strains.

At higher strains, however, not only the dislocation density but also the grain size changes. The grain size is another microstructural parameter that affects the yield strength of a material. The relation between yield stress σ_{ys} of polycrystalline metals (in a recrystallized state) and their average grain diameter d is described by the Hall-Petch relationship (Hall 1951)

$$\sigma_{ys} = \sigma_0 + A \cdot d^{-\frac{1}{2}} \tag{4.2}$$

where σ_0 is the starting stress for dislocation movement (friction stress) and A is a material constant (Hall 1951). It is obvious that a decreasing grain size results in a higher yield strength, causing an interest in grain refinement methods. A typical approach for grain refinement is thermomechanical processing (recrystallization); however, the achievable minimum grain size is limited to a few microns (Valiev and Langdon 2006). Smaller grain sizes (and therefore higher strength) can be realized by severe plastic deformation (SPD) processes. The so-called ultrafine-grained (UFG) materials produced by SPD processes were shown to exhibit superior strength and therefore have been subject of intensive research for more than a decade. There have been various definitions of UFG materials, but the definition by Valiev et al. (2006) has been widely accepted. In their work UFG materials are defined as polycrystals with an average grain size below 1 μm. The microstructure is required to be reasonably equiaxed with a majority of grain boundaries exhibiting high angles of misorientation. The large fraction of high-angle grain boundaries is important to achieve the superior properties of the UFG materials (Valiev et al. 2006).

To produce UFG materials, many different SPD processes such as equal channel angular pressing (ECAP) or high-pressure torsion (HPT) were developed. To achieve grain refinement down to the submicron range very high strains in the order of 7 (Valiev 2004) have to be imposed on the workpiece. In order to prevent failure of the material subjected to high strains, its formability has to be increased by imposing high hydrostatic compressive stresses onto the material. This is only possible if material flow is geometrically constrained during the process. Therefore, severe plastic deformation processes are defined as processes that impose very high strains onto the sample, leading to strong grain refinement in the material and producing a UFG microstructure. An additional requirement is that the high strain has to be imposed without significantly changing the dimensions of the workpiece (Valiev et al. 2006).

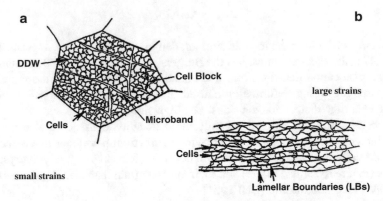

Fig. 4.1 Schematic drawing of deformation microstructures and grain subdivision (reprinted from Hughes and Hansen 1997)

The mechanisms governing grain refinement in severe plastic deformation processes were described by many authors. Hughes and Hansen (1997) as well as Prangnell et al. (2001) describe the formation of deformation-induced substructures within the original grains dividing them into cells. With increasing deformation, the misorientation between the cells increases due to the trapping of dislocations in the cell walls as can be seen in Fig. 4.1. At higher strains the cell boundaries are also starting to reorient from a cellular structure to a more lamellar structure. With increasing strain the amount of high-angle grain boundaries increases and the average distance between the lamellar boundaries decreases up to a strain of approximately $\varepsilon = 5$. Then, the increase in high-angle grain boundary fraction as well as the decrease in the average distance of lamellar boundaries progress more slowly.

The average distance of the lamellar boundaries saturates as soon as it reaches the size of a subgrain, creating thin, elongated "ribbon" grains. (The minimum subgrain size is determined by the rate of recovery and therefore strongly depends on the temperature and the strain rate.)

At this stage further grain refinement is only possible by breaking up the ribbon grains into shorter segments until at $\varepsilon > 10$ a homogeneous submicron grain structure is formed. However, the segmentation process occurs at a very low rate, allowing the ribbon grains to be stable over a wide strain range.

Depending on the SPD process as well as the processed material, the grain refinement results in more or less equiaxed grains in the range of 100 nm to 1 μm as well as in a doubling or even higher increase in yield strength.

High-strain processes linear flow splitting and linear bend splitting: The process principles and characteristics of linear flow splitting and linear bend splitting are detailed in Sect. 3.1. Both introduce high contact stresses between splitting rolls and workpiece yielding high hydrostatic compressive stresses in the forming zone.

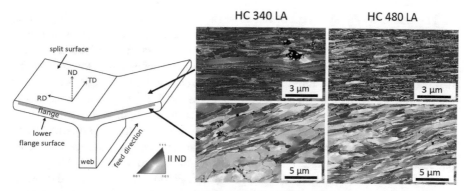

Fig. 4.2 Microstructure in a linear flow split profile (EBSD IPF maps)

The high contact stresses lead to a significant smoothening of the split surface, resulting in a surface comparable to the one of the splitting rolls.

The high hydrostatic stresses induced into the forming zone allow the material to bear the high plastic strains occurring in the process (Sect. 3.1.3). These high plastic strains cause the formation of a fine-grained microstructure with strongly elongated, pancake-shaped grains (short pancake axis normal to the surface) at the flange surfaces, as can be seen in Fig. 4.2 (top) for two low-alloyed steels. The grain boundary spacing perpendicular to the flange surface is below 0.1 μm; parallel to the surface an average grain width of 2 μm is found. To apply the Hall-Petch relation to such non-equiaxed microstructures, an effective grain size based on the average glide length of dislocations in a grain is used (Bruder 2011). In this case, the effective grain size is in the order of 0.2 μm, which is within the ultrafine-grained range. Still, according to the definition of UFG materials mentioned above, this microstructure would not be considered ultrafine grained due to its strong anisotropy in grain shape. However, the mechanisms leading to grain refinement in linear flow splitting are the same as in the standard SPD processes. Moreover, as described above, very anisotropic ribbonlike grains always occur during the grain refinement in severe plastic deformation processes, and since they are stable over wide strain ranges, they can be found in the microstructures resulting from standard SPD processes, e.g., ECAP route A (Langdon et al. 2000; Iwahashi et al. 1998). Because of that, the microstructure produced at the upper flange surface by linear flow splitting will be referred to as UFG microstructure in the following.

By definition, the linear flow splitting process is not an SPD process in a narrower sense, since it changes the geometry of the workpiece. Furthermore, in the case of linear flow splitting and linear bend splitting the formation of UFG microstructures is merely a side effect of the desired change in geometry (i.e., a manufacturing-induced property of the process), whereas it is the main purpose of all SPD processes. However, the large strains and the hydrostatic stress state in the forming zone as well as the produced elongated UFG microstructure with high defect density are all features of SPD processes. Because of that, the linear flow-splitting process can be viewed as an SPD-related process.

The characteristic strain gradient in the linear flow-splitting process is built up during the first few splitting steps. Hence, the first millimeters of the flange exhibit a cold-worked microstructure instead of the UFG microstructure described previously (Müller et al. 2007). However, from a certain geometry-dependent distance from the flange tip onwards (e.g., 4 mm in case of a 2 mm sheet and a 160° angle between the flanges) the characteristic microstructure evolves. Interestingly, this characteristic microstructure is found to be independent of the splitting depth or flange length.

The characteristic microstructure consists of a strong grain size gradient perpendicular to the flange surface. This gradient ranges from a strongly elongated pancake UFG structure below the upper flange surface to a conventionally cold-worked microstructure at the lower flange surface. Figure 4.2 illustrates the location of the UFG microstructure close to the upper surface of the flanges (grey areas) and displays electron backscatter diffraction (EBSD) images of the microstructure of an *HC 320 LA* and an *HC 480 LA* linear flow split profile in a depth of 25 µm as well as 500 µm. It can be seen that the characteristic microstructural features of the linear flow split profiles, being a strong gradient perpendicular to the surface and a constant microstructure parallel to it, are found qualitatively in flanges split from both ferritic steel sheets. We found this to also be the case for other predominantly ferritic steels such as stainless ferritic steels, and mild or other HSLA steels (Kaune 2013; Bödecker 2013).

The same correlation can be found between the microstructures of linear flow split and linear bend split profiles. The inverse pole figure maps in Fig. 4.3 illustrate

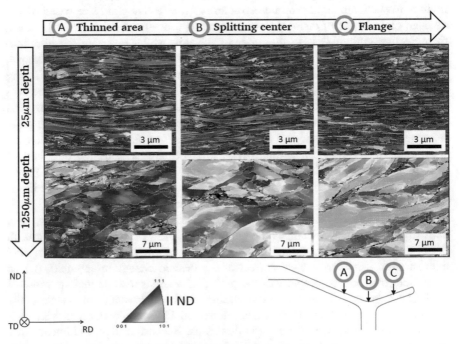

Fig. 4.3 Microstructural gradient found in *DD11* linear bend split profiles

that the microstructures in the flange, the thinned area, and the splitting center exhibit a gradient perpendicular to the surface which is similar to the one in linear flow split profiles. It can also be seen that there is nearly no difference in microstructure between thinned area, splitting center, and flange at a given depth.

From the IPF maps in Figs. 4.2 and 4.3 it is also evident that there is not only a gradient in grain size perpendicular to the split surface, but also in the grain orientations which are represented as different colors in the IPF maps. The presence of predominant orientations means that crystallographic textures are present in the material. Details of these textures as well as their impact on the elastic properties of profiles are described in the following.

The severe deformation during linear flow splitting and linear bend splitting leads to sharp crystallographic textures (Niehuesbernd et al. 2013; Bruder 2012). Due to strain gradients and changes in the deformation mode within the process zone the texture is not constant across the flange thickness. At the split surface a typical rolling texture of body-centered metals with a strong α-fiber ($<110> \parallel RD$) and γ-fiber ($<111> \parallel ND$) is generated (Fig. 4.4, coordinate system according to Fig. 4.2). With increasing distance to the surface the type of texture gradually changes and the intensity of the texture components decreases substantially. At the bottom surface the alpha fiber is reduced to the rotated cube orientation and the gamma fiber has completely vanished. However, additional shear components such as the $\{110\}<100>$ Goss orientation as well as the $\{112\}<111>$ Copper orientation can be observed with increasing distance to the split surface. Consequently, the change in texture can be described as a transition of a rolling fiber texture to a shear texture with discrete orientations.

In materials with significant crystallographic anisotropy such as iron and steel strong crystallographic textures result in anisotropic mechanical properties. The elastic properties at any position within the texture gradient of the flanges can be quite accurately determined using the grain orientation data of EBSD measurements. By averaging single-crystal stiffness tensors rotated according to measured grain orientations it is possible to calculate a local stiffness tensor for the respective position. The geometric mean has proven to be a suitable approximation method

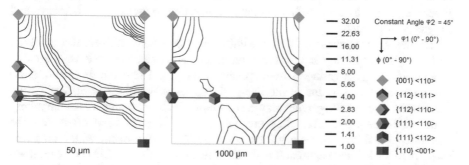

Fig. 4.4 Textures of *HC 480 LA* flanges with a total thickness of 1100 μm, near the split surface (50 μm) and near the bottom surface (1000 μm) represented as orientation distribution functions (Niehuesbernd et al. 2013)

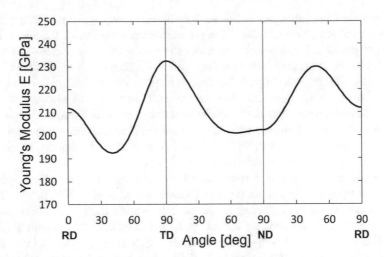

Fig. 4.5 Young's modulus distribution between the main directions RD, TD, and ND of *HC 480 LA* flanges (average over the flange thickness) (Niehuesbernd et al. 2013)

(Niehuesbernd et al. 2013). In low-alloy steels such as *HC 480 LA* and *HC 340 LA* the elastic properties feature no significant gradient in thickness direction despite the considerable changes in texture. In these cases the transition from a fiber texture to discrete orientations compensates the reduction in texture intensity. Therefore, an average stiffness tensor over the whole flange thickness can be calculated. The resulting Young's modulus distribution between the main directions (Fig. 4.5) shows deviations of up to 12% from the isotropic average of 210 GPa. This elastic anisotropy is a manufacturing-induced property of the linear flow splitting process that can be exploited in product development process to increase product performance (Sect. 4.3.1).

Residual stresses: Another aspect that has a considerable impact on product performance is the presence of residual stresses. Depending on their character, residual stresses can have beneficial as well as detrimental effects on a range of properties such as formability, structural stability, dimensional accuracy, and fatigue performance.

The heterogeneous material flow within the process zone of linear flow splitting and linear bend splitting and the resulting strain gradient lead to the development of steep residual stress gradients in the produced flanges. The determination of these stresses is challenging due to the fine-grained, severely deformed microstructure and the strong crystallographic textures. X-ray diffraction methods on a laboratory scale are inadequate, since they are only surface sensitive and often lack the measuring sensitivity necessary for the weak reflexes that occur with strong textures. A more suitable approach is the modification of the established blind hole-drilling method for anisotropic materials (Niehuesbernd et al. 2014). Using the blind hole-drilling method, residual stresses are generally determined by measuring surface strains in dependence of the drilling depth using strain gage rosettes.

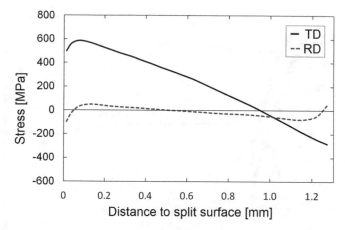

Fig. 4.6 Residual stress distributions in TD and RD of *HC 480 LA* flanges

Conventional approaches to obtain residual stress distributions from those measurements include analytical differential and integral procedures (Schwarz 1996; Schajer 1988). These methods, however, are either not suited for anisotropic materials or are only applicable for residual stresses well below 60% of the yield stress of the material. In order to overcome these effects obviously present in linear flow split flanges, we adopted a finite element (FE) assisted approach to determine the residual stresses in flow split flanges. By comparing measured surface strains to those obtained from an FE model of the drilling process in which the stresses are applied as variable predefined fields the residual stress distributions are determined in an iterative process. Following this approach, it is possible to accurately determine residual stresses even in materials with anisotropic and highly varying properties and stresses close to their yield strength. In the case of HSLA steels the typical residual stress distributions that evolve during the forming process within the flanges comprise of strong stresses in the feed direction (TD) and minor stresses in rolling direction (RD) (Fig. 4.6, coordinate system according to Fig. 4.2). At the split surface tensile stresses of nearly 600 MPa are present in feed direction. With increasing distance to the split surface the stresses decrease and transition into compressive stresses, reaching −300 MPa at the lower flange surface. The stresses in normal direction (ND) are negligible due to the low thickness of the flanges.

4.1.1 Strength

Among the material properties that are changed during metal-forming processes, the strength is certainly the one which is most often taken into account in product development. Considering the severe strains that are imposed to the material as well as the resulting ultrafine-grained microstructure, a strong increase in the strength of linear flow split profiles is to be expected.

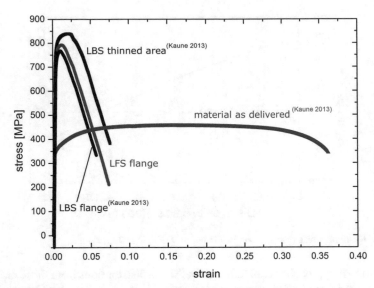

Fig. 4.7 Engineering stress strain curves of *DD11* linear flow split (LFS) and linear bend split (LBS) profiles and as delivered condition of a *DD11* mild steel

To determine the local change in strength caused by linear flow splitting and linear bend splitting, we carried out tensile tests on flat tensile specimens (12 × 4 × 0.8 mm gage dimension) from the material as delivered, the flange and thinned area of a linear bend split profile, as well as the flange of a linear flow split profile. The samples of the flange and thinned area were cut out parallel to the transverse direction (for coordinate system see Fig. 4.2) and milled down starting from the lower flange surface, resulting in samples covering the top 0.8 mm of the flange or thinned area, respectively.

The tensile test results of the flanges (Fig. 4.7) show a strong increase in strength of the material due to linear bend splitting as well as flow splitting. In comparison to the yield stress of the material as delivered of 350 MPa the yield stress of the linear flow split and linear bend split flange is increased by about 115% to 750 MPa. The yield stress of the linear bend split thinned area is even slightly higher than that and amounts to 790 MPa. However, it can be seen that the elongation at fracture of the deformed material is strongly reduced in comparison to the as-delivered material state. The linear bend splitting process reduces the elongation at fracture from 36% in the material as delivered to 5.5% in the flange and 7% in the thinned area. The linear flow split flange exhibits an elongation at fracture of 7%.

However, even though they exhibit a reduced elongation at fracture, the material behavior of the split samples is still ductile. The reason for the reduction in elongation at fracture is the fact that necking occurs already shortly after reaching the yield strength, i.e., a strong reduction in uniform elongation. Hence, a very high amount of elongation is carried by ductile necking behavior of the samples. This early necking combined with a ductile fracture behavior is a characteristic of materials with UFG microstructures and is attributed to their limited work hardening rate (Song et al. 2006; Wang and Ma 2004).

However, due to the strong microstructural gradient in the flange, a strength gradient can be expected from one surface of the tensile specimen to the other. Therefore, the results of tensile tests can only represent an average value of the top 0.8 mm of the gradient. Yet, it is necessary to resolve the mechanical properties on a finer scale for simulation purposes. Since tensile specimens are limited to a certain minimum thickness, other testing methods have to be used. Many studies revealed a linear correlation between the ultimate tensile strength and the hardness of a material with a proportionality factor of approximately 3 (Brooks et al. 2008; Pavlina and Van Tyne 2008; Gaško and Rosenberg 2011). For the highly pre-deformed materials that exhibit only minor subsequent work hardening, this approximation is also applicable to yield strength and hardness. Therefore, Vickers hardness measurements with a load of 50 g will be used in the following as a measure of strength to discuss the microstructural property gradient. The yield strength estimated from hardness measurements will be additionally displayed in the following diagramms.

The hardness profiles of *DD11* linear bend split flange and thinned area as well as of *DD11* linear flow split flange perpendicular to the surface show a very similar behavior (Fig. 4.8). It can be seen that there is a strong decrease in hardness with increasing distance to the surface that saturates above the hardness of the as-delivered condition. The hardness of the thinned area of the linear bend split profile is about 10 HV higher than the hardness of the linear bend split flange and reaches 330 at the split surface. This equals an increase of 109% in comparison to the initial *DD11* hardness of 158 HV.

The hardness characteristics of *HC 480 LA* linear flow split flanges (Fig. 4.9) are comparable to the ones of *DD11* described above. The maximum hardness of 380 HV is found close to the upper flange surface and decreases towards the

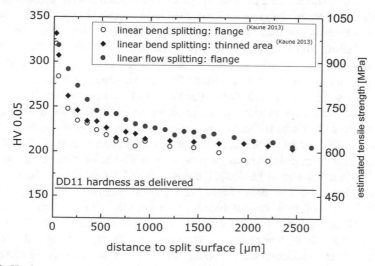

Fig. 4.8 Hardness perpendicular to the flange surface of *DD11* linear flow split and linear bend split profiles

L. Ahmels et al.

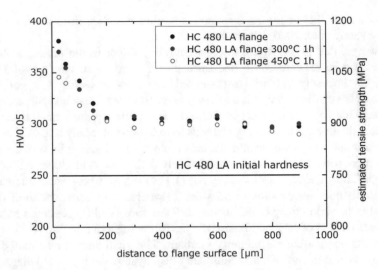

Fig. 4.9 Hardness perpendicular to the flange surface of heat-treated and non-heat-treated *HC 480 LA* linear flow split flanges (Kaune 2013)

lower side over a distance of 200 μm before leveling into a plateau of 300 HV, which is still significantly higher than the initial hardness of the material.

Thermal stability of the UFG microstructure: Microstructures generated by severe plastic deformation are far from thermodynamic equilibrium and therefore prone to grain coarsening, as their high defect density creates a strong driving force for grain growth. Investigations of the thermal stability of UFG microstructures produced by linear flow splitting revealed that at temperatures above 500 °C grain growth occurs, leading to a destruction of the UFG microstructure (Bruder 2011; Bruder 2012). Due to the grain growth, the hardness and strength of the material decrease following the Hall-Petch relationship (Sect. 4.1). However, even at temperatures below 500 °C, where no significant grain growth occurs, a drop in hardness at the upper flange surface is observed (Fig. 4.9). A heat treatment at 300 °C for 1 h causes only an insignificant change in hardness, whereas the decrease at the flange surface amounts to 10% after 1 h at 450 °C. In both cases, the hardness plateau at higher distances to the surface remains unchanged by the heat treatment. The decrease in hardness without a change in grain size cannot be explained by the Hall-Petch relationship, indicating recovery processes as source of the hardness drop. Since recovery processes can only occur in work-hardened materials, it can be concluded that the strong hardness increase caused by the linear flow splitting process is due to both work hardening and grain refinement. This means that the superior strength of the microstructure is not maintained over the same temperature range in which the grain size is thermally stable, since softening due to recovery occurs already at lower temperatures. With regard to manufacturing-induced properties and their consideration in product development, the substantial increase in strength should therefore be treated as metastable.

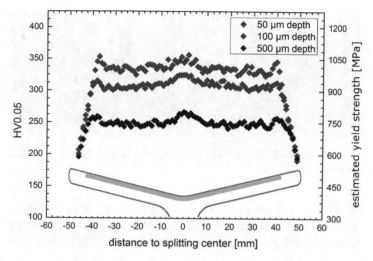

Fig. 4.10 Hardness parallel to the surface of a *DD11* linear flow split profile

Steady state: The hardness parallel to the surface of linear flow split profiles can be divided into two parts (Fig. 4.10): a linear increase from both flange tips towards the splitting center and a constant hardness in the parts of the flange apart from the flange tip area. The hardness of the plateau depends on the measuring depth, i.e., the distance to the upper flange surface. As it was shown in the measurements perpendicular to the surface, the hardness decreases with increasing distance to the splitting center. However, the fact that, aside from the very flange tips, the hardness is independent from the splitting depth implies the evolution of a steady state in the forming zone during linear flow splitting and linear bend splitting. This theory is supported not only by the fact that the process forces also saturate after a certain number of splitting steps (Sect. 3.1.3), but also by the microstructural investigations (Sect. 4.1.1) revealing constant microstructural features after the first few mm at the flange tip. Consequently, the splitting depth in linear flow splitting is not limited by a steady increase in process forces or the formability limit of the material, theoretically allowing infinite splitting depths (Müller et al. 2008).

4.1.2   Subsequent Formability

In the previous chapter the increase in strength during linear flow splitting and linear bend splitting was addressed. This increase is an important manufacturing-induced property which holds the potential to significantly improve the performance of produced parts. However, in order to expand the application range of split profiles further processing might be necessary (Sect. 3.2.1). This also includes subsequent forming operations on the flanges. The increased strength of the flanges, though, is accompanied by considerable limitations in terms of subsequent

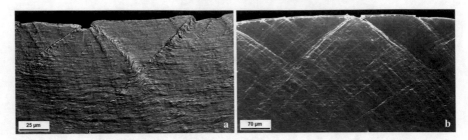

Fig. 4.11 Shear bands on *HC 320 LA* three-point bending samples, with the UFG region subjected to (**a**) compression and (**b**) tension

formability. The flanges are prone to localized deformation through shear bands, which is a well-known phenomenon for materials with ultrafine and nanocrystalline grain structures (Carsley et al. 1998; Joshi and Ramesh 2008). Such shear bands appear in the UFG region of the flanges during bending, independent of the bending direction, i.e., under compressive as well as under tensile loads (Fig. 4.11). They lead to substantial roughening of the surface and even cracking under tensile loads and therefore limit the formability of the UFG side of the flanges. This does not occur in the cold-worked material at the lower flange surface or the initial sheet material.

The most common approach to address the limited formability of cold-worked materials is the conduction of heat treatments. However, the increase in formability is usually accompanied by a decrease in yield strength. Investigations on *HC 480 LA* split profiles revealed a considerable influence of the annealing temperature on the resulting mechanical properties (Bruder et al. 2008; Bruder 2011). Annealing at temperatures below 450 °C appears to have only a minor influence on the strength in the UFG region beneath the split surface and no considerable influence on the ductility (Fig. 4.12). Heat treatments at 550 °C, however, lead to fractioning of the pancake microstructure and to moderate grain growth. Tensile tests reveal only a slight increase in uniform elongation to about 3% in combination with a distinct yield drop effect. This is accompanied by a reduction in yield stress of 15–20% compared to the untreated condition. Annealing at a temperature of 600 °C causes significant grain growth and considerably extends the uniform elongation to about 25%. The yield strength, however, drops below the level of the initial sheet metal.

Based on these results it is clear that conventional heat treatments in furnaces do not offer the possibility to improve the formability of the flanges without sacrificing most of the strength. Another approach which in addition can easily be integrated in the production line is laser annealing. This allows to selectively heat treat the areas where an increased formability is needed. Furthermore, by restricting the heat treatments to the severely deformed split surface it is possible to impede shear band formation in the UFG region while preserving the improved strength in subjacent areas of the flange. In order to assess the capabilities of laser annealing detailed investigations on laser heat-treated *HC 320 LA* flanges were

Fig. 4.12 Engineering stress–strain curves of *HC 480 LA* samples (Bruder 2011)

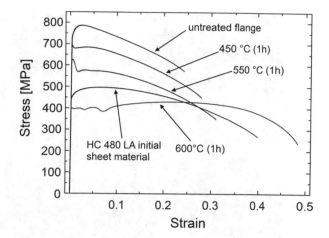

carried out. We were able to show that significant improvements can be achieved by precisely adjusting the annealing parameters (Niehuesbernd et al. 2016). Favorable properties result from heat treatments with a maximum temperature in the center of the heat-treated zone of approx. 690 °C for a fraction of a second. It was shown that such annealing conditions lead to recrystallization up to a distance of 15–20 μm beneath the split surface, resulting in an equiaxed microstructure with an average grain size of about 500 nm (Fig. 4.13). This grain size is still in the UFG regime, since the high heating rates in combination with the self-quenching effect restrict the time for grain growth. The microstructure in subjacent areas solely results from grain growth, which is accompanied by a reduction in grain aspect ratio. Hardness measurements revealed that these heat treatments effectively level out the gradient within the flange, resulting in a constant hardness of 255 HV over the flange thickness in the central area of the heat-affected zone (Fig. 4.14).

The loss in strength under these conditions is comparatively low. Tensile tests on samples with a thickness of 200 μm cut out from the heat-affected zone show a reduction in yield stress and ultimate tensile strength of approx. 20% compared to the as-processed condition (Fig. 4.15). However, since the sample only represents 20% of the total flange thickness the overall loss is considerably lower. Additionally, due to the fact that the strength exhibits virtually no gradient over the flange thickness, the plastic deformation is much more homogeneous after the heat treatment, resulting in a symmetrical necking behavior. The elongation at fracture is also increased considerably, as well as the reduction of area at fracture, which is 79% after the heat treatment compared to 34% in the as-processed condition.

The heat treatment allows flanges to be bent to a 90° angle with a bending radius of 1 mm without any cracks, which in the untreated state led to excessive cracking and severe surface roughening (Fig. 4.16). From a product development point of view, it is of particular relevance that laser annealing allows to locally overcome the manufacturing-induced limitation in formability without sacrificing the important

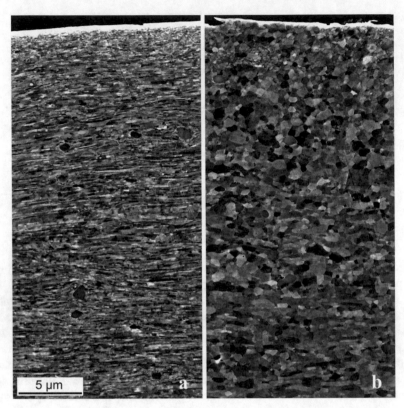

Fig. 4.13 Channeling contrast images of microstructures at the split surface of *HC 320 LA* flanges, (**a**) in untreated condition and (**b**) after laser heat treatment (Niehuesbernd et al. 2016)

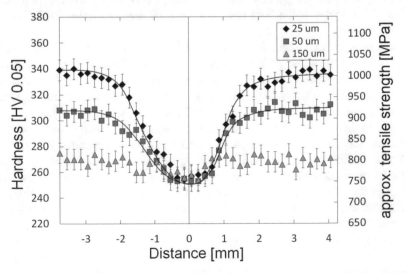

Fig. 4.14 Hardness distributions and corresponding approximated tensile strength along the heat-treated zone of a laser-annealed *HC 320 LA* flange, measured in 25, 50, and 150 μm beneath the split surface (Niehuesbernd et al. 2016)

Fig. 4.15 Engineering
stress–strain curves of *HC
320 LA* flanges in the
untreated state and
after laser annealing
(Niehuesbernd et al. 2016)

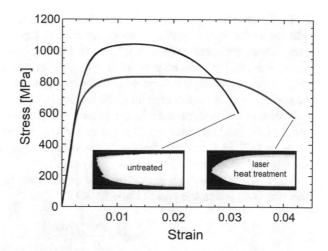

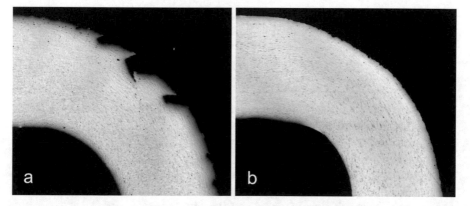

Fig. 4.16 *HC 320 LA* flanges bent to an angle of 90°, (**a**) in the untreated state and (**b**) after the laser heat treatment (bending radius 1 mm)

manufacturing-induced improvement in strength in adjacent regions. The improved formability of those laser heat-treated flanges enables the manufacturing of diverse shapes and thus expands the application range of split profiles.

4.1.3 Durability

The loading conditions of components in service are manifold. Therefore, the manufacturing-induced properties of linear flow split profiles not only under monotonic loading conditions but also under cyclic loading have to be considered during the product design and are discussed in the following.

To be able to include the manufacturing-induced properties of linear flow splitting under cyclic loading into component design, a method to estimate the durability of the components under service loading conditions from the material- and manufacturing-induced properties is needed. In case of cyclic loading, this can be achieved by using the local strain concept. It allows the determination of the fatigue strength of the component if the cyclic stress–strain behavior and the strain S–N curve of component material are known. In the case of linear flow split profiles under cyclic loading, however, it is necessary to determine the cyclic stress–strain curve as well as the strain–fatigue life correlation not only for the as-delivered material, but also for the linear flow split flanges.

The cyclic stress–strain curve can be expressed by the Ramberg-Osgood equation (Eq. (4.3)) (Ramberg and Osgood 1943):

$$\varepsilon_{a,t} = \frac{\sigma_a}{E} + \left(\frac{\sigma_a}{K'}\right)^{\frac{1}{n'}} \tag{4.3}$$

whereas the Manson-Coffin-Basquin equation (Eq. (4.4)) (Manson 1965) is used to characterize the strain S–N curve:

$$\varepsilon_{a,t} = \frac{\sigma_f'}{E} \cdot \left(2N_f\right)^b + \varepsilon_f' \cdot \left(2N_f\right)^c. \tag{4.4}$$

Both equations assess the total strain amplitude by a sum of elastic and plastic strains. The cyclic material properties ($\sigma_f', \varepsilon_f', b, c, K'$, and n') which are used to describe the cyclic material behavior are summarized in Table 4.1.

The unknown cyclic parameters of Eq. (4.3) and Eq. (4.4) are needed to describe the fatigue behavior of the material or component. They can be determined by performing strain-controlled fatigue tests with constant total strain amplitude to obtain stress–strain number of cycles to failure triples of the form

Table 4.1 Abbreviations used for the description of cyclic and monotonic material behavior

Sym.	Description	Sym.	Description
σ_a	Stress amplitude	$R_{p0,2}$	Yield limit for 0.2%
$\varepsilon_{a,t}$	Strain amplitude		Residual elongation
σ_f'	Fatigue ductility coefficient	R_m	Tensile strength
ε_f'	Fatigue strength coefficient	φ_v	Degree of deformation
b	Fatigue ductility exponent	RD	Rolling direction
c	Fatigue strength exponent	E	Young's modulus
K'	Cyclic strength coefficient	K	Static strength coefficient
n'	Cyclic strain hardening exponent	n	Static strain hardening
$\sigma_{a,k}$	Stress amplitude at knee-point		Exponent
N_k	Number of cycles at knee-point	K_f	Fatigue notch factor
N_f, N_i	Number of cycles to fracture/crack initiation	K_t	Theoretical stress–concentration factor

$$\left(\varepsilon_{a,t,1}, \left(N_{f,1}, \sigma_{a,1}\right)\right), \ldots, \left(\varepsilon_{a,t,N}, \left(N_{f,N}, \sigma_{a,N}\right)\right). \tag{4.5}$$

The cyclic material properties can be deduced using linear regressions for the elastic and plastic parts of the strain–life curve and the compatibility conditions to derive the stress–strain curve (Eqs. (4.6) and (4.7)):

$$n' = \frac{b}{c} \tag{4.6}$$

$$K' = \frac{\sigma'_f}{\left(\varepsilon'_f\right)^{n'}} \tag{4.7}$$

Experimental results and cyclic material behavior: The cyclic stress–strain curve of *HC 340 LA* was obtained by the analysis of the stabilized hysteresis loops during cyclic loading under constant amplitudes. It can be seen that the cyclic proof stress of the material is increased by the linear flow-splitting process from 338 MPa in the as-received state to 667 MPa for the flange specimens (Fig. 4.17). The difference in the cyclic Young's modulus between as-received ($E = 176$ GPa) and flange specimens ($E = 207$ GPa) can be attributed to the elastic anisotropy that

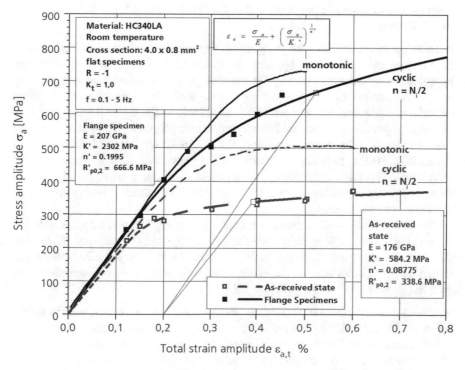

Fig. 4.17 Cyclic stress–strain curves and initial loading curves for *HC 340 LA* in as-received state and for linear flow split flange specimens

evolves during linear flow splitting (Sect. 4.1) and potentially also a stronger anelastic effect in the as-received material, considering the pronounced decrease from monotonic to cyclic curve. By further comparing the static stress–strain curves to the cyclic ones, the cyclic softening of the flange is found to be lower than the softening of the as-delivered material. For strain amplitudes exceeding 0.30% the cyclic softening in as-received state amounts to about 40%, whereas the strength reduction of the flange material is in the range of 10%.

In accordance with their higher cyclic strength, the flanges are able to endure more loading cycles when subjected to the same stress amplitude or higher stress amplitudes for a given fatigue life than the as-received material. At $N = 10^5$ cycles the as-received material endures a stress amplitude of 300 MPa and the flange material 400 MPa till fracture. The increase of 33% is somewhat lower than what is typically observed in UFG steels processed by severe plastic deformation (Okayasu et al. 2008). This is not surprising with regard to the gradient microstructure according to which only a third of the sample cross section is actually UFG, whereas the majority contains a cold-worked microstructure (Fig. 4.18). When comparing the load-controlled tests for $R = -1$ of unnotched ($K_t = 1.0$) as well as notched ($K_t = 2.3$) samples cut from flanges and from as-delivered material, it can be seen that the fatigue notch factor of the flow split material ($K_f = 1.74$) is 6% less than of one of the as-received material states ($K_f = 1.84$). This finding stands in contrast to the common experience that the notch sensitivity increases with the material strength.

Estimation of cyclic material properties: The fatigue assessment of formed parts or components relies on appropriate cyclic material properties. Different approaches for estimating the cyclic material properties have been suggested to reduce the number of time-consuming fatigue tests. One method is the uniform material law (Bäumel and Seeger 1990), which allows the estimation of the cyclic material properties of a material from its monotonic properties. In order to consider the influence of cold forming on the cyclic material behavior the material law of steel sheet (MLSS) (Masendorf 2000) and the method of variable slopes (MVS) (Hatscher 2004) have been developed. In addition to the static material behavior these two methods use the equivalent plastic strain to consider the manufacturing influences of cold working.

The linear flow-splitting process introduces very high strains into the material, leading not only to cold working but also to a strong grain refinement. Because of that, both estimations (MLSS and MVS) are limited in the usage for linear flow split materials. Thus, other methods to estimate the fatigue behavior of linear flow split material under cyclic loading are needed.

Artificial Neural Networks (ANNs): Aside from the equivalent plastic strain-based MLSS and MVS approaches, there are other methods to estimate the cyclic material parameters from monotonic material properties and manufacturing parameters, which use artificial neural networks (ANNs) (Artymiak et al. 1997; Marquardt and Zenner 2003; Marquardt et al. 2006; el Dsoki et al. 2012). An ANN consists of a network of transfer functions, which simulates the behavior of

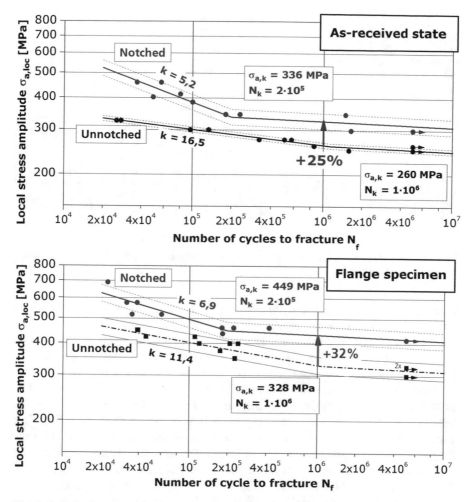

Fig. 4.18 Influence of notches on fatigue life for *HC 340 LA* in as-received state and for linear flow split flange specimens with $R = -1$

biological neurons Fig. 4.19 (Zell 1994; Scherer 1997). The transfer function consists of a weighted input and an activation function. Parameters for the configuration of an ANN are the network topology, choice of activation functions, and different models and algorithms for the training process. They are able to identify ("learn") nonlinear, complex, and hidden relationships between data and data representations (Wilamowski 2009).

A very useful capability of ANNs is the evaluation of binary variables, which facilitates a case differentiation and difficult-to-quantify parameters. In the fatigue strength lifetime estimation, this allows the inclusion of further input parameters, such as certain manufacturing parameters (thermal treatment, surface finishing), environmental conditions (temperature, ambient air, salt, water, gasoline, etc.), or

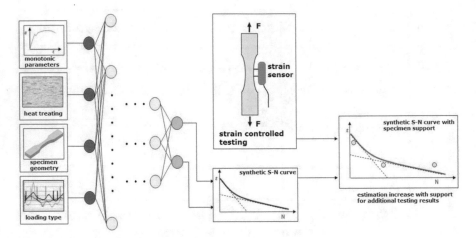

Fig. 4.19 ANN Overview: Input parameters, topology, output, and support for additional testing results (from left to right)

specimen geometry (round, flat, dimensions). ANNs offer the possibility to improve the estimation accuracy by adding additional experimental results. This method allows with just a few specimens to correct uncertainties in the estimation (el Dsoki et al. 2009). Furthermore, ANNs have the ability to adapt or improve well-known existing correlations, and mathematical or physical models. For example, this can be used for a description of the known elastic plastic material behavior and to extend it with the description of cyclic hardening/softening. Another example for the use of ANNs is the FlowSplitStrength-Toolbox (Landersheim et al. 2010). This Toolbox uses FE models based on the local elastic plastic material behavior for the fatigue strength analysis of flow split materials. The influence of the manufacturing process is considered by the residual stress field and the local work hardening (Heinrietz et al. 2011). Besides interpolation of cyclic material parameters or hardness and uniform material law (UML), ANNs are one method to take account of the history of a comparable equivalent plastic strain of flow split profiles for sheet metal and low-alloyed steels (Landersheim 2013; el Dsoki et al. 2009).

Disadvantages of ANNs are the fact that the recognized correlation or model is hidden in a black box and that ANNs work very well for interpolation problems, but only have a modest accuracy for extrapolation problems. This can lead to a very poor estimation accuracy for dataset out of the covered range of the ANN training dataset. Another challenge in the creation process of an ANN is the finding of an appropriate topology and configuration, which can be a time-consuming task, yet increasing computing power and improved algorithms can compensate this aspect. Regarding the application to innovative manufacturing technologies such as linear flow splitting, the benefit of ANNs for the estimation of the cyclic material behavior hinges on the question whether or not sufficient datasets are available for training. In view of the high equivalent plastic strains and strain gradients in linear flow split flanges (Sect. 6.3.2) most of the metal-forming processes are out of the question as references. However, there's a lot of literature on fatigue of metals processed by severe plastic deformation (Höppel et al. 2009) which could serve as potential input.

Nonparametric density and quantile estimation of the number of cycles till failure: Current approaches are only focusing on the estimation of the mean value of the number of cycles till failure as a function of the strain amplitude, based on Eqs. (4.3) and (4.4) (Williams et al. 2002). Since the real behavior can vary extremely around this mean, it is useful for the lifetime assessment to gain more information about the underlying probability distribution. This information can for example be obtained by estimating the density and the α-quantiles of the number of cycles until failure. A density of a real-valued random quantity X is a nonnegative function f, which integrates to one over $\mathbb{R}$, such that the probability that X takes on a value in an interval $[a, b]$ is given by

$$P(X \in [a, b]) = \int_a^b f(x)dx \qquad (4.8)$$

for all $a, b \in \mathbb{R}$. Even though the number of cycles until failure is actually a discrete quantity, its stochastic behavior will be approximated by a density estimate, which can be used to estimate the probability of a breakdown in a certain range.

For $\alpha \in (0, 1)$, an α-quantile is defined as the minimal value, such that a random quantity X falls below this value with a probability of at least α. Thus, for a given strain amplitude an α-quantile of the number of cycles until failure is the minimal number of cycles such that a failure occurs with a probability of (at least) α. In other words, an α-quantile is the number of cycles such that no failure occurs with a probability of approximately $(1 - \alpha)$.

In the following a description is given how the density and the α-quantile can be approximated for each material depending on the strain amplitude. To do this, data from "Materials Database For Cyclic Loading" (Boller et al. 2008) is used. In addition to results of the strain-controlled fatigue tests, static material properties are also needed. Since the experiments for obtaining the data sets

$$D^{(m)} = \left\{ \left(\varepsilon_{a,t,1}^{(m)}, \left(N_{f,1}^{(m)}, \sigma_{a,1}^{(m)} \right) \right), \ldots, \left(\varepsilon_{a,t,l_m}^{(m)}, \left(N_{f,l_m}^{(m)}, \sigma_{a,l_m}^{(m)} \right) \right) \right\}, \qquad (4.9)$$

for each of the $T = 132$ materials are very time consuming, the amount of data per material m is very limited and thus insufficient for a nonparametric approach. Therefore, our idea is to combine the data sets of related materials in order to improve the statistical power of the estimation. This is suggestive, because it can be assumed that similar materials also show similar fatigue behavior. The similarity of materials is measured by the closeness of five of their static material properties, namely Young's modulus E, the yield limit for the 0.2% residual elongation $R_{p0,2}$, the tensile strength R_m, the static strength coefficient K, and the static strain-hardening exponent n.

In our model the main assumption is that the numbers of cycles N_f until failure are given by

$$N_f^{(m)}(\varepsilon_{a,t}) = \mu^{(m)}(\varepsilon_{a,t}) + \sigma^{(m)}(\varepsilon_{a,t}) \cdot \delta^{(m)}, \qquad (4.10)$$

for each material $m = 1, \ldots, 132$, where $\mu^{(m)}(\varepsilon_{a,t})$ is the expected value, and $\sigma^{(m)}(\varepsilon_{a,t})$ the standard deviation, and both depend on the material m and the strain amplitude $\varepsilon_{a,t}$. $\delta^{(m)}$ is a random error term that depends on the material m and has expected value 0 and variance 1. Assume that $f^{(m)}$ is a density of $\delta^{(m)}$; the assumption in Eq. (4.10) directly implies that

$$g^{(m,\varepsilon_{a,t})}(\cdot) = \frac{f^{(m)}\left(\dfrac{\cdot - \mu^{(m)}(\varepsilon_{a,t})}{\sigma^m(\varepsilon_{a,t})}\right)}{\sigma^m(\varepsilon_{a,t})} \qquad (4.11)$$

is a density estimate of $N_f^{(m)}$. Analogously, a conditional quantile of $N_f^{(m)}$ is given by

$$q_{N_f^{(m)},\alpha}(\varepsilon_{a,t}) = \sigma^{(m)}(\varepsilon_{a,t}) \cdot q_{\delta^{(m)},\alpha} + \mu^{(m)}(\varepsilon_{a,t}), \qquad (4.12)$$

where $q_{\delta^{(m)},\alpha}$ is the α-quantile of $\delta^{(m)}$. Due to the transformations of Eqs. (4.11) and (4.12), the purpose of estimating the density and the α-quantile of the number of cycles $N_f^{(m)}$ until failure for each material m reduces to estimating the density and the α-quantile of $\delta^{(m)}$. Since μ and σ are unknown, estimates $\widehat{\mu}^{(m)}(\varepsilon_{a,t})$ of $\mu^{(m)}(\varepsilon_{a,t})$ and $\widehat{\sigma}^{(m)}(\varepsilon_{a,t})$ of $\sigma^{(m)}(\varepsilon_{a,t})$ are used to construct approximate data

$$\widehat{\delta}_i^{(m)} = \frac{N_{f,i}^{(m)}(\varepsilon_{a,t,i}) - \widehat{\mu}^{(m)}(\varepsilon_{a,t,i})}{\widehat{\sigma}^{(m)}(\varepsilon_{a,t,i})}, \quad i = 1, \ldots, l_m, \qquad (4.13)$$

which will be utilized as samples of $\delta^{(m)}$ to estimate the density and the α-quantile of $\delta^{(m)}$.

In order to estimate $\mu^{(m)}(\varepsilon_{a,t})$, the parametric estimator of Williams et al. (2002) is applied. It is assumed that the mean behavior of $N_f^{(m)}$ is given by the strain S–N curve of Eq. (4.4) and consequently the aim is to approximate the cyclic material parameters that determine this curve. By considering compatibility to the cyclic stress–strain curve of Eq. (4.3) a linearization happens and thus a linear least squares estimate can be applied (Williams et al. 2002).

The standard deviation $\sigma^{(m)}(\varepsilon_{a,t})$ will be estimated indirectly by approximating the variance. Therefore estimated squared deviations

$$Y_i^{(m)} = \left(N_{f,i}^{(m)} - \widehat{\mu}^{(m)}(\varepsilon_{a,t,i})\right)^2 \qquad (4.14)$$

are considered. Due to the lack of a parametric model, the method is much more complex compared to the expected value. In order to increase the expressiveness, we construct 100 artificial data points with the method described in Furer et al. (2013) and Furer and Kohler (2015). More details on this approach can be found in Bott (2015), Bott and Kohler (2016b), or Hansmann and Kohler (2016a). In this process again the data of related materials are taken into account.

With the help of the estimates of μ and σ, the estimated samples $\widehat{\delta}_1^{(m)}, \cdots, \widehat{\delta}_{l_m}^{(m)}$ of $\delta^{(m)}$ can be obtained for each material m, using Eq. (4.13). For each sample of $\delta^{(m)}$ the corresponding density can be estimated by the Rosenblatt-Parzen kernel density estimate

$$\widetilde{f}^{(m)}(y) = \frac{1}{l_m \cdot H} \cdot \sum_{k=1}^{l_m} \widetilde{K}\left(\frac{y - \widehat{\delta}_k^{(m)}}{H}\right) \tag{4.15}$$

(Parzen 1962; Rosenblatt 1956), where $H > 0$ is the bandwidth and $\widetilde{K}(x) = 3/4 \cdot (1 - x^2) \cdot I_{\{|x| \le 1\}}$ the so-called Epanechnikov kernel. Since these are not enough data points per material, we combine the density estimates by a regression estimate (Györfi et al. 2002):

$$\widehat{f}^{(m)}(y) = \frac{\sum_{i=1}^{T} I_{\{\|X^{(m)} - X^{(i)}\| \le h_1\}} \sum_{k=1}^{l_i} \widetilde{K}\left(\frac{y - \delta_k^{(i)}}{H}\right)}{H \cdot \sum_{i=1}^{T} l_i \cdot I_{\{\|X^{(m)} - X^{(i)}\| \le h_1\}}}, \tag{4.16}$$

where $h_1, H > 0$ are bandwidths. Here $X^{(i)}$ for $i = 1, \ldots, T$ is the vector of the five static material properties of the material i that have been mentioned above. In this way the data of related materials are also taken into account. If the distance (measured in the Euclidian norm) between the static material properties is smaller than the bandwidth h_1, the corresponding data is included in the density estimate. Here, the consideration of these additional data points increases the number of considered samples and therefore the power of the estimation. This approach is based on the above-mentioned idea that materials with similar static material properties show similar fatigue behavior.

The bandwidths h_1 and H of the density estimate are chosen by a modification of the combinatorial method by Devroye and Lugosi (1996) and Devroye and Lugosi (2001). Details on this method can be found in Bott (2015) and Bott and Kohler (2016a).

This approach of considering samples of similar materials can analogously be transferred to the estimation of the α-quantile by choosing

$$\widehat{q}_{\delta^{(m)}, \alpha} = \min\left\{z \in \mathbb{R}: \widehat{F}_{\delta^{(m)}}\left(z, X^{(m)}\right) \ge \alpha\right\}, \tag{4.17}$$

where

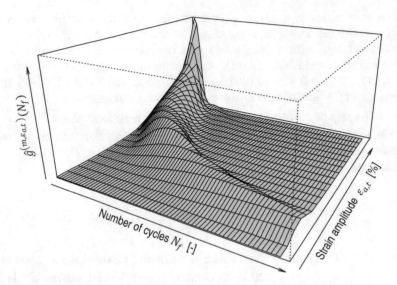

Fig. 4.20 Density estimate of $N_f^{(m)}$ for the linear flow split material of *HC 480 LA* as a function of the strain amplitude $\varepsilon_{a,t}$

$$\widehat{F}_{\delta^{(m)}}\left(z, X^{(m)}\right) = \frac{\sum_{i=1}^{T} I_{\left\{\|X^{(m)} - X^{(i)}\| \leq h_2\right\}} \sum_{k=1}^{l_i} I_{\left\{\delta_k^{(i)} \leq z\right\}}}{\sum_{i=1}^{T} l_i \cdot I_{\left\{\|X^{(m)} - X^{(i)}\| \leq h_2\right\}}} \tag{4.18}$$

is a kernel estimate of the distribution function of $\delta^{(m)}$, conditional on $X = X^{(m)}$, with a bandwidth $h_2 > 0$ (Nadaraya 1964; Watson 1964). The bandwidth h_2 of the quantile estimate is chosen by a cross-validation (Györfi et al. 2002).

The density and quantile estimates $\widehat{g}^{\left(m, \varepsilon_{a,t}\right)}$ and $\widehat{q}_{N_f^{(m)}, \alpha}$ of the number of cycles until failure can now be obtained from the estimates $\widehat{f}^{(m)}$ and $\widehat{q}_{\delta^{(m)}, \alpha}$ using the transformations of Eqs. (4.11) and (4.12).

In Fig. 4.20 the resulting density estimate is presented for the linear flow split material of *HC 480 LA*. The estimator $\widehat{g}^{\left(m, \varepsilon_{a,t}\right)}$ is shown in dependence of the strain amplitude $\varepsilon_{a,t}$. Due to the fact that the Epanechnikov kernel $\widetilde{K}$, used for the density estimate, is monotonically decreasing in the absolute value of x, the maxima of each curve for a fixed $\varepsilon_{a,t}$ describe the cyclic stress–strain curve with the estimated parameters as before. Thus, Fig. 4.20 shows how the numbers of cycles till failure $N_f^{(m)}$ vary around their expected values. As intuitively expectable, the variance of the number of cycles $N_f^{(m)}$ decreases for increasing values of the strain amplitude $\varepsilon_{a,t}$.

In Fig. 4.21 the resulting density estimates and the 5%-quantile estimates for the as-received and the linear flow split material of *HC 480 LA* are presented for the strain amplitude $\varepsilon_{a,t} = 0.12\%$. Obviously, the maximum point of the density of the linear flow split material (which as above corresponds to the expected number of

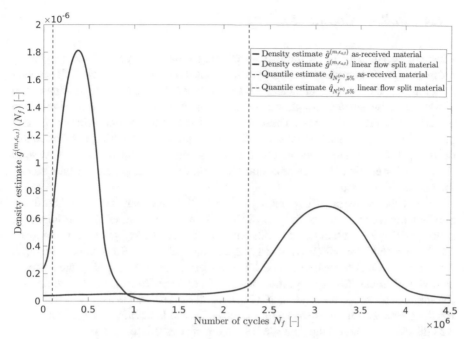

Fig. 4.21 Comparison of the estimated density $\widehat{g}^{(m,\varepsilon_{a,t})}$ and the estimated 5%-quantiles $\widehat{q}_{N_f^{(m)},5\%}$ of $N_f^{(m)}$ from the as-received and the linear flow split material of *HC 480 LA* for the strain amplitude $\varepsilon_{a,t} = 0.12\%$

cycles) appears at a higher number of cycles than the maximum point of the as-received material. This is expectable since the *S–N* curve of the flow split material (with ultrafine grain size) shows a higher fatigue life than the one of the as-received material (conventional grain size) for this strain amplitude. Due to the fact that the density of the linear flow split material has a higher variance, the number of cycles until failure is more difficult to predict and therefore the 5%-quantile of the number of cycles shows a higher deviation from the expected number than the one of the as-received material. Thus, the advantage of the higher expected number of cycles in case of the flow split material is somewhat reduced when estimating values for the number of cycles, such that no failure occurs with a probability of approximately 95%.

As a small inconsistency, both densities take positive values for negative numbers of cycles. This can be attributed to the fact that the estimates of the variances are obviously too high, which can occur since we also used data sets of related materials in our estimation. However, as the 5%-quantile estimates show, the probability that the number of cycles gets negative according to this density estimates is smaller than 5% and therefore negligible.

4.1.4 Rolling Contact Fatigue

Linear flow splitting is an innovative forming process, which allows producing bifurcated profiles with graded properties. This process leads to an UFG microstructure in the forming area and as a consequence to increased hardness and potentially low surface roughness of the flanges compared to the as-received material (Groche et al. 2007). These manufacturing-induced properties are perfect preconditions to use such structures for linear guides under rolling contact conditions (Sect. 8.1.2). However in order to utilize the potential of linear flow split structures for applications under contact loading, the rolling contact fatigue behavior has to be assessed.

Due to the characteristic geometry of linear flow split structures, standard test benches for rolling contact fatigue (RCF) investigations, which are designed for rolling motion of contact bodies both with round surfaces, e.g., bearings and gears (Hoffmann and Lipp 2002; Glover 1982; Ehrlenspiel et al. 1968), cannot be used. Therefore, we had to design a new RCF test bench for investigating the material behavior of linear flow split profiles (flat surface) under rolling contact conditions (Nommel et al. 2012; Karin et al. 2013b). In order to reduce the experimental effort for the characterization of the rolling contact fatigue behavior, a numerical model was developed, which takes the gradient microstructure into account.

Rolling Contact Fatigue Test Bench for Linear Flow Split Flanges: The test principle is based on a translational movement of a linear specimen and a rotatory movement of a rolled body, Fig. 4.22. The flow split flanges (specimen) are fixed on the sliding carriage, which is moved by linear motors. Contact pressure is realized by two crowned loading rolls, which are pressed on the linear specimen via electric motors with a maximum load capacity of 10 kN per roll in vertical direction. Thus, different slip values can be adjusted. During the test the loaded area of the rolled track is periodically monitored, e.g., with an optical camera device for detailed investigation, thus allowing the analysis of progressing surface damage.

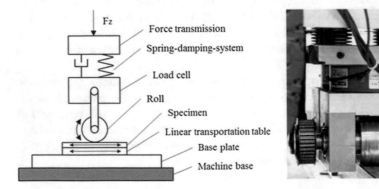

Fig. 4.22 Rolling contact fatigue test bench, configuration, and contact area (Karin 2016)

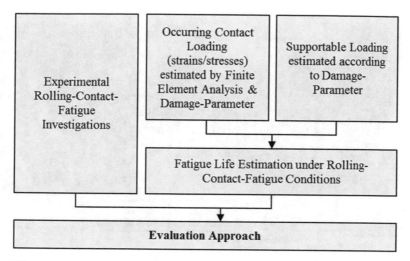

Fig. 4.23 Local strain approach for the estimation of rolling contact fatigue behavior of linear flow split flanges

Evaluation of the Rolling Contact Fatigue Behavior of Linear Flow Split Flanges by Using the Local Strain Approach: An estimation of the rolling contact fatigue behavior being exclusively based on experiments would be very elaborate. Hence, more efficient methods with predictive capabilities such as finite-element-based approaches are necessary. Some experiments are still required though to validate the numerical results. The examination of mechanical behavior of the flow split surface (higher hardness due to the UFG microstructure) is required when considering rolling contact fatigue of linear flow split components. A possible method for the numerical evaluation of the rolling contact fatigue behavior is based on the local strain approach, which takes the features at the initiated fatigue damage into account. The occurring contact loading considering the local nonlinear elasto-plastic material behavior in the contact area was simulated by finite-element calculations and used for subsequent fatigue life estimations, Fig. 4.23.

The estimation is based on a comparison of the occurring contact loading and the supportable loading. The occurring contact loading (strains/stresses) is estimated by finite-element analysis and the damage parameter according to Bergmann (1983), which considers mean-stress effects. The number of cycles to crack initiation however is estimated with the damage parameter according to Smith, Watson, and Topper (Smith et al. 1970). The coefficients used in this damage parameter are obtained from cyclic stress–strain curves from axial experiments. We have applied the approach for the as-received material state and the linear flow split flanges separately for an implementation into the finite-element model (Karin et al. 2013a; Karin et al. 2012). The cyclic material parameters of the as-received material state were taken directly from experimental tests on corresponding specimens. The particular material behavior of the linear flow split flanges depending on the graded properties, however, cannot be tested directly, as it is impossible to cut

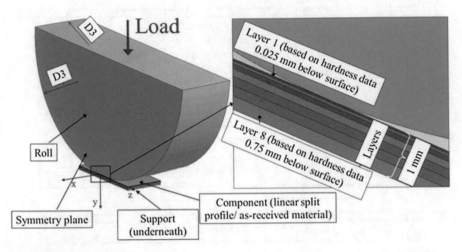

Fig. 4.24 Finite-element model with layers based on measured hardness values (Karin 2016)

sufficiently thin specimens from the flanges. Therefore, the strength gradient in direction perpendicular to the deformed surface has to be taken into account, numerically. The Hertzian theory is only valid for homogeneous materials and can't be directly applied to the flanges with a gradient microstructure. Hence, an accurate modeling of the material is required to take the gradient in the mechanical properties into account. To implement a reliable material model, we have modeled the steep gradient of the linear flow split material in the finite-element model of the contact area by layers comprising the individual mechanical properties. Figure 4.24 shows the numerical model with the implemented layers in direction perpendicular to the contact surface. The modeled geometry was realized according to the contact conditions which are included in the rolling contact fatigue test bench considering the symmetry.

The cyclic material properties describing the cyclic stress–strain curves as well as the strain–life curves with flange specimen are therefore an average value of the different layers that constitute the steep gradient. The estimation of the cyclic material parameters of the flow split flanges is based on measured hardness values, which are correlated to the ultimate tensile strength of the material, according to the median method (Meggiolaro and Castro 2004).

The calculated results (estimated strains/stresses) of the approach are compared to experimental rolling contact fatigue tests (supportable strains/stresses) under the assumption of rolling with sliding and by that considering the effect of surface roughness and residual stresses. For the linear flow split surface condition a higher endurable contact pressure was achieved. For a detailed description of the evaluation approach, which also includes residual stresses that are induced during the linear flow-splitting process, we refer to Karin et al. (2013a) and Karin (2016).

Damage Modes Under Rolling Contact Fatigue: To validate the calculated results a correlation with the damage appearance of the experimental results is

Table 4.2 Detected damage appearance under rolling contact fatigue loading depending on the material state

HC 480 LA material state	Damage appearance
As-received material	Particle formation, plastic deformation, pitting, surface flaking, groove
Linear flow split flanges	Particle formation, surface flaking

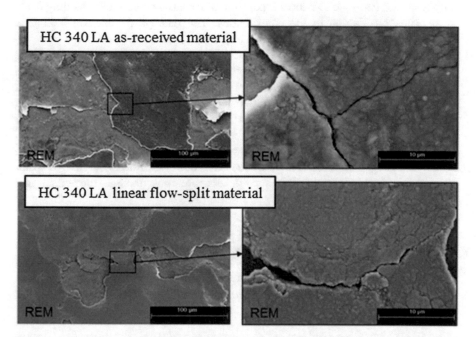

Fig. 4.25 Damage in the form of surface flaking, caused by rolling contact loading after 5×10^5 loading cycles for the *HC 340 LA*. Top: as-received material state, contact load of 600 N. Bottom: linear flow split flange, contact load of 3.000 N (Karin 2016)

necessary. We have performed the experimental rolling contact fatigue tests under constant loading amplitudes, without lubrication and with sliding using slip conditions between 0.4 and 1.2% at a frequency of 3 Hz. Within the experimental rolling contact fatigue investigations on linear flow split flanges and material in the as-received state (*HC 480 LA*), the damage types particle formation and microstructural change were identified for both material states. The occurred damages are summarized in Table 4.2 (Karin 2016).

For the as-received material state at the beginning of the loading plastic deformation occurs followed by pitting in a later stage. Pitting results from surface fatigue when the local strength of the material is exceeded. The damage mechanism surface flaking (Fig. 4.25) was detected for both material states.

In the case of linear flow split flanges no plastic deformation and pitting were observed and flaking occurred at later stages of loading. For the specimens in the as-received state the damage from flaking was observed after a certain number of loading cycles. By contrast, the flanges of the flow split material show no significant damage in former loading cycles. It is estimated that the flaking results from slip-afflicted rolling contact due to microscopic cracks on the contact surface because of surface spalling and wear. Thus, the occurring microscopic cracks on the surface propagate and coalesce to a local separation of material and ongoing loading leads to an extensive flaking in the contact area. The damage mechanism flaking is similar for both material states; see Karin (2016).

However, the plastic deformation due to the loading leads in the case of the as-received material to a hardening in both lubricated and unlubricated state. In the case of the linear flow split flanges a softening of the surface in the unlubricated state and a hardening in lubricated state can be detected. These results are based on the hardness measurement after $N_f = 5105$ loading cycles. Compressive residual stresses after the loading were detected in both cases. Furthermore a pronounced grain refinement for as-received material was observed in the rolling contact area in lateral and longitudinal direction. In the case of linear flow split flanges grain segmentation was observed. For more details regarding damage appearance and mechanisms see Karin et al. (2014) and Karin (2016).

The results of the experimental rolling contact fatigue investigation of the flow split material were evaluated in the form of Wohler curves with the Hertz pressure as the relevant loading parameter. For the considered damage criterion in the form of surface flaking, we have compared the determined curves for the as-received and flow split material in Fig. 4.26. The *HC 480 LA* flange shows 20% higher endurable Hertz pressure in comparison to the as-received material state under rolling contact

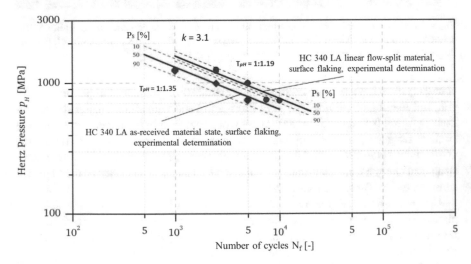

Fig. 4.26 Experimental Wohler curves for the *HC 340 LA* as-received material state and the linear flow split flanges for the considered failure damage surface flaking (Karin 2016)

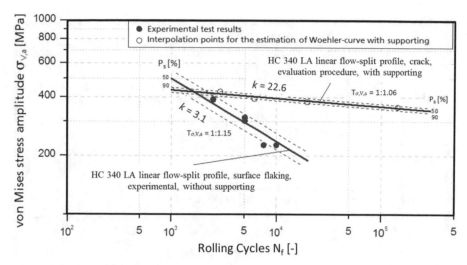

Fig. 4.27 Comparison of experimental results and numerical estimation for the damage criterion flaking for the flow split material state (Karin 2016)

fatigue resulting in about two times higher lifetime than the as-received material state. These results illustrate the relevance of manufacturing-induced properties regarding potential benefits of linear flow split flanges for applications under rolling contact fatigue.

A direct comparison of the Wohler curves in Karin (2016) for damage criterion flaking derived from experiments and numerical assessment using local strain approach shows a strong deviation from each other (Fig. 4.27). Therefore no estimation of rolling contact fatigue is possible for the criteria flaking. The same behavior can be noted for the as-received material. We suspect that a possible reason for the deviation in slope lies in the assumption of the fatigue-based damage mechanism. The experiments were conducted under non-lubricated conditions and therefore other damage mechanisms like wear can appear predominantly. In this regard, further investigations are needed to clarify the nature of the occurring damage mechanisms.

A better comparison of experimental and numerical results in Karin (2016) can be detected if pitting as damage criteria, which occurs only in case of as-received material, is considered, Fig. 4.28. The damage criteria pitting is caused by fatigue and occurs in the former loading cycles.

The damage determined for the flow split material under rolling contact fatigue results in particle formation, microstructural changes, and finally flaking; see Karin (2016). Compared to the as-received material state no plastic deformation and pitting were observed as a consequence of cyclic fatigue damage due to advantageous material properties of the flow split flanges, Table 4.2. Flaking, which is the damage appearance of the flow split material, is mainly based on the wear behavior than on fatigue. Therefore, the approach of the numerical fatigue life estimation

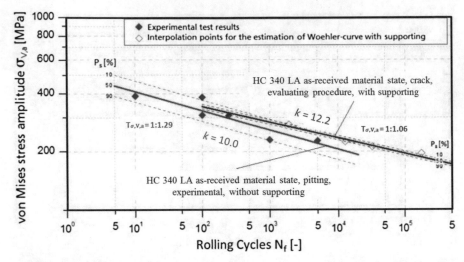

Fig. 4.28 Comparison of experimental results and numerical estimation for the damage criterion pitting for the as-received material state (Karin 2016)

does not result in an acceptable correlation with the experimental results for damage mechanism flaking, Fig. 4.27.

However, considering the damage criterion pitting, which occurs in the as-received material state, a good correlation between the experimentally determined results and the numerical fatigue life estimation could be achieved, Fig. 4.28.

4.2 Processing of Manufacturing-Induced Properties for Product Development

A huge amount of comprehensive insights into linear flow splitting is made available by examinations, analyses, and simulations, as shown in previous chapters. This comprises not only insights into the manufacturing processes but also into achievable geometrical and material properties. By applying these insights manufacturability can be ensured and the manufacturing potential of lightweight design, reliability, and functional benefits can be realized. Insights regarding the content and formalism of documentation are very heterogenic. This is all the more challenging when applying these insights during product development. Systematic identification, selection, and uniform and formalized preparation of manufacturing insights are indispensable.

The first step of a systematic procedure for processing manufacturing insights is represented by the acquisition of manufacturing-induced properties (Fig. 4.29 and Sect. 4.1).

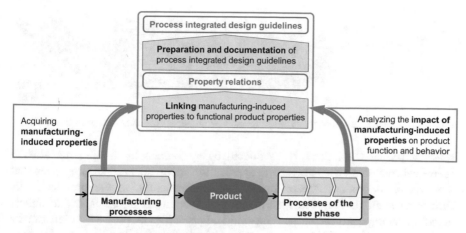

Fig. 4.29 Processing of manufacturing-induced properties for product development

Manufacturing-induced properties distinctly characterize manufacturing technologics by describing the final state of the manufactured product in terms of its geometrical, material, and mechanical properties (Gramlich 2013). The increased strength of the linear flow split flanges, for example, allows lightweight potential and offers opportunities for improved product reliability. Manufacturing-induced properties are the intersection of information relevant to product development and manufacturing. In addition to the acquisition of manufacturing-induced properties their impact on product function and behavior is analyzed with respect to processes of the product's use phase (Fig. 4.29). Insights into properties, such as rolling contact fatigue, are gathered and prepared for product development purposes.

Further steps in the systematic processing of manufacturing insights comprise the linking of manufacturing-induced properties to functional product properties to identify relevant property relations. To improve accessibility of these property relations for the product designer, preparation and documentation in the form of *process-integrated design guidelines* (PIDGs) are conducted (Wagner et al. 2014; Wagner et al. 2016) (Fig. 4.29).

4.2.1   Linking Manufacturing-Induced Properties to Functional Product Properties

The identification of relevant property relations is one of the essential aims of the procedure. The importance of property relations derives from product modeling: Property relations describe the designer's influence on the product in its functional context while determining key properties for product design. Identification of

rolling contact   notches in linear
                                   flow split flange

Fig. 4.30 Implementation of manufacturing-induced properties in linear flow split products

relevant property relations is conducted by systematically linking the already identified manufacturing-induced properties to functional product properties that are crucial to product concretization and, consequently, to the product in its functional context. Targeted identification of property relations is key to understanding the impact of manufacturing-induced properties on product design and for systematic application of manufacturing insights during product development (Wagner et al. 2016). The identified property relations are not specific to particular development projects and can be universally applied to a chosen manufacturing technology.

Increased hardness, for example, as a manufacturing-induced property of linear flow split flanges, and reduced rolling contact fatigue lead to an increased suitability of these flanges as rolling contact areas. This relates to functional properties of linear guides, where the properties of rolling contacts are crucial to realizing product function. Thus, the application of linear flow split flanges for rolling contact is advantageous. The high fatigue resistance of linear flow split flanges also matches products with areas of high stress. Arrangement of the flanges in terms of this should be considered by the designer. The decreased notch sensitivity also fits holes with a noticeable notch effect. Therefore, using flanges within these areas of the product reduces this adverse effect. Figure 4.30 provides an overview of the advantageous implementation of manufacturing-induced properties in linear flow split products.

Figure 4.31a illustrates the hardness decrease towards the end of the flanges, caused by the linear flow-splitting process. Additionally, welding at the linear flow split flanges leads to a significant reduction in hardness, which is caused by destruction of the UFG microstructure due to melting and solidification in the weld zone as well as recovery and grain growth in the heat-affected zone (Sect. 4.1.2). The combination of these two aspects results in a product design where welding seams have to be placed at flange ends to ensure the functional properties of closed chambered profile structures (Fig. 4.31b)

As a result of this processing step, property relations are identified to make their integration into product development possible. Thus, concretization of the product in its functional context with regard to manufacturing-induced properties becomes feasible.

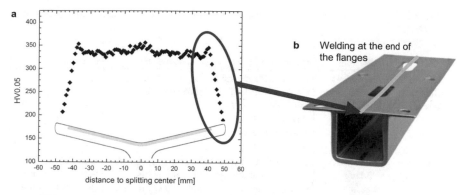

Fig. 4.31 (**a**) Hardness decrease towards the end of the flanges, (**b**) welding seam positioning for linear flow split flanges

4.2.2 Preparation and Documentation of Process-Integrated Design Guidelines

To support the designers in their decisions, the processed property relations have to be prepared and documented in such a way that the designer can apply manufacturing-induced properties in actual development projects. For this purpose, design guidelines are chosen. They are easily applicable, widely adaptable, and established in various fields, such as Design for Manufacture (DfM). This tool for documenting property relations, with a focus on suitable product solutions, is further extended to a more manufacturing process-oriented view by defining process-integrated design guidelines (Wagner et al. 2016).

The core of a PIDG is a generally valid design recommendation that supports the finding of a possible solution to a specific design problem, to realize manufacturing potential. The design recommendation defines manufacturing-induced design elements (described by manufacturing-induced properties) for realizing the product function (described by functional properties) that can be used as solution elements for product design (Wagner et al. 2016).

To ensure applicability, design guidelines consist of: an identifier, comprising information on the underlying design problem; a design recommendation and its consequence, which can also be understood as the actual guideline based on property relations; and an explanation of the guideline for when it is provided to someone without a manufacturing background (Wagner et al. 2016). Table 4.3 is an excerpt from various PIDGs that are based on processed manufacturing-induced properties for linear flow splitting. The goal of PIDGs is to support the integration of manufacturing-induced properties into product design to concretize the product with the help of manufacturing-initiated solution elements. On this basis, innovative products whose manufacturing potential has been realized can be developed systematically.

Table 4.3 Excerpt from process-integrated design guidelines (PIDGs) according to Gramlich et al. (2015)

Problem	Identifier	Design recommendation	Consequence	Explanation
Guided longitudinal motion	PIDG 4.1: Linear flow split flanges as rolling contact area	Use **linear flow split flanges** with *UFG microstructure* to realize a ***rolling contact area*** of a linear guide	High longevity and surface quality	UFG microstructure with high hardness and low surface roughness
Welding	PIDG 4.2: Placing welding seams at the end of flanges	Use **welding seams** positioned at the *end of the flanges* to assemble ***structural elements***	Minimizing effects of welding on material within the UFG microstructure of the bifurcated profile	Hardness of the flanges decreases towards their end. Thus, welding leads to a less significant reduction in UFG material properties
Transfer forces	PIDG 4.3: Linear flow split flanges in highly stressed areas	Place **linear flow split flanges** with *UFG microstructure* in ***highly stressed components*** of the profile, which are relevant for transferring forces	Increased capacity to withstand stresses	UFG microstructure with increased capacity to withstand stresses (static, cyclic)
Notch effect	PIDG 4.4: Positioning of form elements in the flange area	Use the **flange area** with *low notch sensitivity* for adding ***functional elements*** to the profile	Increased hold capacity of the form element	UFG microstructure is characterized by a lower notch sensitivity, which can be seen by comparing Wohler curves

4.3 Benefits: Some Examples

4.3.1 Mathematical Optimization of Stringer Sheets

The linear bend-splitting process offers the possibility to realize stiffening ribs at virtually any position of a sheet metal without additional joining operations (Sect. 3.1.1). Therefore, it stands to reason to apply this technology for producing stringer-stiffened sheets. This offers the advantage of avoiding weak spots that would be introduced by welding and could significantly reduce lifetime, particularly under cyclic loads. Furthermore, the manufacturing-induced properties described in Sect. 4.1 allow additional improvements in terms of stiffness and stability. By utilizing the improved strength and the increased Young's modulus in longitudinal direction in a synergetic manner, it is possible to develop stringer sheets with significantly higher stiffness compared to those produced differentially by joining separate sheets.

The primary failure mode of stringer sheets which defines the maximum load capacity is buckling at the free ends of the stringers under compressive bending loads. This limits the height of the stringers and therefore the achievable stiffness. An increase in Young's modulus and strength in the areas that bear the highest loads counteracts this issue and enables advantageous stringer configurations.

To objectively assess the improvement capabilities, structurally optimized stringer sheets produced conventionally and by linear bend splitting need to be compared under the same loading conditions. In the present case this was done by comparing the behavior of stringer sheet FE models with different material properties subjected to equal bending loads. The cross sections of the sheets were optimized mathematically for maximum stiffness. Sheets with preassigned dimensions $l = 600$ mm, $w = 300$ mm, and $t = 1.85$ mm having two stringers with variable height h, width b, and distance d from the edge of the sheet were modeled using *ABAQUS* (Fig. 4.32). The cross sections of the stringers were kept at a constant value of 22.5 mm^2 so that h and b are dependent variables. The two stringers were modeled symmetrically to the centerline of the base sheet. All implemented properties were measured on split profiles produced from *HC 320 LA* sheet. Three different cases with different material properties were investigated:

- **Conventional model** with homogeneous material properties throughout the whole model including the stringers and an isotropic Young's modulus of 210 GPa. A single flow curve with a yield stress of 330 MPa was applied (green in Fig. 4.32). This model represents a differentially produced part, generated by welding stringers to a sheet but ignoring changes in properties within the weld seam.
- **Intermediate model** with increased strength for the stringers but the same isotropic Young's modulus of 210 GPa as in the conventional model. The stringers and adjacent regions were assigned additional flow curves with yield stresses of approx. 570 MPa and 900 MPa, respectively (red and blue in Fig. 4.32). This model is meant to demonstrate the sole impact of locally increased strength after linear bend splitting.
- **Linear bend split model** with increased strength and anisotropic elastic properties for the stringers. The implemented flow curves are equal to the intermediate model. For the elastic properties of the stringers and adjacent regions (red and blue in Fig. 4.32), the orthotropic stiffness tensor of linear bend split material was used (Sect. 4.1.1).

In the FE model, movement in Z-direction was prohibited by boundary conditions at one end of the stringer sheet but the stringers were allowed to move laterally to avoid buckling restriction. At the other end a line load of 300 N was applied in Y-direction. To prevent stress concentrations, the force was applied through a small undeformable segment with a width of 8 mm. The optimization process was done

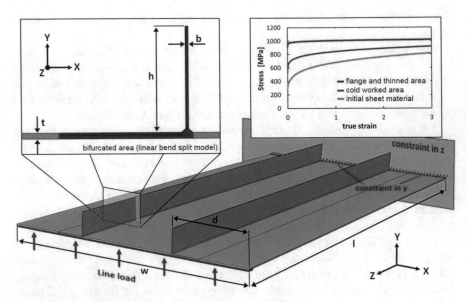

Fig. 4.32 Geometry of the FE models, including the definition of the stringer dimensions (*top left*) and the implemented flow curves for the three defined regions (*top right*)

by coupling the *ABAQUS* FE model with the NLopt library using a Python script. The derivative-free method COBYLA was used to solve the optimization problem. The position *d* and the width *b* were defined to be the optimization variables. The stringer height *h* resulted from the volume constancy of the stringers. The maximum displacement of the sheet in *Y*-direction at the point of the applied force was defined as objective function and was minimized iteratively with the additional constraint of prohibited buckling. In this case buckling was defined as a displacement of the stringer tip of more than 50 μm in *X*-direction at any given point. This value was chosen after being found to be a tripping point in the model, after which the deformation becomes unstable and the stringers buckle rapidly. All three geometries feature the same load capacity after the optimization process, since the applied force is the maximum applicable without buckling.

The optimization results reveal a clear advantage of the stringer sheet produced by linear bend splitting compared to the one produced differentially. The optimized cross sections of the stringer sheets are shown in Fig. 4.33. For the conventional model the optimized combination of stringer height *h* and position *d* results in a minimized maximum displacement in *Y*-direction of about 6.95 mm. The increase in strength alone appears to have only a marginal effect on the stiffness, since the decrease in maximum displacement in the intermediate model is less than 2% compared to the conventional model. It is worth noting that the optimization towards maximum stiffness results in a slight decrease in stringer height in the intermediate model compared to the conventional model. In the linear bend split

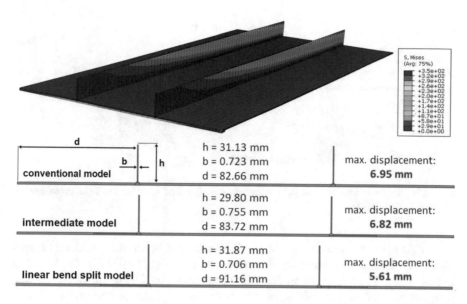

conventional model		h = 31.13 mm b = 0.723 mm d = 82.66 mm	max. displacement: **6.95 mm**
intermediate model		h = 29.80 mm b = 0.755 mm d = 83.72 mm	max. displacement: **6.82 mm**
linear bend split model		h = 31.87 mm b = 0.706 mm d = 91.16 mm	max. displacement: **5.61 mm**

Fig. 4.33 Resulting cross sections for the three cases after the optimization procedure and stress distribution in optimized stringer sheet (linear bend split model)

model the optimization results in a shift of the position d towards the centerline of the sheet and in a small increase of the stringer height h, which allows a reduction of the maximum displacement to 5.61 mm. This corresponds to an increase in bending stiffness of more than 19% compared to the conventional model. This is particularly noticeable, since the increase in Young's modulus is only about 11% in the stringers in load direction (Z-direction). The synergetic effect between the increased strength and Young's modulus allows parameter combinations of stringer width, height, and position which result in a further increased stiffness as compared to what would be possible with improved strength or Young's modulus alone.

The example demonstrates that by taking manufacturing-induced properties into account and integrating them into the design process to optimize the shape of the part accordingly, products with significantly improved properties can be developed.

4.3.2   Light Crane System

The full potential of a specific manufacturing technology such as linear flow splitting cannot be characterized by only comparing its manufacturing-induced properties to the as-received material (Sect. 4.1). Instead, product designers have to comprehensively consider how the provided manufacturing-induced properties can be beneficially utilized in product design, illustrated by the following example of a light crane system.

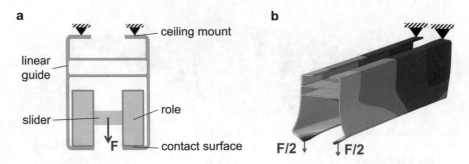

Fig. 4.34 (**a**) Reference for the light crane system, (**b**) load case for the light crane system (Gramlich et al. 2015)

A light crane system consists of a ceiling-mounted linear guide with an internal slider. A reference system made of an aluminum alloy (*EN AW-6061-T1*) is shown in Fig. 4.34a. The most relevant load case is approximated by a four-point bending setup comprised of the ceiling mount at the ends of the profile and the slider positioned in the middle of the profile where the deflection, induced by force F, is most significant. As shown in Fig. 4.34b, the highest strain is in the area of rolling contact (Gramlich et al. 2015).

The light crane system can be realized by linear flow splitting in the context of a continuous flow production. Linear flow split profiles (material: *HC 340 LA*), characterized by structural bifurcations and high strength and hardness at the flanges, are especially suitable for realizing the light crane system's linear guide and its rolling contact areas (Gramlich et al. 2015). Stepwise integration based on PIDGs supports a comprehensive realization of linear flow splitting's design possibilities (Sect. 4.2 and Table 4.3). The impact of each guideline on product design is shown in Table 4.4.

The final design of the linear flow split linear guide is shown in Table 4.4, Step 4, where the slider and the ceiling mount are already attached to the linear guide. Regarding the aforementioned load case, the final design of the profile is characterized by a 2.5 times greater longevity than the reference (Fig. 4.35), which also means increased reliability of the linear guide. The load per mass (load: external load *F* on the slider; mass: mass of the profile) could be increased by about 67%; the stiffness of the structure could be increased by about 100%; and the deflection is reduced by about 50% (Fig. 4.35). Thus, the potential of linear flow splitting is realized within the light crane system's linear guide by systematically integrating manufacturing-induced properties, resulting in an improved lightweight design. Additional functional benefits are realized in the form of integrated fastening elements with increased capacity to secure the ceiling mount.

Table 4.4 Stepwise integration of manufacturing-induced properties (Gramlich et al. 2015)

Process-integrated design guidelines and impact on the design of a light crane system	Design integration
Step 1—Integration of PIDG 4.1 The linear flow split flanges are used to realize the highly stressed rolling contact area at the bottom of the profile *Thus, the properties of linear flow split flanges regarding reduced rolling contact fatigue are used to increase the longevity of the profile*	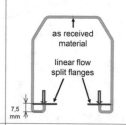
Step 2—Integration of PIDG 4.2 To increase stiffness of the profile in the highly strained area at the bottom of the profile, the profile structure is locally closed using welding seams. Seams are positioned at the end of the flanges to minimize the effects of welding on the UFG microstructure *Thus, the stiffness of the profile is increased and material properties, such as high hardness and high rolling contact fatigue of the flanges, are preserved*	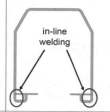
Step 3—Integration of PIDG 4.3 Additional linear bend split flanges are placed at the top of the profile that is highly stressed in the four-point bending load case *Thus, the increased capacity of the linear flow split flanges to withstand stress (static and cyclic) is used to increase the load capacity of the whole profile without increasing its mass*	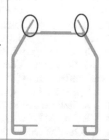
Step 4—Integration of PIDG 4.4 Existing flanges are used to place fastening elements for the intended ceiling mount because of the increased hold capacity of form elements placed at the flanges *Thus, additional benefits in resisting prevailing stresses are generated when mounting the profile*	

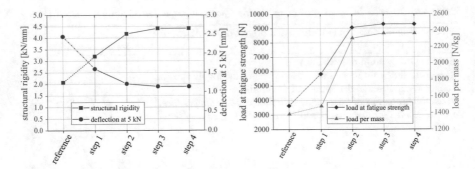

Fig. 4.35 Benefits of the stepwise integration of manufacturing-induced properties compared to the reference system (Gramlich et al. 2015)

References

Artymiak P, Bukowski L, Feliks J (1997) Forecasting of durability of machin components using artificial neural network. In: 3rd International Conference on Neural Networks, Kule 430, Oct 1997

Bäumel A, Seeger T (1990) Materials data for cyclic loading, Mater Sci Monog 61, Elsevier

Bergmann JW (1983) Zur Betriebsfestigkeit gekerbter Bauteile auf der Grundlage der örtlichen Beanspruchungen. Dissertation, TU Darmstadt

Bödecker J (2013) Randschichtmodifikation von integral verzweigten Blechprofilen mit UFG Gradientengefügen. Dissertation, TU Darmstadt

Boller C, Seeger T, Vormwald M (2008) Materials database for cyclic loading. Fachgebiet Werkstoffmechanik, TU Darmstadt

Bott A (2015) Nichtparametrische Schätzung einer (bedingten) Verteilung ausgehend von Daten mit zusätzlichen Messfehlern. Dissertation, TU Darmstadt

Bott A, Kohler M (2016a) Adaptive estimation of a conditional density. Int Stat Rev 84:291–316

Bott A, Kohler M (2016b) Nonparametric estimation of a conditional density. (To appear in Ann I Stat Math)

Brooks I, Lin P, Palumbo G et al (2008) Analysis of hardness-tensile strength relationships for electroformed nanocrystalline materials. Mat Sci Eng A 491:412–419

Bruder E, Bohn T, Müller C (2008) Properties of UFG HSLA steel profiles produced by linear flow splitting. Mater Sci Forum 584–586:661–666

Bruder E (2011) Thermische Stabilität von Stählen mit ultrafeinkörnigen Gradientengefügen und deren mechanische Eigenschaften. Dissertation, TU Darmstadt

Bruder E (2012) The effect of deformation texture on the thermal stability of UFG HSLA steel. J Mater Sci 47:7751–7758

Carsley JE, Fisher A, Milligan WW, Aifantis EC (1998) Mechanical behavior of a bulk nano-structured iron alloy. Metall Mater Trans A 29A:2261–2271

Devroye L, Lugosi G (1996) A universally acceptable smoothing factor for kernel density estimation. Ann Stat 24:2499–2512

Devroye L, Lugosi G (2001) Combinatorial methods in density estimation. Springer, New York

Ehrlenspiel K, Kugelfischer Georg Schäfer & Co. (1968) Vorrichtung zur Bestimmung der Wälzfestigkeit von rotationssymmetrischen Körpern. Patentschrift Nr. 1 260 824

el Dsoki C, Hanselka H, Kaufmann H et al (2009) Das ANSLC Programm zur Abschätzung zyklischer Werkstoffkennwerte. Materialwissenschaften und Werkstofftechnik 40(8):612–617

el Dsoki C, Lohmann F, Hanselka H et al (2012) Influence of the topology of an artificial neural network on the results for estimating the cyclic material proper-ties. Mater Sci Eng 43 (8):681–686

Furer D, Kohler M, Krzyżak A (2013) Fixed-design regression estimation based on real and artificial data. J Nonparametr Stat 25:223–241

Furer D, Kohler M (2015) Smoothing spline regression estimation based on real and artificial data. Metrika 78:711–746

Gaško M, Rosenberg G (2011) Correlation between hardness and tensile properties in ultra-high strength dual phase steels—short communication. Mater Eng 18:155–159

Glover D (1982) A ball-rod rolling contact fatigue tester, rolling contact fatigue testing of bearing steels. In: Hoo JJC (ed) ASTM STP 771, pp 107–124

Gramlich S (2013) Vom fertigungsgerechten Konstruieren zum produktionsintegrierenden Entwickeln: Durchgängige Modelle und Methoden im Produktlebenszyklus. Dissertation, TU Darmstadt

Gramlich S, Roos M, Ahmels L, Kaune V, Müller C, Bauer O, Karin I, Tomasella A, Melz T (2015) Ein wissensbasierter fertigungsintegrierender Produktentwicklungsansatz. In: Binz H, Bertsche B, Bauer W, Roth D (eds) Stuttgarter Symposium für Produktentwicklung: Entwicklung smarter Produkte für die Zukunft, Stuttgart, 19 June 2015

Groche P, Vucic D, Jöckel M (2007) Basics of linear flow splitting. J Mater Process Technol 183:249–255

Györfi L, Kohler M, Krzyżak A, Walk H (2002) A distribution-free theory of nonparametric regression. Springer, New York

Hall EO (1951) The deformation and ageing of mild steel: III discussion of results. Proc Phys Soc Lond 64:747–753

Hansmann M, Kohler M (2016a) Estimation of quantiles from data with additional measurement errors. (To appear in Stat Sinica)

Hansmann M, Kohler M (2016b) Estimation of conditional quantiles from data with additional measurement errors

Hatscher A (2004) Abschätzung zyklischer Kennwerte von Stählen. Dissertation, TU Clausthal

Heinrietz A, Diefenbach C, Landersheim V et al. (2011) Potential innovativer Methoden für den Betriebsfestigkeitsnachweis unter Berücksichtigung von Werkstoff und Fertigung. Presented at DVM Tagung Betriebsfestigkeit, 2011

Hoffmann G, Lipp K (2002) Design for rolling contact fatigue 2002 advances in powder metal-lurgy & particulate materials. In: Conference Proceedings 2002 P/M World Congress, MPIF, Orlando, 2002

Höppel HW, Mughrabi H, Vinogradov A (2009) Fatigue Properties of Bulk Nanostructured Materials. In: Zehetbauer MJ, Zhu YT (eds) Bulk nanostructured materials. Wiley, Weinhein, pp 481–500

Hornbogen E (2006) Werkstoffe—Aufbau und Eigenschaften von Keramik-, Metall-, Polymer-, und Verbundwerkstoffen. Springer, Heidelberg

Hughes D, Hansen N (1997) High angle boundaries formed by grain subdivision mechanisms. Acta Mater 45(9):3871–3886

Iwahashi Y, Furukawa M, Horita Z et al (1998) Microstructural characteristics of ultrafine-grained aluminum produced using equal-channel angular pressing. Metall Mater Trans A 29 (9):2245–2252

Joshi SP, Ramesh KT (2008) Grain size dependent shear instabilities in body centered and face-centered cubic materials. Mat Sci Eng A 493:65–70

Karin I, Lommatzsch N, Lipp K et al. (2012) Gestaltung von Wälzkontakten mit spaltprofilierten Flanschen. In: Tagungsband des 4. Zwischenkolloquiums vom SFB666, Darmstadt, 14–15 Nov 2012, pp 123–130

Karin I, Tomasella A, Landersheim V et al (2013a) Application of the local strain approach on a rolling point contact model. Int J Fatigue 47:351–360

Karin I, Hößbacher J, Lipp K et al (2013b) Prüfung linearer Bauteile auf Wälzfestigkeit. Mater Test 55(1):12–16

Karin I, Wagner C, Lipp K et al. (2014) Innovative Linearsysteme mit spaltprofilierten Wälzkontaktflächen—Nutzung der Möglichkeiten der kontinuierlichen Fließfertigung zur Funktionsintegration. In: Tagungsband des 5. Zwischenkolloquiums vom SFB666, Mörfelden-Walldorf, 19–20 Nov 2014, pp 123–132

Karin I (2016) Zur Verwendung von durch Spaltprofilieren hergestellten Blechprofilen als wälzbeanspruchte Oberflächen im Vergleich zum Ausgangszustand. Dissertation, TU Darmstadt

Kaune V (2013) Entstehung und Eigenschaften von UFG Gradientengefügen durch Spaltprofilieren und Spaltbiegen höherfester Stähle. Dissertation, TU Darmstadt

Landersheim V, Rullbaum F, Jöckel M et al. (2010) Untersuchung schädigungsmechanischer Ansätze an UFG Gefügebauteilen sowie Bewertung ihrer Schwingfestigkeit mit Hilfe der FEM, SFB 666 3. Zwischenkolloquium des Sonderforschungsbereichs 666, Darmstadt, Sept 2010

Landersheim V (2013) Numerische Schwingfestigkeitsbewertung inhomogener Spaltprofile mit dem örtlichen Dehnungskonzept. Dissertation, TU Darmstadt

Langdon T, Furukawa M, Nemoto M et al (2000) Using equal-channel angular pressing for refining grain size. JOM 52(4):30–33

Manson SS (1965) Fatigue: a complex subject—some simple approximation. Exp Mech 5:193–226

Marquardt C, Zenner H (2003) Neurolebensdauer-Vorhaben Nr. 346, Forschungskuratorium Maschinenbau e.V., Frankfurt am Main

Marquardt C, Bacher-Höchst M, Zenner H (2006) Prognose von Werkstoff- Bauteilwöhlerelinien mit künstlich neuronalen Netzen. DVM-Bericht 673:145–154

Masendorf R (2000) Einfluss der Umformung auf die zyklischen Werkstoffkennwerte von Feinblech. Dissertation, TU Clausthal

Meggiolaro MA, Castro JTP (2004) Statistical evaluation of strain life fatigue crack initiation predictions. Int J Fatigue 26:463–476

Müller C, Bohn B, Bruder E (2007) Severe plastic deformation by linear flow splitting. Mater Sci Eng Tech 38(10):842–854

Müller C, Bohn T, Bruder E et al (2008) UFG microstructures by linear flow splitting. Mater Sci Forum 584–586:68–73

Nadaraya EA (1964) On estimating regression. Theor Probab Appl 9:141–142

Niehuesbernd J, Müller C, Pantleon W (2013) Quantification of local and global elastic anisotropy in ultrafine grained gradient microstructures, produced by linear flow splitting. Mater Sci Eng A 560:273–277

Niehuesbernd J, Bruder E, Müller C (2014) Influence of gradients in the elastic anisotropy on the reliability of residual stresses determined by the hole drilling method. Adv Mater Res 996:289–294

Niehuesbernd J, Monnerjahn V, Bruder E, Groche P, Müller C (2016) Improving the formability of linear flow split profiles by laser annealing. Material Wiss Werkst (Submitted for publication)

Nommel A, Karin I, Hößbacher J (2012) Ein wahrer Antriebs-Kraftakt. Der Konstrukteur 2012 (11):10–12

Okayasu M, Sato K, Mizuno M et al (2008) Fatigue properties of ultra-fine grained dual phase ferrite/martensite low carbon steel. Int J Fatigue 30:1358–1365

Parzen E (1962) On the estimation of a probability density function and the mode. Ann Math Stat 33:1065–1076

Pavlina EV, Van Tyne C (2008) Correlation of yield strength and tensile strength with hardness for steels. JMEPEG 17:888–893

Prangnell P, Bowen J, Gholinia A (2001) The formation of submicron and nanocrystalline grain structures by severe plastic deformation. In: Dinesen et al. (ed.) Proceedings of the 22nd Risø International Symposium on Materials Science, Roskilde, 2001

Ramberg W, Osgood WR (1943) Description of stress-strain curves by three parameters. Technical Note No. 902. National Advisory Committee for Aeronautics, Washington DC

Rösler J, Harders H, Bäker M (2008) Mechanisches Verhalten der Werkstoffe. Vieweg+Teubner, Wiesbaden

Rosenblatt M (1956) Remarks on some nonparametric estimates of a density function. Ann Math Stat 27:832–837

Schajer GS (1988) Measurement of non-uniform residual stresses using the hole drilling method. J Eng Mater Technol 110:338–343

Scherer A (1997) Neuronale Netze—Grundlagen und Anwendungen, Vieweg-Verlag

Schwarz T (1996) Beitrag zur Eigenspannungsermittlung an isotropen, anisotropen sowie inhomogenen, schichtweise aufgebauten Werkstoffen mittels Bohrlochmethode und Ringkernverfahren. Dissertation, Staatliche Materialprüfanstalt (MPA), Universität Stuttgart

Smith KN, Watson P, Topper TH (1970) A stress-strain function for the fatigue of metals. J Mater 5(4):767–778

Song R, Ponge D, Raabe D et al (2006) Overview of processing, microstructure and mechanical properties of ultrafine grained bcc steels. Mater Sci Eng A 441:1–17

Valiev R (2004) Nanostructuring of metals by severe plastic deformation for advanced properties. Nat Mater 3:511–516

Valiev R, Langdon T (2006) Principles of equal-channel angular pressing as a processing tool for grain refinement. Prog Mater Sci 51:881–981

Valiev R, Estrin Y et al (2006) Producing bulk ultrafine-grained materials by severe plastic deformation. JOM 58(4):33–39

Wagner C, Gramlich S, Kloberdanz H (2014) Entwicklung innovativer Produkte durch Verknüpfung von Funktionsintegration und Fertigungsprozessintegration. In: Krause D, Paetzold K, Wartzack S (eds) Proceedings of the 24th Symposium Design for X, Bamberg, 1–2 Oct 2014, pp 361–372

Wagner C, Roos M, Gramlich S, Kloberdanz H (2016) Process integrated design guidelines: systematically linking manufacturing processes to product design. In: Marjanovic D, Storga M, Pavkovic N, Bojcetic N, Skec S (eds) Proceedings of the DESIGN 2016 14th International Design Conference, Dubrovnik, pp 739–748

Wang YM, Ma E (2004) Strain hardening, strain rate sensitivity, and ductility of nanostructured metals. Mater Sci Eng A 375–377:46–52

Watson GS (1964) Smooth regression analysis. Sankhya Series A 26:359–372

Wilamowski BM (2009) Neural network architectures and learning algorithms. IEEE Ind Electron Mag 3(4):56–63

Williams CR, Lee YL, Rilly JT (2002) A practical method for statistical analysis of strain-life fatigue data. Int J Fatigue 25:427–436

Zell A (1994) Simulation Neuronaler Netze, Oldenbourg-Verlag

Chapter 5
Finding the Best: Mathematical Optimization Based on Product and Process Requirements

H. Lüthen, S. Gramlich, B. Horn, I. Mattmann, M. Pfetsch, M. Roos, S. Ulbrich, C. Wagner, and A. Walter

The challenge of finding the best solution for a given problem plays a central role in many fields and disciplines. In mathematics, best solutions can be found by formulating and solving optimization problems. An optimization problem consists of an objective function, optimization variables, and optimization constraints, all of which define the solution space. Finding the optimal solution within this space means minimizing or maximizing the objective function by finding the optimal variables of the solution. Problems, such as geometry optimization of profiles (Hess and Ulbrich 2012), process control for stringer sheet forming (Bäcker et al. 2015), and optimization of the production sequence for branched sheet metal products (Günther and Martin 2006), are solved using mathematical optimization methods (Sects. 5.2 and 5.3). A variety of mathematical optimization methods is comprised within the field of engineering design optimization (EDO) (Roy et al. 2008).

Finding the optimal solution in the context of product and process development means finding appropriate product properties and process parameters. An analytical solution to this product and process optimization problem is not possible in many cases; numerical methods have to be used. *Algorithm-based product and process development* processes provide a promising way to deal with the

H. Lüthen (✉) • M. Pfetsch
Research Group Discrete Optimization (DOpt), Technische Universität Darmstadt, Darmstadt, Germany
e-mail: luethen@mathematik.tu-darmstadt.de

S. Gramlich • I. Mattmann • M. Roos • C. Wagner
Institute for Product Development and Machine Elements (pmd), Technische Universität Darmstadt, Darmstadt, Germany

B. Horn • S. Ulbrich • A. Walter
Research Group Nonlinear Optimization (NOpt), Technische Universität Darmstadt, Darmstadt, Germany

© Springer International Publishing AG 2017
P. Groche et al. (eds.), *Manufacturing Integrated Design*,
DOI 10.1007/978-3-319-52377-4_5

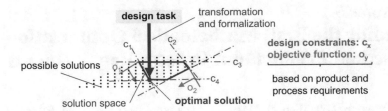

Fig. 5.1 Algorithm-based solution finding according to Birkhofer (2005) and Birkhofer and Wäldele (2005)

complexity caused by the huge number of correlating product and process requirements (Groche et al. 2012). Applying specialized algorithms allows a fast and efficient solution finding for all problems mentioned above. By determining variables, like, for example, process parameters, the arrangement of nodes and edges within a profile cross section, or the dimensions of bifurcations and stringers, an optimal solution with regard to specific properties, like, for example, stiffness, can be found.

Focusing product development, the challenges are to develop innovative products that have maximum value for the customer, generate economic benefits, and satisfy stakeholder expectations. Hence, product and process requirements have to be considered equally during solution finding. Product design carried out in accordance with established approaches, such as VDI 2221 (1993) and Axiomatic Design (Suh 1998), consist of a step-by-step procedure to concretize product properties by repeating variation and selection procedures (Birkhofer and Wäldele 2005). Ensuring manufacturability often results in time-consuming iterations. By contrast, algorithm-based solution finding starts from the possible solution space, then delimits the solution space, based on elaborated product and (manufacturing) process requirements in the form of constraints (Birkhofer and Wäldele 2005) (Fig. 5.1 red-framed border). The solution space comprises conventional solutions, characterized by established designs, as well as unconventional solutions, which realize manufacturing potential, in the form of innovative solutions. Implemented objective functions based on product and process requirements define where the optimal solution can be found within the solution space (Fig. 5.1). Appropriate optimization methods have to be applied to efficiently generate the optimal product solution. To ensure the applicability of mathematical optimization methods, the design task has to be systematically formalized.

The consideration of process requirements is relevant for developing and realizing the intended product. Manufacturing processes in particular provide additional potential for product design. Thus, the consideration of manufacturing processes is essential to formalize the design task by extending the solution space with further *manufacturing-initiated solutions*. Specific solutions are generated that realize the potential of the determined manufacturing technologies to optimally fulfill customer and stakeholder requirements of product use and function. Before

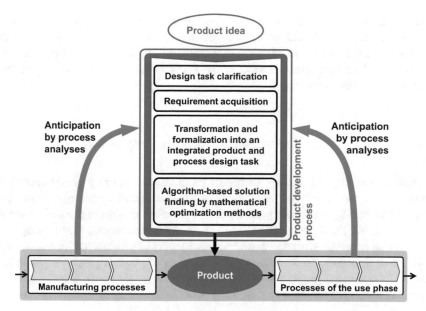

Fig. 5.2 Manufacturing-integrated algorithm-based product design approach

mathematical optimization is carried out, manufacturing technologies and processes have to be analyzed and appropriate *manufacturing-induced properties* need to be integrated into the design task during the early phases of product design. The *manufacturing-integrated algorithm-based product design approach* (Fig. 5.2) consists of four steps which are classified as follows:

1. Design task clarification
2. Requirement acquisition
3. Transformation and formalization into an integrated product and process design task, with concurrent integration of manufacturing insights, supported by prior identification and determination of appropriate manufacturing technologies
4. Algorithm-based product and process solution finding using mathematical optimization methods

5.1 Formalization of the Design Task

Finding the best solution using mathematical optimization methods requires a proper formulation of the optimization problem (Fig. 5.1). Solving design problems is especially demanding since they are often characterized by constraints that are not explicit or only vaguely formulated (Birkhofer 2005; Franke 1976). Formalizing and comprehensively preparing the design task is necessary: Consistent and uniform modeling of technical systems according to their properties allows comprehensive acquisition and processing of product and process constraints and objectives as well as the integration of manufacturing-initiated solutions into the design task. This formalization of the design task is the basis for applying mathematical optimization methods.

5.1.1 Modeling of Technical Systems by Properties

Technical systems, particularly technical products, are entirely characterized by product properties. This is the basis for a purpose-specific description and modeling of products during the whole product design process. Design approaches such as VDI 2221 (1993) and Axiomatic Design (Suh 1998) recommend step-by-step procedures that correlate with decreasing abstract levels of product modeling. At each step, the designer has to determine a combination of product properties in such a way that the desired properties of the previous step (at the more abstract product modeling level) are realized so that the product becomes more and more concrete (Gramlich 2013).

Axiomatic design theory describes the designer's task as structured mapping that spans four dimensions and asks at each mapping "what do we want to achieve?" and "how do we want to achieve it?". The functional requirements (functional domain) have to be determined in a way that enables the customer's attributes (customer domain) to be achieved. Design parameters (physical domain) have to be defined to fulfill functional requirements that characterize product functions. In the last step, the process variables (process domain) are determined, considering the defined design parameters (Suh 2001) (Sect. 1.2).

Pahl et al. (2007), Ehrlenspiel (2009), and other authors describe levels of product models that correlate with appropriate design approaches. The aim of a technical product is to meet customer needs. This is achieved by realizing its product function. The product function expresses the intended causal and solution-neutral relationship between the inputs and the outputs of the product (Ehrlenspiel 2009). Starting from the product function, product attributes are formed by desired properties. Designers have to define an applicable working principle with corresponding product properties. At the next step, they have to determine appropriate design elements, with corresponding geometric and material product properties, to implement the working principle. Hence, product design can be understood as modeling solutions or solution elements by determining product properties at each level of product modeling.

While the designer determines some product properties directly, the customer perceives different aspects of the product through its properties. According to this differentiation, there are several views of product properties that lead to a categorization. Even though different authors such as Hubka (1973), Ehrlenspiel (2009), Lindemann (2009), DIN 4000 (2012), Birkhofer (1980), and Birkhofer and Wäldele (2008) define different categories for product properties, there are two central ideas

Independent product properties		Dependent product properties
• length of coil spring • diameter of coil • diameter of wire • material of wire		• stiffness • damping/dissipation • mass • volume

Fig. 5.3 Independent and dependent properties of a coil spring

addressed by each author. Firstly, there are properties, which are directly influenced and determined by the designer. Secondly, there are properties that can be perceived from outside the product, by the customer, for example. However, these properties cannot be directly determined by the designer. Both ideas result from a comprehensive consideration of the design process itself, where product properties are determined to satisfy requirements (Birkhofer 1980).

This leads to a categorization of *independent* and *dependent properties* (Birkhofer and Wäldele 2008). Independent properties are used for an entire and independent description of the product to fully determine product geometry and material. By determining independent properties, the product is fully determined in its function context. These independent properties can also be utilized as design variables in the context of algorithm-based solution finding. Figure 5.3 shows a coil spring that is characterized by a combination of independent and dependent product properties. The dependent product properties, such as mass and stiffness, depend on the independent product properties, such as material (density) and length of the coil spring (Mattmann et al. 2014). Consequently, dependent properties provide a description of the product that depends on independent product properties. The product in its function context can mostly be comprehended only by dependent properties. Dependent properties can only be indirectly determined by determining independent properties.

The differentiation of independent and dependent properties allows a systematic modeling of dependencies in the form of property relations. This leads to a better understanding of the impacts of design decisions on the product in its function context. To determine independent properties, dependent properties have to be broken down to the independent properties. This is based on specific models from various disciplines that describe the underlying effects and principles. Examples of such models are finite element models, analytical models, and physical models. Property relations can be illustrated in the form of *property networks*, where each property is represented by a node related to further nodes (Wäldele 2012).

Figure 5.4 shows a property network for determining the stiffness of a coil spring. The relation of the stiffness to independent properties is also given as formula. In this case, stiffness gives the relation between displacement as external input and the resulting force.

From a designer's perspective, it is necessary to model the product in its function context to provide customer value by realizing the product function (Pahl et al. 2007). Determination of independent properties, modeling of property relations, and structuring into design elements are carried out to realize the product function

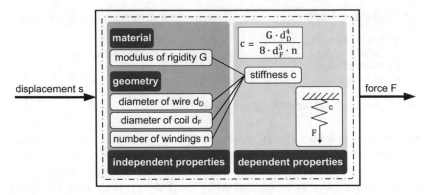

Fig. 5.4 Property network of a coil spring according to Wäldele (2012) and Gramlich (2013)

according to the acquired requirements. A coil spring, for example, has to provide a certain force depending on its displacement. Accordingly, the stiffness of the coil spring has to be determined by the designer to ensure the product function for the product's use (Fig. 5.4) to fulfill stakeholder expectations.

To apply the potential of manufacturing-integrated product design, manufacturing processes have to be formalized and modeled in a process-related context. Technical processes transform technical objects (defined as work-pieces in manufacturing processes) from an initial state into a defined final state (Heidemann 2001). The technical object can be consistently described by properties in all states of the transformation process (Gramlich 2013) and is comprised of geometric, mechanical, and material properties. This consistent description according to product properties is the key to consolidating product and process modeling. The process-related transformation of the technical object is realized by the chosen manufacturing technology. Consequently, the manufacturing technology significantly affects the final state of the technical object. Properties that are transformed by a specific manufacturing technology are summarized as manufacturing-induced properties (Groche et al. 2012) (Chap. 2). The final state of the technical object is composed of manufacturing-induced design elements, characterized by a technology-specific combination of manufacturing-induced properties.

In metal forming processes, geometric and mechanical properties of the technical object are transformed and affected by the mechanisms of plastic deformation. In case of linear flow splitting, the final state is characterized by modifications in hardness, surface roughness, and microstructural conditions (Groche et al. 2012; Tekkaya et al. 2015). Figure 5.5 summarizes the characterization of the manufacturing technology linear flow splitting by *manufacturing-induced design elements* and the corresponding technology-specific combination of *manufacturing-induced properties*.

Linear flow splitting allows the manufacturing of bifurcated structures in integral style, with two thin-walled flanges at the band edge of sheet metals (Chap. 3). An UFG microstructure develops on the bifurcation surfaces during the forming

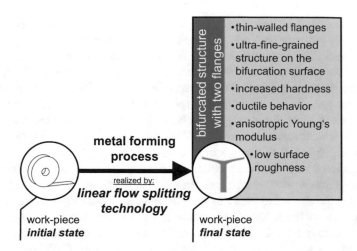

Fig. 5.5 Characterization of linear flow splitting technology by manufacturing-induced design elements (*dark blue*) and the technology-specific combination of manufacturing-induced properties (*light blue*) (Gramlich 2013)

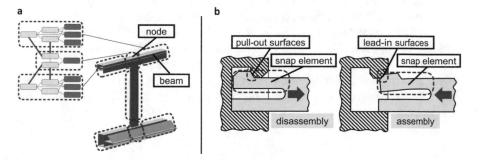

Fig. 5.6 (**a**) Function carriers of a poster stand (Gramlich et al. 2011), (**b**) assembly and disassembly process-relevant elements of a snap-fit

process. This shifting in microstructural conditions is accompanied by increased hardness and ductile behavior. The linear flow split flanges are further characterized by an anisotropic Young's modulus (Chap. 4) and minimal surface roughness (Chap. 3) on the upper side of the flanges.

The complexity of modeling products by properties in function and process contexts significantly increases with the number of functions and processes that have to be considered. To reduce complexity, products can be structured into topological elements according to context and underlying models, where the granularity of the structure can be adjusted contextually. This context and model-based structuring of the product into elements clusters specific property combinations. The underlying model can cause the properties to interrelate, which leads to specific modules of property relations and networks (Fig. 5.6a). These property network modules are linked by relations between the included properties. This leads to a superordinate property network with manageable complexity.

The elements of the structured product are *function carriers* in the function context of the product. Function carriers define elements of the product, such as parts, principle solutions and solution principles, or the entire product, which fulfills one or more specific functions (Ehrlenspiel 2009; Pahl et al. 2007; Wagner et al. 2016). A poster stand, for example, can be structured into beams and nodes, where the beam conducts forces to the nodes. The nodes distribute the loads to the profile structure (Fig. 5.6a), fulfilling the function of transmitting forces.

In the process context of the product, the elements of the product structure can be defined as *process-relevant elements*. In the case of the product's use phase, only processes with noticeable value to the customer are considered. An example for such processes of the product's use phase is assembly processes conducted by the customer. A snap-fit, for example, comprises elements relevant to the assembly process, such as lead-in surfaces and yielding snap-elements. Both have significant influence on the assembly force (Fig. 5.6b). The disassembly process is characterized by pull-out surfaces and the snap element, which both influence the disassembly and holding forces. This shows that the chosen structure does not necessarily correlate with the construction structure. Elements can be defined across components and more than one element can be assigned to one component (Fig. 5.6b).

Consistent modeling of technical systems by their properties in their function and process contexts forms the basis for formulating mathematical optimization problems, with integrated manufacturing characteristics in the form of manufacturing-induced properties. This is also the basis for structuring product and process requirement acquisition and transformation to formalize the design task.

5.1.2   Requirement Acquisition and Transformation into Properties

Mathematical optimization methods are applied to find optimal solutions within the solution space by constraining the objective function. For applying mathematical optimization methods within the integrated product and process design, the inputs have to be appropriately preprocessed. Stakeholder expectations are often vague or not explicitly documented. Thus, stakeholder expectations cannot be directly used as input in mathematical optimization methods due to their vast scope for interpretation.

The analysis of relevant product life cycle processes, like use and disposal, contributes a large number of highly relevant, process-related requirements to product development. Process-related requirements contribute valuable information about the product to be developed and its functions (Mattmann et al. 2016a). Requirements have to be considered across the entire product and process design to

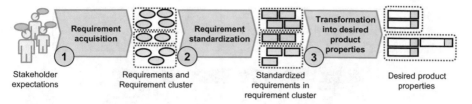

Stakeholder expectations   Requirements and Requirement cluster   Standardized requirements in requirement cluster   Desired product properties

Fig. 5.7 Procedure of requirement acquisition and transformation into properties

ensure decision-making is aligned to marketability of the product (Mattmann et al. 2016b).

To formalize the vague and often not explicitly documented input for mathematical optimization methods, stakeholder expectations have to be systematically transformed into *desired product properties*, which are assigned to function carriers (Mattmann et al. 2016a). The following procedure (Fig. 5.7) has to be applied.

1. Requirements are systematically acquired from different sources to ascertain the completeness of requirements and their explicit documentation (requirement acquisition).
2. Requirements are standardized for their content and formal structure (requirement standardization) to ensure a precise documentation that is appropriate for the transformation of requirements into properties.
3. Standardized requirements are transformed into desired product properties that are formalized for applying them in mathematical optimization.

Requirement Acquisition

The use of mathematical optimization methods assists in the avoidance of multiple iterations during early design phases. The best product geometry and processes are found much faster than during gradual iteration. The use of mathematical optimization methods necessitates complete acquisition of requirements from various sources, such as stakeholders, the intended product life cycle processes, and restrictive conditions that result from legislation and norms (Pahl et al. 2007). Missing and possibly self-evident requirements cannot be compensated with experiential designer knowledge.

By focusing on the product function and according properties the differing goals and contexts of product and process requirements (arranged in clusters based on the requirement source) can be handled. Requirement clusters are used to systematically complement semantically related requirements. Unfortunately, complexity increases massively with the volume of requirements acquired in pursuit of completeness. To counter these challenges, methodological support is provided by the *Smart Requirement Configuration System* (SRCS) (Fig. 5.8) (Röder 2014).8

SRCS structurally integrates process requirements acquired from product life cycle processes and requirements from use processes in a database. Each requirement that is acquired from stakeholders and product sources, e.g., its surroundings and intended use processes, is documented in the form of product and process

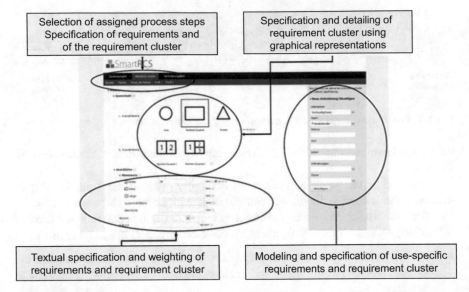

Fig. 5.8 Smart Requirement Configuration System (SRCS) according to Röder (2014)

defining descriptions with appropriate quantitative and qualitative values. The requirements are then prioritized by their importance in fulfilling stakeholder expectations.

Stakeholder expectations from multiple sources have been documented using the SRCS during the exemplary development of an information stand. For presentations using multimedia devices, many people use devices like a tablet, which has to be connected to a power source and a projector. However, most information stands do not have appropriate interfaces for the connection of multimedia devices. This leads to the idea of an information stand with integrated energy storage for multimedia devices (such as a 10-in. tablet), to present multimedia content. The idea of the information stand is concretized in the form of essential function carriers: information stand socket, socket to power tablet pcs, profile. The profile connects each element to ensure easy assembly and disassembly of the information stand by defining process-relevant elements: connection between tablet socket and profile and connection between profile and information stand socket (structured in Fig. 5.9).

For the information stand, requirements are acquired from different sources such as intended use processes. Demands such as easy transportability of the information stand, fast assembly and disassembly, load of typical multimedia device weights, and energy supply have to be considered. The information stand should provide sufficient holding force to ensure stability even during misuse. For easy assembly of the information stand socket, the elements should be connectable and separable using a flexible connection. The capacity to withstand stresses is limited. Cables

Fig. 5.9 Function carriers
and process-relevant
elements of an information
stand

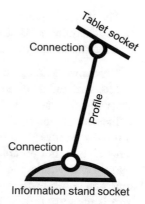

Connection

Tablet socket

Profile

Connection

Information stand socket

should be integrally lead through the information stand to fulfill aesthetic requirements. The requirements above are mainly the result of analyses in terms of the use
processes of the information stand. However, to achieve a high degree of completeness regarding requirement acquisition, also other requirement clusters have to be
considered. The identified function carriers and process-relevant elements (socket,
profile, tablet socket, connections) activate corresponding requirement clusters
within SRCS, leading to a higher degree of completeness.

Requirement Standardization

Requirements that are formulated with implicit information (Lindemann 2009)
within conventional product development processes cannot be used in mathematical optimization methods as they are too open to interpretation (Mattmann et al.
2015b; Mattmann et al. 2016b). Mathematical optimization methods require highly
standardized information within a formalized structure. According to the dimensions of requirements the standardization of requirements falls into two types:
standardization of requirement content and requirement formal structure
(Mattmann et al. 2015b).

The standardization of requirement content in the context of an integrated
product and process design approach necessitates the transcription of all process-
related requirements to product-related requirements. In the example of the information stand, the profile structure is described by its number of chambers, dimensions, stiffness, sheet thickness, and bending angle. Process-related information
within requirements, resulting from product life cycle processes, is broken down to
a product-specific terminology. Thus, a minimally tolerable load (resulting from
use processes) is broken down to a description of the smallest acceptable sheet
profile stiffness, e.g., during misuse when a user leans on the information stand with
his entire body weight.

Formal aspects of requirements are guaranteed using specific documentation
patterns that fulfill quality criteria for the formal and correct documentation of
requirements (Mattmann et al. 2015b). This leads to standardized requirement data
sets within the SRCS. Standardized data sets allow the derivation of standardized
requirement lists.

The standardization of requirements by transforming process requirements into product requirements that satisfy quality criteria is the basis for the following transformation into properties (Röder et al. 2011).

Transformation into Properties

Standardized requirements have to be transformed into a representation suitable for mathematical optimization methods. The direct use of requirements within mathematical optimization methods does not work since their implementation into the optimization problem (objective function, design variables, and design constraints) is not yet defined. Requirements need to be represented in a form that is manageable within the integrated algorithm-based product and process design approach with consistent models for product and process description. The common base is property theory, as shown in Sect. 5.1.1, which allows a semantically and formally standardized representation of requirements in the form of properties. Product and process requirements are broken down to desired product properties (Mattmann et al. 2015a). Desired product properties characterize the optimal solution within the solution space. The set of desired product properties is reduced to the most important design-determining ones that are highly relevant to functional fulfillment of the technical product. The values of desired product properties are extracted from the related requirements. In this way, permitted value ranges that constrain the solution space for optimization are specified (Mattmann et al. 2016a). The most relevant sets of desired product properties are linked to function carriers.

Using the example of the information stand, it is structured into function carriers and process-relevant elements that are each characterized by a minimal set of desired product properties: the socket of the information stand, the profile element, the socket of the tablet, and connections between the information stand base, profile element, and tablet socket, as shown in Fig. 5.10. The sheet metal profile has to have a minimum stiffness to carry the load of the multimedia unit into the base, requiring high stiffness and a thin-walled, lightweight structure in which the cables of the multimedia unit are guided. The minimum number of chambers is predefined by the cable guiding function and the required stiffness of the sheet profile. The maximum length of the sheet profile results from ergonomic requirements gathered through analysis of operation by the user. The connection element has to ensure specific assembly, disassembly, and holding forces, which are necessitated by requirements from the intended assembly, disassembly, and use processes. To provide stability, the socket of the information stand must have a minimum weight. However, the socket must be realized within a maximum installation space to provide for the integration of a battery to supply additional energy for the tablet during use. The connection may be stressed during use due to bending forces that result from misuse of the information stand, which can occur through leaning on the tablet socket with a large portion of the user's bodyweight.

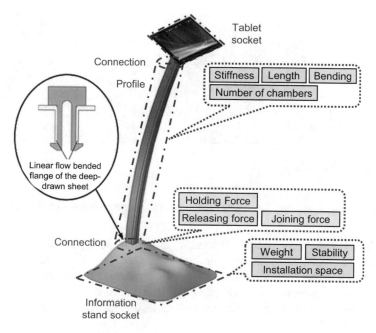

Fig. 5.10 Assigning desired product properties to function carriers and process-relevant elements of the information stand according to Röder et al. (2012)

5.1.3 Integration of Manufacturing Technologies

The formalized modeling of technical systems and products by their properties in function and process contexts provides a promising basis for systematically linking manufacturing characteristics and product design. There are two essential steps to incorporating the possibilities and characteristics of manufacturing technologies into the product design process:

1. Identification of applicable manufacturing technologies, including manufacturing-induced design elements and properties at an early stage, based on systematic *property matching* with desired product properties.
2. Integration of manufacturing-induced design elements and their specific combinations of manufacturing-induced properties into solution finding, based on property relations (*property linking*).

Identification and Determination of Applicable Manufacturing Technologies

Identification of appropriate manufacturing technologies at an early stage of the design process enables the integration of manufacturing characteristics into the process of solution finding. The product has to have the attributes required by customers and stakeholders in the form of desired product properties. The wide range of available manufacturing technologies provides a huge variety of possible

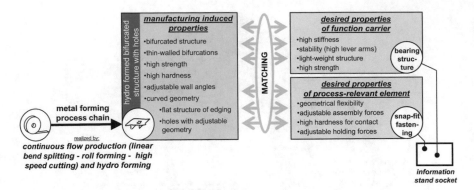

Fig. 5.11 Property matching for an information stand socket according to Gramlich (2013)

manufacturing-induced design elements, characterized by technology-specific combinations of manufacturing-induced properties. By matching these technology-specific combinations of manufacturing-induced properties with desired product properties, manufacturing-induced design elements can be identified that are predestined to realize the product in its function and process contexts (Gramlich 2013).

Desired product properties are assigned to function carriers and process-relevant elements. Manufacturing-induced design elements and function carriers or process-relevant elements are matched with the corresponding combinations of properties. By applying systematic property matching, applicable manufacturing technologies and manufacturing process chains can be identified for subsequent integration of manufacturing-induced design elements, manufacturing-induced properties, and property relations into the design task.

In the example case, the socket of an information stand comprises a bearing structure (function carrier) and a detachable snap-fit fastening (assembly and disassembly process-relevant element). Both elements are characterized by a combination of desired product properties in the underlying working elements, like high stiffness, stability, and lightweight structure (Fig. 5.11). Linear bend splitting, in combination with roll forming and high-speed cutting, as part of continuous flow production with subsequent hydroforming provides a combination of manufacturing-induced properties, like high stiffness and a thin-walled bifurcated structure with curved geometry (Chap. 3). The matching of both properties reveals that there is beneficial applicability of linear bend splitting within continuous flow production for the design of an information stand socket: The high strength and high stiffness of the linear flow split, hydroformed profiles are promising for realizing a bearing structure that has high stiffness and can bear high forces. The thin-walled bifurcated structure guarantees a lightweight design, where the geometric flexibility of the hydroformed, bifurcated structure can be used to ensure stability of the information stand socket. The snap-fit fastening benefits from the adjustable, high-speed cut geometry and the roll-formed, linear bend split flanges whose angle can be flexibly adjusted. Consequently, assembly and holding forces of this fastening can be adjusted up to material strength limits. The chosen combination of manufacturing technologies seems to be beneficial for realizing the information stand socket.

Integration of Manufacturing-Induced Design Elements and Properties

The feasibility of a product solution highly depends on the identified and determined manufacturing technologies. Consideration of manufacturing process requirements is crucial to ensure manufacturability of a product and has to be prepared and integrated into the design task.

While requirements originating from product use processes and manufacturing processes constrain the possible solution space, manufacturing characteristics create potential for unconventional product solutions. The manufacturing potential leads to further design opportunities that result in additional solutions within the solution space, increasing the chance of finding a better product solution. To comprehensively realize manufacturing potential for product design, manufacturing-induced design elements and properties of the chosen manufacturing technologies have to be integrated into the design task. This can only be done by applying identified property relations between manufacturing-induced properties and desired product properties. By integrating manufacturing-induced design elements function carriers and process-relevant elements are realized. This step also involves the determination of product topology in terms of structural properties, such as number and arrangement of manufacturing-induced design elements.

To support integration, the designer can use documented, *process-integrated design guidelines* (PIDGs) (Sect. 4.2), where the relation of function and process information to the product is captured (Wagner et al. 2016). Using process-integrated design guidelines, *manufacturing-initiated solution elements* can be added to the product to concretize the design task. Manufacturing-initiated solution elements are key to realizing manufacturing potential within solution finding. Solution elements are represented by concretized, manufacturing-induced design elements, which are applied to realize function carriers or process-relevant elements. They comprise generally valid statements about how the manufacturing-induced properties can be applied to realize the product in its function and process contexts. The generation of these solution elements is not part of integration but has to be conducted in terms of processing manufacturing-induced properties (Sect. 4.2). Solution elements can be further applied during mathematical solution finding to ensure comprehensively realized manufacturing potential.

An information stand socket comprises a bearing structure and a snap-fit fastening (Fig. 5.12). Linear bend splitting and roll forming have been identified and determined to realize both elements of the socket. Manufacturing-induced properties have to be applied to realize the product while conflicting design goals are adequately handled, like in case of snap-fit fastening assembly and disassembly forces, which both depend on each other, based on geometric undercut (Roos et al. 2016). To adequately handle the conflicting design goals of assembly and disassembly forces, property relations are modeled to understand the impact of manufacturing-induced properties. As shown in Fig. 5.12a, the length and angle

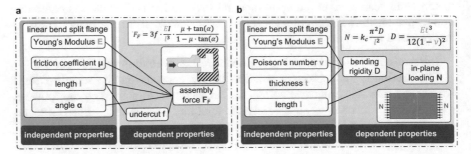

Fig. 5.12 Property networks according to Wäldele (2012) and Gramlich (2013): (**a**) Assembly force of a snap-fit, (**b**) buckling of a linear bend split structure

Problem	Identifier	Design Recommendation	Consequence	Explanation	Solution elements
Detachable fastenings	PIDG 5.1: Linear flow split flanges as snap elements	Use the linear flow split flanges or linear bend split flanges with increased strength as snap element of detachable fastenings.	realizing increased elastic disassembly displacements	The linear flow split flanges have an increased strength which allows higher displacement at the same Young's Modulus.	
Detachable fastenings	PIDG 5.2: Linear flow split flanges as lead-in/pull-out surfaces	Use the upper surface of linear flow split flanges or linear bend split flanges with increased hardness as lead-in or pull-out surfaces of detachable fastenings.	realizing line contact	The upper surface of linear flow split flanges have an increased hardness.	
Locking degrees of freedom	PIDG 5.3: Roll formed linear flow split flanges to lock degrees of freedom	Use roll formed linear flow split flanges or roll formed linear bend split flanges with specific bending radii (> 1 mm) for providing multiple working surfaces for detachable fastenings.	lock additional degrees of freedom	Bending radii at heat treated linear flow split flanges have to be > 1 mm to avoid cracks at the upper surface of the linear flow split flanges.	

Fig. 5.13 Process-integrated design guidelines (PIDGs) with solution elements for a linear bend split snap-fit according to Roos et al. (2016)

of the flange, as well as Young's Modulus and friction coefficient at the flanges, have the biggest influence on assembly force, while length also influences the undercut. The angle, friction coefficient, and Young's Modulus have to be considered during solution finding to define low assembly forces separately to disassembly forces.

In the bearing structure of the information stand socket, the linear bend split flanges can be used to increase the stiffness of the structure for the use processes of the product, as already identified by property matching. A higher flange length increases the risk of buckling during the manufacturing process of hydroforming (Chap. 3). To embed the crucial dependencies between manufacturing-induced properties and the risk of buckling in optimization models, a property network is set up to process and integrate the relevant relations (Fig. 5.12b) into the design task to solve conflicting design goals during solution finding.

To additionally realize manufacturing potential, several process-integrated design guidelines are generated in a prior processing step, as described in Sect. 4.2 (Fig. 5.13). These process-integrated design guidelines are applied to integrate the corresponding manufacturing-induced properties into product design. This concerns, for example, the flanges of the linear bend split profile (manufacturing-induced design elements) that can be applied as pull-out surfaces (disassembly process-relevant element) to a snap-fit fastening. The increased strength of the flanges and the increased hardness of the flange's upper surface are beneficial in realizing increased holding forces while allowing increased displacement of the snap element.

5.1.4 Formalized Integrated Product and Process Design Task

The integrated product and process design task, consistently formalized by product properties in function and process contexts, is the basis for a mathematical formulation of the *optimization problem*. The formalized design task comprises all prior processed information in early development steps to formulate the optimization problem. This includes specific information about product use processes, such as external forces, signals, and energies, that pass across system boundaries. Accordingly, working surfaces and working elements are determined and positioned within the geometric boundaries of the solution space. This information is documented in the form of a *design space diagram* (Fig. 5.14). Requirements acquired while considering the use phase result in constraints and objective functions that structure the solution space to find the optimal solution while considering the predefined working surfaces.

During use processes in the case study, the information stand is subject to various uses that affect formulation of the mathematical optimization problem. For example, the information stand has to resist much higher forces in various directions when the user leans on the information stand with a portion of his body weight. Assembly and disassembly processes lead to forces required within the fastening that have to be considered during positioning of working surfaces within the fastening (Fig. 5.14). These important requirements contribute information about position and characteristics of essential working surfaces.

Requirements of the product limit the permissible design space. The information stand socket has to provide enough stability for the entire product, even if the user accidentally bumps the tablet socket (misuse). Dimensions of the information stand socket are limited by the ergonomics of the user standing in front of the tablet and the weight of the information stand socket, which increases with its dimensions. The information stand socket has to provide enough space for the integration of an additional energy supply in the form of a battery. These design constraints restrict the design space, as illustrated in the design space diagram (Fig. 5.14: design space—white, design constraints—red, working surfaces with applied forces—red arrows).

By integrating manufacturing-induced properties in the form of manufacturing-initiated solution elements, function carriers or process-relevant elements of the product can be realized (Fig. 5.13). Manufacturing-induced properties act as design variables in subsequent solution finding. There are additional manufacturing

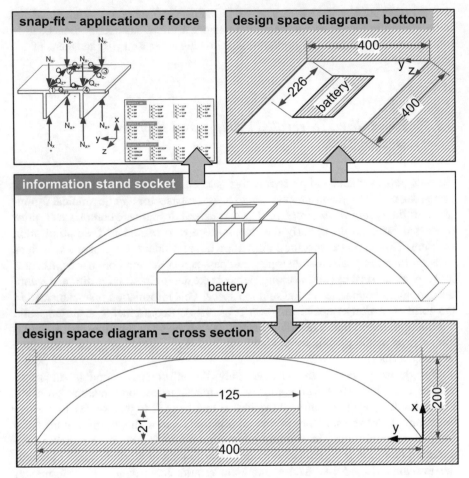

Fig. 5.14 Design space diagram for an information stand socket

requirements in the form of design constraints or objective functions that have to be considered during solution finding. Most important for the formalized design task is information about the realization of function carriers and process-relevant elements by manufacturing-induced design elements. Manufacturing-initiated solution elements are applied to generate a starting solution for subsequent algorithmic optimization. Specific property relations, for example, property networks, link product and process information. These are implemented into algorithm-based solution finding to model the function and process contexts of the product (Fig. 5.12).

The stiffness of the sheet metal structure used to realize an information stand socket can be increased by applying linear bend split flanges. These flanges provide additional benefit when realizing an integrated snap-fit fastening. Because of their geometric and material properties, they are also predestined to be solution elements in a detachable snap-fit fastening (property relations, Fig. 5.12). These solution

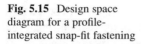

Fig. 5.15 Design space diagram for a profile-integrated snap-fit fastening

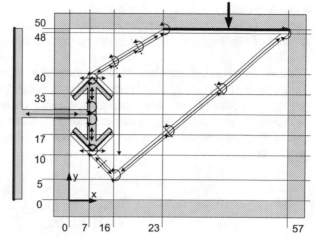

elements are integrated into a starting point for further steps in solution finding, so that manufacturing potential is comprehensively realized (Fig. 5.14). Consideration of the flanges also influences the application of forces in the form of additional, clearly defined working surfaces to transmit external forces. Additional, non-geometric constraints, resulting from linear bend split processes, are documented separately to the design space diagram.

A second example is a snap-fit fastening within a linear flow split profile structure (Fig. 5.15) for bearing lateral forces. The design space diagram illustrates the manufacturing-initiated solution elements as part of the starting geometry within a restricted design space. This is extended by giving information about the degrees of freedom, represented by independent properties, in bending points and adjustable lengths. The focus for subsequent optimization is on increasing the stiffness of the profile structure while considering assembly and disassembly process requirements of the linear flow split snap-fit fastening.

5.2   Optimization of Product Design

Many engineering tasks related to lightweight design or manufacturing of branched sheet metal products are characterized by a high degree of complexity due to a large number of controllable parameters and dependencies. Thus, finding the optimal parameter setting manually is not the most efficient approach or even possible. A promising technique to tackle these problems is mathematical optimization. In this and the following section different optimization methods are applied to find the optimal product and process design as introduced in Sect. 2.4. The first application is the geometry optimization of branched sheet metal products with the aim of maximizing the structural stiffness. In the next section we consider another application which is the control of deep drawing processes where the aim is to avoid material failures such as stringer buckling or cracking. Finally, we optimize the number of retooling operations of the assembly line and the production sequence of profiles.

As presented in the introduction of this chapter, to solve problems resulting from the formalized design task explained in Sect. 5.1, an algorithmic solution finding process is applied. In our case, this process relies on mathematical optimization problems. To abstractly state these problems, the mixed integer nonlinear problem (MINLP) is introduced which forms the basis of all the following optimization problems

$$\begin{aligned}
\min \ & c(x) \hspace{4cm} \text{(MINLP)}\\
s.t. \ & e_{in}(x) \leq 0,\\
& e_{eq}(x) = 0,\\
& x \in \mathbb{R}^{n-p} \times \mathbb{Z}^p.
\end{aligned}$$

The aim of this optimization problem is to minimize the objective function $c : \mathbb{R}^n \to \mathbb{R}$ subject to different constraints. These constraints are defined by functions $e_{in} : \mathbb{R}^n \to \mathbb{R}^m$, $e_{eq} : \mathbb{R}^n \to \mathbb{R}^l$ and an additional integer constraint for a subset of the optimization variables.

If $p = 0$, the optimization problem is called nonlinear program (NLP). The shape optimization problem described in this section and the optimal control problem in Sect. 5.3.1 can be formulated in terms of NLPs. This type of mathematical program can be solved, for example, by interior point or sequential quadratic programming (SQP) methods (Nocedal and Wright 2006).

The MINLP is called mixed integer program (MIP) for $0 < p \leq n$ and linear objective function c and constraints e_{in}, e_{eq}. MIPs are addressed in discrete optimization and can be solved, for example, by a branch and cut method, see, e.g., Nemhauser and Wolsey (1988). These programs appear in Sect. 5.2.2 within the combined topology and geometry optimization approach, in Sect. 5.3.2 in form of the minimization of the retooling time, and in Sect. 5.3.3 as the optimization of the production sequence.

To find the optimal design for a branched sheet metal product as illustrated in Fig. 5.16a, we follow the approach of the algorithmic solution finding process. For this purpose, the requirements and restrictions concentrated within the formalized design task are translated into a mathematical optimization problem. The aim is to find the optimal geometry for a specific load scenario and a given starting geometry under certain design constraints. Hence, the stiffness of the geometry is evaluated by an objective function, for example, the compliance (Eq. (5.2)) or the mean quadratic displacement (Eq. (5.3)).

The geometry is represented by $\overline{\Omega} = \Omega \cup \partial\Omega \subset \mathbb{R}^3$, the interior Ω is an open set, see Fig. 5.16b. The boundary $\partial\Omega$ is divided into the Dirichlet boundary Γ_D, where the geometry is fixed and the Neumann boundary Γ_N, where the surface load q is applied. The volume load f, for example, gravity, acts on the whole domain Ω. The mechanical behavior for a geometry $\overline{\Omega}$ is determined by the linear elasticity equations.

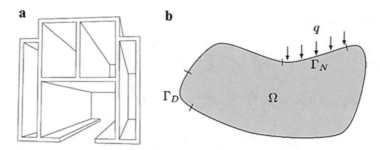

Fig. 5.16 (**a**) Draft of a multi-chambered profile, (**b**) abstract mathematical setting for the elasticity problem

Problem 5.1 (Elasticity Problem)

Find the displacement field $y : \Omega \to \mathbb{R}^3, y \in Y := [H^1_D(\Omega)]^3$, such that for all test functions $v \in Y$

$$\int_\Omega \varepsilon(v) : C\varepsilon(y)\, dx - \left(\int_\Omega v \cdot f\, dx + \int_{\Gamma_N} v \cdot q\, dS(x) \right) = 0$$

holds, where $H^1_D(\Omega) := \{ v \in H^1(\Omega) : v|_{\Gamma_D} = 0 \}$.

The linearized strain tensor is defined as $\varepsilon = \frac{1}{2}(\nabla y + \nabla y^T)$ and is linked to the stress tensor $\sigma : \mathbb{R}^3 \to \mathbb{R}^3 \times \mathbb{R}^3$ through Hooke's law $\sigma(y) = C\varepsilon(y)$. In our case the material tensor C is characterized by the Poisson ratio and Young's modulus.

For the numerical simulation of the physical behavior, Problem 5.1 is solved by the finite element method (FEM). For this purpose, the function space Y is substituted by a suitable finite dimensional subspace $Y_h \subset Y$. The finite dimensional version of y can be written as

$$y_h = \sum_{i=1}^{\hat{n}} \phi_i \cdot (\vec{y}_{\to h})_i, \tag{5.1}$$

where $\Phi := \{ \phi_1, \ldots, \phi_{\hat{n}} \}$ is a basis of Y_h.

The coefficients are denoted by $(\vec{y}_h)_i \in \mathbb{R}$. The coefficient vector $\vec{y}_{\to h} \in \mathbb{R}^{\hat{n}}$ is defined as the collection of all coefficients of y_h in a single vector. It is sufficient to use the basis functions ϕ_i, $i = 1, \ldots, \hat{n}$ as test functions v_h. The finite dimensional formulation of Problem 5.1 is shown in Problem 5.2.

Problem 5.2 (Finite Dimensional Formulation of the Elasticity Problem)

Find $\vec{y}_{\to h} \in \mathbb{R}^{\hat{n}}$, such that

$$A_h \vec{y}_h - \underbrace{\left(M_h^\Omega \vec{f}_h + M_h^{\Gamma_n} \vec{q}_h \right)}_{F_h} = 0,$$

with $\hat{n} \in \mathbb{N}$ denoting the number of vertices in the finite element mesh.

The stiffness matrix $A_h \in \mathbb{R}^{\hat{n} \times \hat{n}}$ is defined as

$$(A_h)_{i,j} = \int_\Omega \varepsilon(\phi_i) : C\varepsilon(\phi_j) dx$$

and the mass matrices $M_h^\Omega \in \mathbb{R}^{\hat{n} \times \hat{n}}$, $M_h^{\Gamma_N} \in \mathbb{R}^{\hat{n} \times \hat{n}}$ as

$$(M_h^\Omega)_{ij} = \int_\Omega \phi_i \cdot \phi_j dx, \quad (M_h^{\Gamma_n})_{ij} = \int_{\Gamma_N} \phi_i \cdot \phi_j ds.$$

For further details on the derivation of the discrete formulation, we refer to Braess (2007), Hess (2010), and Brenner and Scott (2007). Next we formulate the finite dimensional shape optimization problem in linear elasticity. In the following, two different ways of measuring the structural stiffness are considered as objective functions. The design variables are denoted by $u \in \mathbb{R}^n$, with $n \in \mathbb{N}$ denoting the number of design variables. We use the notation $\Omega(u)$ to express the design dependency of Ω on u. An introduction to shape optimization is given in Haslinger and Mäkinen (2003). The considered objective functions are the compliance

$$J_{c,h}(u, \vec{y}_h) = \vec{y}_h^T M_h^{\Omega(u)} \vec{f}_h + \vec{y}_h^T M_h^{\Gamma_N(u)} \vec{q}_h \tag{5.2}$$

and the mean quadratic displacement

$$J_{d,h}(u, \vec{y}_h) = \frac{\vec{y}_h^T M_h^{\Omega(u)} \vec{y}_h}{vol(\Omega(u))}. \tag{5.3}$$

Since the elasticity problem has a unique solution, a Lipschitz continuous operator $\mathbb{R}^n \ni u \mapsto y_h(u) \in Y$ can be defined such that the reduced form of the functions Eqs. (5.2) and (5.3)

$$j_{k,h}(u) := J_{k,h}\left(u, \vec{y} \to_h(u)\right), \quad k \in \{c, d\}$$

can be defined.

Given an objective function Eqs. (5.2) or (5.3), the finite dimensional shape optimization problem is stated as in Problem 5.3.

Problem 5.3 (Discretized Shape Optimization Problem in Linear Elasticity)

$$\min J_{k,h}(u, \vec{y}_h), \quad k \in \{c, d\}$$
$$\text{s.t. } A_h \vec{y}_h - F_h = 0$$
$$u = U_{ad}.$$

The set of admissible designs $U_{ad} \subset \mathbb{R}^n$ is defined by design constraints, for example, angle or length restrictions, which are provided by the formalized design task presented in Sect. 5.1. The design u can be interpreted as an independent variable and the displacement y as a dependent variable, see Fig. 5.4. In the following, optimization methods for different types of geometries as multi-chambered profiles (Sects. 5.2.1 and 5.2.2), hydroformed branched sheet metal products (Sect. 5.2.3), and mechanical connections (Sect. 5.2.4) are presented.

5.2.1   Geometry Optimization of Multi-Chambered Profiles

In this section, the geometry optimization of linear profiles with constant cross section and fixed length in the three-dimensional space is considered. The following is based on Hess (2010). Profiles are natural candidates for continuously manufactured and highly integrated products. We assume that the topology of the profile is already known and fixed. An extension where the topology is also part of the optimization is presented in Sect. 5.2.2. The design of multi-chambered profiles is determined by the sheet thickness and the chamber size. The resulting shape optimization problem is shown in Problem 5.3. Since this is an NLP, an SQP method, see Algorithm 5.1, is applied to find the optimal design. Within this iterative method quadratic subproblems are solved to determine the next step s^k which can be written as

$$
\begin{aligned}
&\min_{s \in \mathbb{R}^n} \nabla j(u^k)s^k + \rho \sum_{i=1}^{m} \xi_i + \frac{1}{2}s^{k,\mathrm{T}}H_k s^k \\
&\text{s.t. } h_i(u^k) + \nabla h_i(u^k)s^k \leq \xi_i \quad \forall i = 1 \ldots m \\
&\qquad \xi_i \in \mathbb{R}, \quad \xi_i \geq 0 \quad \forall i = 1 \ldots m.
\end{aligned}
\tag{5.4}
$$

The matrix H_k represents the Hesse matrix of the corresponding Lagrange function or a suitable approximation of it. The functions $h_i(u)$, $i = 1, \ldots, m$ represent design constraints, for example, maximal sheet thickness or minimal chamber size. These subproblems can easily be solved, for example, with an interior point method or an active set strategy. To ensure global convergence, an Armijo step size rule for the ℓ_1-penalty function is used. The resulting penalty function P_1 is defined as

$$
P_1(\mathbf{u}) := j_{k,h}(\mathbf{u}) + \rho \sum_{i=1}^{m} \max\left(0, h_i(\mathbf{u})\right), \quad k \in \{c, d\}.
$$

The constraints are satisfied for a sufficiently large penalty parameter $\rho > 0$.

Algorithm 5.1 (SQP with l_1-penalty globalization)
Input: Problem 5.3 with start solution $\boldsymbol{u}^0$ and constants $\beta, \sigma \in (0, 1), \rho > 0$
Output: Optimal solution $\boldsymbol{u}^* \in U_{ad}$ of (5.4)
For $k = 0, 1, \ldots$ **do**
> Determine step $\boldsymbol{s}^k$ as a solution of the quadratic subproblem (5.4)
> **If** $s = 0$
>> **return** optimal solution $\boldsymbol{u}^*$
> **Else**
>> Find maximal step length $\delta_k \in \{\beta^0, \beta^1, \ldots\}$ with
>> $$P_1(\boldsymbol{u}^k + \delta_k \boldsymbol{s}^k) \leq P_1(\boldsymbol{u}^k) - \sigma \delta_k \boldsymbol{s}^{k,\mathrm{T}} H_k \boldsymbol{s}^k$$
>> Set $\boldsymbol{u}^{k+1} = \boldsymbol{u}^k + \delta_k \boldsymbol{s}^k$

For convergence results of the SQP method we refer to Hess and Ulbrich (2012). For efficiency reasons several improvements have been implemented such as a multigrid method to solve the elasticity problem (Braess 1986; Yserentant 1993). This method reduces the computational time significantly, but results in inexact evaluations of the objective function and the corresponding gradient. The algorithm has been extended to handle this inexactness, for a detailed description see Curtis et al. (2013), Heinkenschloss and Vicente (2001), and Hess and Ulbrich (2012). A multilevel approach for the optimization problem has also been added. The main idea is to start on a coarse mesh, where the simulations are less computationally expensive and then refine the mesh until a desired accuracy is reached. The optimal solution of the former level is used as starting point for the current level to reduce the number of iterations needed on the finer meshes. The refinement of the meshes is controlled through an error estimator and is applied in an adaptive manner. For further details on the adaptive refinement controlled by an error estimator, we refer to Bartels and Carstensen (2002), Bornemann et al. (1993), Hess (2010), and Ziems and Ulbrich (2011).

As an example we apply the inexact version of Algorithm 5.1 to a multi-chambered profile. The considered profile is 50 mm long and 52 mm high and wide. It is fixed at the back and in addition there are forces q_1 acting downwards on the bottom of the upper right chamber, see Fig. 5.17a. Beside constraints on the sheet thickness, there is also a constraint on the volume. For further details on the numerical result, see Hess (2010) and Hess and Ulbrich (2012). We use eight mesh levels, denoted by k in Table 5.1, for the optimization. The progress of the algorithm is shown in Table 5.1. For the approximate solution of the elasticity problems a preconditioned conjugate gradient method (PCG) with multigrid precondition is used. On the first level a direct solver has been applied. The mean quadratic displacement of the optimal solution (Fig. 5.17c) could be reduced by about 40% compared to the starting solution (Fig. 5.17b).

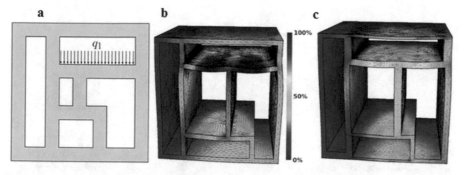

Fig. 5.17 (**a**) The load scenario is defined as follows. The surface load q_1 is uniformly applied to the profile and acts *downwards* on the *bottom* of the *upper* chamber, (**b**) starting, (**c**) optimized geometry. The color represents the norm of the displacement, ranging from *dark blue* (=0%) to *dark red* for the maximal displacement of the starting solution (=100%)

Table 5.1 Numerical results of the inexact SQP method

k	Iter.	Objective value	Constraint violation	PCG its. (avg.)	Elements
1	99	4.568×10^{-4}	1.250×10^{-6}	–	7696
2	51	4.680×10^{-4}	6.474×10^{-7}	1.765	12,666
3	21	4.761×10^{-4}	9.911×10^{-7}	3.646	24,034
4	6	4.814×10^{-4}	7.265×10^{-7}	4.880	56,605
5	9	4.815×10^{-4}	6.755×10^{-7}	6.091	132,569
6	2	4.862×10^{-4}	4.995×10^{-7}	7.000	311,538
7	1	4.870×10^{-4}	2.455×10^{-7}	8.500	694,214
8	1	4.874×10^{-4}	1.079×10^{-7}	9.000	1,553,730

5.2.2   Topology and Geometry Optimization: A Combined Approach

The following section extends the geometry optimization from Sect. 5.2.1 to deal with the generation of the optimal topology of multi-chambered profiles with a fixed length and an upper bound on the number of 90° bends, compare Fig. 5.18. Given a load case for an enclosed profile as well as minimal and maximal sizes of chambers, e.g., Figs. 5.19 and 5.20, the aim is to find the optimal structure satisfying the constraints and minimizing a given objective function. To this end, a geometry optimization problem and a topology optimization problem is solved in sequence based on the general form already shown in Problem 5.3. The main difference between these are the different definitions of the control or optimization variable u. In the topology optimization we use density values in contrast to the geometry optimization where the sheet thickness and length is used as control variables.

   To circumvent the problem of enumerating all possible topologies, the algorithm relies on a branch and bound framework. In every additional level of the branch and bound tree, a further chamber is introduced such that the leaves of this tree contain

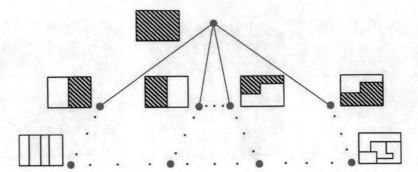

Fig. 5.18 Schematic figure representing the branch and bound idea. Hatched area means that the corresponding chamber is to be divided further in the child nodes

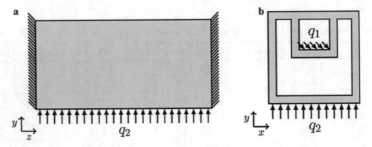

Fig. 5.19 (**a**) Load case scenario side view, (**b**) load case scenario cross-sectional view. The profile is fixed on both ends. The surface loads q_1 and q_2 are uniformly applied to the profile, on the *bottom* of the *upper* chamber a load of 1980 N (q_1) acts with an angle of 45°. On the lower boundary of the profile a load of 1400 N (q_2) acts in upside direction

the topologies with the requested number of chambers, see Fig. 5.18. The child nodes are generated such that all feasible subdivisions of the current chamber are created. In every node of the tree, first a relaxation of the discrete topology optimization problem based on a pixel model is solved. The next step is to find a chamber that best resembles the solution of the relaxation to generate a feasible topology. Lastly a geometry optimization process on the found topology is started. Since the feasible solutions are only found in the leaves of the branch and bound tree, it is traversed in classical depth-first order.

The node that is selected next is the one with the maximal fitness value. The calculation of the fitness will be explained below. We also tried different node selection rules, but no speed-up of the algorithm could be obtained in our experiments. The idea of this framework is based on Göllner et al. (2010) and Göllner et al. (2012). In the next paragraphs the different steps are explained in more detail.

In every node of the branch and bound tree, first a relaxation of the discrete topology subproblem is solved. The whole design space, which is bounded by the outer borders of the profile, is divided into hexahedral elements. Since we are interested in a constant cross section, we are able to reduce the number of

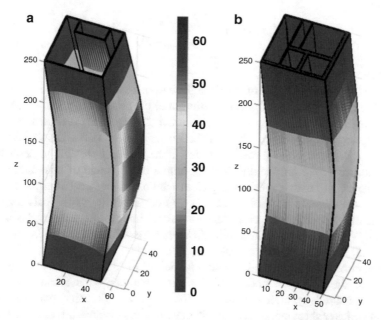

Fig. 5.20 Comparison of (**a**) the start and (**b**) the optimal profile design for the example. The displacement is multiplied by 5000 for better visualization. The colors represent the norm of the displacement

optimization variables by considering only a cross section of the profile. The finite element simulation to determine the displacement y is still performed in the three-dimensional space. The density of each element is used as an optimization variable. The real valued densities are in the interval $[0, 1]$. Fixing the lower bound to one enforces material for some elements. Analogously setting the upper bound to zero enforces space containing no material. These constraints are used only to compute the topology optimization in one chamber in each node. The already optimized chambers remain fixed in the topology optimization. There are different methods available to solve the relaxation of the topology optimization subproblem. One possible method is to formulate the subproblem as a MIP which has been presented in Fügenschuh et al. (2008). Another method is to use the SIMP approach (Solid Isotropic Material with Penalization). Following the latter, a modified version of the topology optimization code presented in Andreassen et al. (2011) is used. The solution of this topology optimization subproblem is given by a density matrix containing entries between 0 and 1. Therefore, it is necessary to find a feasible topology with discrete densities, which approximates the given solution matrix as good as possible.

This approximation is computed such that one new chamber with less than the maximal number of 90° bends is introduced which best describes the solution of the topology optimization subproblem. In order to define the degree of similarity between this solution and the discrete topology decision the fitness function

$$K(d) = 1 - \frac{\sum_{p \in Q} A_p |d_p - s_p|}{\sum_{p \in Q} A_p} \qquad (5.5)$$

is introduced and maximized.

In this formula Q is the set of pixels in the current chamber, A_p is the area of pixel p, $s_p \in [0, 1]$ is the density of p as the solution of the relaxation, and $d := [d_p]_{p \in Q}$, $d_p \in \{0, 1\}$ is the optimization variable. In order to maximize the fitness value for a given topology Q and a density matrix $s := [s_p]_{p \in Q}$ a MIP is formulated. The topology is represented as a matrix, which is the smallest possible to describe the profile structure. We are looking for an optimal enlargement of this matrix to the size of the density matrix maximizing K. Instead of solving the corresponding MIP directly, an algorithm based on successive shortest paths calculation is implemented which works considerably faster, but results still in the optimal solution. We also experimented with different fitness functions but the one given in Eq. (5.5) worked best in our cases.

Furthermore, we introduced a parameter describing a minimal distance between the rim of the "old" chambers and the new chamber. This circumvents small chambers, caused by high densities in the topology relaxation next to the rim implying thicker sheet metal. The densities within this distance are ignored because the optimization of the sheet thickness is part of the geometry optimization, which is applied later.

Since the enlarged matrix also contains information about length and width of the sheet metal surrounding the new chamber, this information is used as a starting point for the geometry optimization subproblem. Within the geometry optimization the shape of the cross section is optimized with fixed topology through changing the sheet thickness and length. The topology is characterized by the number of chambers and the arrangement of the different edges to each other. The optimization variables are the length and the width of each edge. The meshes used for the topology subproblem and geometry optimization are decoupled. This allows us to use different element sizes for each subproblem. The geometry optimization is restricted by both linear, for example, minimal and maximal sheet thickness, and nonlinear constraints such as bounds on the size of individual chambers. The geometry subproblem is solved with an SQP algorithm, see Algorithm 5.1, or alternatively an interior point algorithm.

To illustrate the presented approach, we apply the presented optimization scheme to a specific asymmetric load case. This load case is chosen to illustrate the possible complexity of the optimal inner profile structure for certain load scenarios and is not motived by a practical use case. Figure 5.19 shows the longitudinal and the cross section of the profile and the applied loads. The profile is fixed on both ends and a uniform load of 1980 N (q_1) acts on the upper chamber and another uniform load of 1400 N (q_2) acts on the whole profile from below. There are also restrictions on the maximal and minimal cross-sectional area of each chamber. The sheet thickness is restricted by 1 mm and 3 mm. The number of required chambers is set to five and the maximal number of bends for the additional chambers to zero. Figure 5.20 shows the solution of the algorithm which reduced the objective function by about 63.27% compared to the starting solution.

5.2.3 Geometry Optimization of Hydroformed Branched Sheet Metal Products

In many applications deep drawn metal products, as introduced in Sect. 3.2.4, play an important role, for example, as casing bodies or cladding. Based on the shape optimization in Sect. 5.2.1, we consider surface structures with additional stringers on one side of the sheet. These products can, for example, be manufactured by hydroforming. The optimization of the hydroforming process itself is considered in Sect. 5.3.1. The design of such hydroformed branched sheet metal products can be described through a tensor product surface with B-splines. We use such a B-spline surface as a parametrization of a given domain. The following section is based on Göllner (2014).

A B-spline surface $b(s, t) : [0, 1]^2 \to \mathbb{R}^3$ of degree $(n_1, n_2) \in \mathbb{N} \times \mathbb{N}$ is defined by $m_1 \cdot m_2$ control points $c_{ij} \in \mathbb{R}^3, (i,j) \in \{1, \dots, m_1\} \times \{1, \dots, m_2\}$, with and corresponding basis functions $N_i^p : [0, 1] \to \mathbb{R}_{\geq 0}$, with $p \in \{n_1, n_2\}$ and can be written as

$$b(s, t) = \sum_{i=1}^{m_1} \sum_{i=1}^{m_2} c_{ij} N_i^{n_1}(s) N_j^{n_2}(t).$$

In our case $n_1 = n_2 = 3$ for a bicubic surface is considered. The basis functions N_i^p are recursively defined by the Cox-de Boor algorithm (Piegl and Tiller 1997). The spline surface is used to parameterize the shape of the sheet. We assume that the B-spline surface is the separating plane between the sheet and the stringers. The whole geometry is determined by this separating plane. Since the sheet metal products considered here are characterized by a low ratio of thickness to length and width, it is efficient to parameterize the thickness of the sheet as a distance in the direction of the B-spline surface normal. For this purpose, the thickness is added at each point of the structure, where the thickness change is modeled by an additional B-spline surface. The stringer height is handled analogously, with one B-spline curve for each stringer. For an example of the parameterization see Fig. 5.21. The geometry of the whole surface structure can be changed through the control points c_{ij}.

We consider the shape optimization problem defined in Problem 5.3. The aim of the considered problem is to increase the stiffness of the branched sheet metal product under different restrictions on the geometry. Therefore, the objective functions Eqs. (5.2) or (5.3) are used. To solve the shape optimization problem, a modification of the SQP method, called *ArcEx*, based on Cartis et al. (2011a) and Cartis et al. (2011b), is applied. Instead of solving a quadratic subproblem, an NLP is solved. As a result, in every iteration feasible designs are found. The resulting

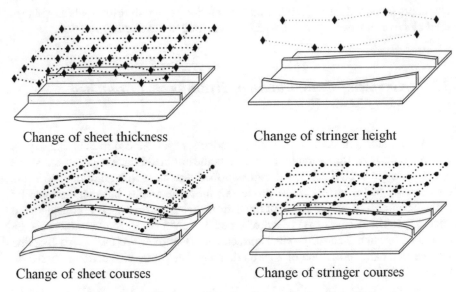

Change of sheet thickness | Change of stringer height

Change of sheet courses | Change of stringer courses

Fig. 5.21 Parameterization of a surface with stringers by cubic B-splines

subproblem with an additional cubic regularization as globalization strategy is defined as

$$\min \nabla j\left(\mathbf{u}^k\right)^T \mathbf{s} + \frac{1}{2}\mathbf{s}^T \mathbf{H}^k \mathbf{s} + \frac{1}{3}\sigma_k \|\mathbf{s}\|^3$$
$$\text{s.t.} \quad h(\boldsymbol{u}^k + \boldsymbol{s}) \leq 0.$$

Under certain assumptions it can be shown that *ArcEx* is globally convergent, see Göllner (2014) for details. The algorithm can be extended to handle inexactness of the objective function and gradient evaluation, which occurs if the elasticity problem is solved inexactly by using fast multigrid solvers, see Sect. 5.2.1 and Göllner (2014). The algorithm has also been extended to profile structures. A numerical result for this case is shown in Sect. 9.2.3.

Now we apply the method described above for the optimization of a foot of an information presenter, see Fig. 5.10. The starting geometry, necessary constraints, and the load case are provided by the formalized design task, see Fig. 5.14. We focus on the optimization of the sheet shape. Other choices of optimization variables, for example, sheet thickness or stringer courses, are presented in Göllner (2014). The starting geometry, depicted in Fig. 5.22a, is a square with a side length of 400 mm and 2 mm thickness.

The number of design variables can be reduced due to symmetry and restrictions on the edges from 412 to 32. As the objective function the mean quadratic displacement (Eq. (5.3)) is chosen. There are a total of 513 constraints. Since a multilevel method is used, see Sect. 5.2.1, based on Hess (2010), the optimization

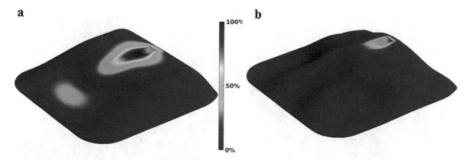

Fig. 5.22 Displacement of (**a**) the starting and (**b**) the optimized geometry. The color represents the norm of the displacement, ranging from *dark blue* (=0%) to *dark red* for the maximal displacement of the starting solution (=100%)

Table 5.2 Number of iterations on different meshes

Level	Elements	Iter.
1	16,996	20
2	31,834	1
3	63,717	1
4	130,974	3
5	282,050	5

problem is solved on different mesh levels. The size of these meshes and the number of optimization iterations performed on the different levels are shown in Table 5.2. The optimized geometry is shown in Fig. 5.22b. The objective value is decreased by 87.29%. The maximal displacement is reduced by 41.32% compared to the starting geometry.

5.2.4 Shape Optimization of Mechanical Connections

A natural extension of profiles are form-fit connections with a profile-like structure. They play an important role in the context of highly integrated products, for example, in rack systems or other module systems. Here, we address the shape optimization of mechanical connections; for an illustration, see Fig. 5.23a. An overview of connections with flow split flanges is shown in Sect. 7.2.2. We consider the two-dimensional cross section of a connection. The general abstract setting is shown in Fig. 5.23b. To distinguish the different parts of the connection, they are denoted by the subscripts $\cdot_s$ for slave and $\cdot_m$ for master. This assignment is arbitrary but fixed.

The physical behavior of such a connection can be formulated mathematically as a multibody contact problem. The multibody contact problem can be formulated analogously to Problem 5.1 for $\Omega = \Omega_s \cup \Omega_m$, except that only displacement fields y,

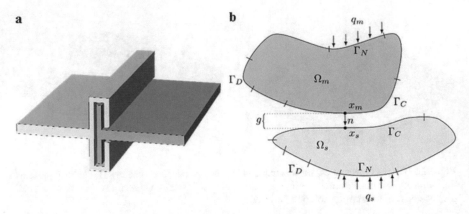

Fig. 5.23 (**a**) Draft of a plug-in connection, (**b**) general abstract setting of the contact problem between two elastic bodies Ω_s and Ω_m. The bodies are fixed on Γ_D and surface loads q_s and q_m act on Γ_N. The potential contact zone Γ_C includes the a priori unknown contact zone

which do not violate a non-penetration condition are allowed. This non-penetration condition is defined by

$$\tau_n^C(y) \leq \bar{g}, \tag{5.6}$$

where $\tau_n^C : Y \to H^{1/2}(\Gamma_C)$ is the normal trace operator, i.e., $\tau_n^C(y)(x) = n^T y(x)$ for all $x \in \Gamma_C$ and n denoting the outer unit normal on Γ_C. The normal distance between the different bodies in the initial configuration is denoted by a sufficiently smooth function $\bar{g}$. Condition (Eq. (5.6)) ensures that the displacement on the contact boundary is not greater than the gap between master and slave. Similar to Problem 5.2 a finite formulation of the multibody contact problem can be derived. For this purpose the multibody contact problem, which can be written as Problem 5.1 with additional non-penetration constraint, can be reformulated as a cone constrained optimization problem, see Problem 5.4, since (Eq. (5.6)) is equivalent to

$$\bar{g} - \tau_n^C(y) \in \mathcal{K} := \left\{ v \in H^{\frac{1}{2}}(\Gamma_C) : v \geq 0 \text{ on } \Gamma_C \right\},$$

where $\mathcal{K}$ is a closed convex cone.

Problem 5.4 (Multibody Contact Problem in Linear Elasticity)

$$\min \frac{1}{2} \int_\Omega \varepsilon(y) : C\varepsilon(y) dx - \left(\int_\Omega y \cdot f dx + \int_{\Gamma_N} y \cdot q dS(x) \right)$$

$$\text{s.t.} \quad \bar{g} - \tau_n^C(y) \in \mathcal{K}, \ y \in Y.$$

The Karush-Kuhn-Tucker (KKT) conditions for Problem 5.4 can be reformulated as a nonsmooth system of equations. After introducing an L^2-

regularization (Ulbrich et al. 2016) the finite dimensional optimality system for Problem 5.4 can be written as

$$H_h\left(\vec{y}_h, \vec{\lambda}_h\right) := \left(\begin{array}{c} A_h \vec{y}_h + B_h^* \vec{\lambda}_h - F_h \\ \vec{\lambda}_h - \max\left(0, \vec{\mu}_{r,h} + \alpha^{-1}(B_h \vec{y}_h - \vec{g}_h)\right) \end{array}\right) = 0, \qquad (5.7)$$

with B_h denoting the finite dimensional counterpart of the normal trace operator τ_n^C. The L^2-regularization has been introduced to allow for a convergence theory of the semismooth Newton method in function space and results in a mesh independent convergence behavior.

Following the mortar approach, Wohlmuth (2000), we chose a biorthogonal dual basis $\Psi = \{\psi_1, \ldots, \psi_{n_c}\}$, i.e.,

$$\int_{\Gamma_C} \phi_i \cdot \psi_j dx = \delta_{ij} \int_{\Gamma_C} \phi_i dx, \;\; i,j \in \{1, \ldots, n_C\},$$

for the finite dimensional Lagrange multiplier subspace $\Lambda_h \subset \Lambda := H^{-1/2}(\Gamma_C)$, with $n_C \in \mathbb{N}$ denoting the number of nodes on the contact boundary Γ_C. The finite dimensional Lagrange multiplier λ_h is defined as in Eq. (5.1) and can be interpreted as the normal contact pressure.

The optimality system (Eq. (5.7)) can be solved, for example, by a semismooth Newton method. For a detailed derivation of Eq. (5.7), we refer to Kikuchi and Oden (1988), Laursen (2002), Ulbrich et al. (2016), and Wriggers and Laursen (2006) and for the semismooth Newton method to Bratzke (2015), Ulbrich (2002), and Ulbrich et al. (2016). The resulting finite dimensional shape optimization problem is defined analogously to Problem 5.3, but instead of the state equation for the linear elasticity problem $A_h \vec{y}_h - F_h = 0$, the state equation for the contact problem $H_h\left(\vec{y}_h, \vec{\lambda}_h\right) = 0$ is used.

For a sufficiently smooth contact boundary Γ_C the existence of a solution operator $\mathbb{R}^n \ni u \mapsto (y_h(u), \lambda_h(u)) \in Y_h \times \Lambda_h$ can be shown. The operator is nonsmooth due to the properties of the multibody contact problem and so is the reduced objective function. The objective function can be chosen, for example, as Eqs. (5.2) or (5.3), respectively. A suitable algorithm for solving this problem is the Bundle Trust Region algorithm (BTR) (Outrata et al. 1998; Schramm and Zowe 1992), since both, the nonconvexity and the nonsmoothness, can be handled. The general approach is to approximate the objective function iteratively through cutting planes and optimize over the cutting plane model. The subproblem to calculate the step in the k-th iteration can be written as

$$\min_{\xi \in \mathbb{R}, s^k \in \mathbb{R}^n} \xi + \frac{1}{2\gamma} \| s^k \|^2$$

$$\text{s.t. } \xi \geq j'(\boldsymbol{u}^i)^{\mathrm{T}} \boldsymbol{s}^k - \beta_i^k \ \forall i \in J_k,$$

$$\boldsymbol{u}^k + \boldsymbol{s}^k \in X. \tag{5.8}$$

The regularization parameter $\gamma \in \mathbb{R}_+$ can be interpreted as a trust region radius. A large parameter γ results in smaller steps $\boldsymbol{s}^k$. The linearization error β_i^k is a measure for the approximation of the objective function through the cutting plane model. In contrast to the original algorithm, we add the design constraints to the subproblem to get feasible designs in each iteration.

Generating the cutting planes requires subgradients (Outrata et al. 1998), which we determine by the adjoint approach (Hinze et al. 2009). The shape subgradients are determined in points of strict complementarity, i.e., if the contact pressure does not vanish at all contact points, by

$$j'(\boldsymbol{u}) = \frac{\partial}{\partial \boldsymbol{u}} J(\boldsymbol{u}, y, \lambda) + \frac{\partial}{\partial \boldsymbol{u}} H(\boldsymbol{u}, y, \lambda)^* p, \tag{5.9}$$

where p denotes the adjoint determined by solving the adjoint equation

$$-\frac{\partial}{\partial (y, \lambda)} J(\boldsymbol{u}, y, \lambda) = \frac{\partial}{\partial (y, \lambda)} H(\boldsymbol{u}, y, \lambda)^* p.$$

If strict complementarity does not hold, it can be shown using a perturbation argument that Eq. (5.9) is an approximate subgradient.

Now we apply the presented method to an exemplary fastening of two different profiles, see Fig. 5.24. The input data for the optimization are provided by the formalized design task as presented in Sect. 5.1.4, see Fig. 5.15. The whole connection has the size of 50 mm × 65 mm. The right profile is fixed at the top. On the left profile a force acts to the left pulling the connection apart. The left profile is additionally guided by slide bearings. The optimization is performed under the constraints of constant volume and different restrictions on the flanges

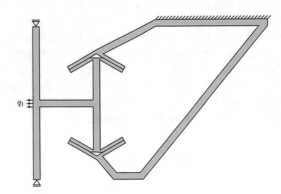

Fig. 5.24 The surface load q_1 is uniformly applied to the *middle* of the *left* profile

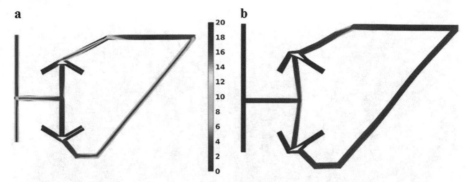

Fig. 5.25 (**a**) Starting and (**b**) optimized geometry with respect to the load case shown in Fig. 5.24. The color represents the von Mises stress (MPa)

such that all degrees of freedom are locked with respect to the motion during the use case. The sheet thickness is assumed to be 2 mm. To measure the stiffness of the connection we used Eq. (5.2) as objective function. The objective value was reduced by about 93.76% within 10 iterations. The optimized geometry is shown in Fig. 5.25.

5.3 Process Optimization

The following section addresses different problems arising in process optimization. As its general idea was already motivated in Sect. 2.4, the focus lies on the mathematical description for which the general overview can be found in Sect. 5.2. In Sect. 5.3.1 the deep drawing process introduced in Sect. 3.2.4 is considered. Here, the aim is to find the optimal process controls such that the resulting sheet metal product approximates the desired shape as good as possible. In addition, the optimal control shall ensure an error-free production process. In Sects. 5.3.2 and 5.3.3 we inspect the production of profiles on integrated flow production lines such as the one introduced in Sect. 3.2.1. First we deal with the problem of minimizing the number of very time-consuming retooling processes. Second the optimal sequence of production steps to manufacture a given profile is calculated.

5.3.1 Optimal Control of Deep Drawing Processes

The hydroforming process of sheet metals with additional stringers as presented in Sect. 3.2.4 can lead to failures during the production if forces like blank holder force or the internal pressure are too low or too high. Furthermore, it is favorable to control the energy consumption due to environmental aspects or costs. We will

Fig. 5.26 FE simulation of the stringer sheet at the end of the forming process with (**a**) initial controls and (**b**) optimal controls

present two different approaches to improve the hydroforming process of sheet metals.

The main content of this section is based on Bratzke (2015).

Derivative-free Approach. This section deals with an optimization approach to reach a desired shape of the stringer sheet and avoid failure modes of the process by determining optimal time-varying process controls. In addition, the geometry of the stringer is optimized so that the stringer buckling is minimal. As already stated in Sect. 3.2.4, this leads to an increased stiffness of the sheet.

The deep drawing process can be modeled as a constrained optimal control problem. The process controls are time-dependent functions $u(t)$ with $t \in [0, T]$, where T is the processing time. The displacement of the stringer sheet is denoted by $y(u(t))$ and is calculated by an FE simulation. It is assumed that the stringer is located in the xy-plane and is orthogonal to the z-axis, see Fig. 5.26.

The deep drawing process is simulated by the software *ABAQUS* using the "Dynamic Implicit" solver. Further process relevant constraints, which characterize the set of admissible displacements will be described in the following. To treat the resulting nonlinear inequalities and equalities, we use a penalty approach. The appropriate penalty terms of the constraints are discretized in time and space based on the FE simulation. The control variable u consists of the blank holder force F_b and the fluid pressure p_I. The controls are parameterized by continuous piecewise linear functions for each interval $[t_j, t_{j+1}]$ of the M subintervals of the process time. We use the amplitude values for the relative blank holder force $\bar{a}_j$ and the relative internal pressure $\hat{a}_j$ as optimization variables, which are between 0 and 1. The control vector is defined by $\boldsymbol{u} := (\bar{a}_0, \cdots, \bar{a}_M, \hat{a}_0, \cdots, \hat{a}_M)$. In order to avoid the opening of the tool system, the blank holder force has to be at least as large as a counterforce generated by the internal pressure during the entire forming process. Therefore, a linear constraint $h_j(\boldsymbol{u})$ is used, which couples the blank holder force and the internal pressure. The buckling of the stringer is measured by the deviation of a stringer face from a best fitting plane at the final time of the forming process. Due to the small stringer thickness, compared to the length and width, we are able to use one face of the stringer to determine the buckling of the stringer. The following penalty term:

$$S_1\Big(y(T)\Big) = \frac{1}{|N_S|} \sum_{i=1}^{|N_S|} \Big(\bar{z}(n_i, T) - z(n_i)\Big)^2$$

states a measure for the buckling, where $n_i \in N_S$ are the mesh nodes of the grid of the stringer face N_S. The value $\bar{z}(n_i)$ denotes the z-coordinate of node n_i on the best fitting plane and $z(n_i)$ the z-coordinate on the stringer. In order to prevent a leaking of the tool system, the flange region is not allowed to be entirely drawn in, the corresponding penalty term is denoted by $S_2(y)$ (Bäcker et al. 2015). In addition, we have to restrict the change of fluid pressure. The change of the fluid pressure influences the relative growth of the dome height $\Delta_k h$ per time increment $\Delta_k t$ which can be restricted by the parameter ν_{rel}. A quadratic penalty term, denoted by $S_3(y)$, is used to handle the inequality $\frac{\Delta_k h}{\Delta_k t} \leq \nu_{rel}$. Together with a regularization term for the blank holder force and the internal pressure $R(u)$ to prevent high energy consumption, we obtain the objective function with additional penalty terms

$$\tilde{J}(y, u) := \text{vol}(\Omega_z)\Big(y(u, T)\Big) + w_1 S_1(y(T)) + w_2 S_2(y) + w_3 S_3(y) + w_4 R(u),$$

where $w_1, w_2, w_3, w_4 > 0$ are given penalty parameters. The difference between the sheet metal at the end of the forming process and the desired shape can be measured as the remaining volume $\text{vol}(\Omega_z)$ between the target shape and the formed sheet metal.

The simulation is coupled through the ABAQUS scripting interface in Python with the optimization algorithm. The optimization problem is solved by using the implementation of COBYLA (Constrained Optimization BY Linear Approximations) from the NLopt library (Johnson 2016). This method is a model-based derivative-free optimization method. It is based on a trust region (TR) method which uses a linear model to approximate the objective function.

To demonstrate our concept, the optimization approach is applied to the hydroforming process of a stringer sheet with a stringer height of 15 mm and thickness of 1 mm. Figure 5.26 shows the stringer sheet at the end of the hydroforming process in the FE simulation. Figure 5.26a shows the result performed with the initial controls and Fig. 5.26b performed with optimal controls. The related control curves can be seen in Fig. 5.27. After 79 iterations, the objective function value is reduced by 93.3%. The simulation of the deep drawing process with *ABAQUS* takes 13 h each which constitutes the main part of the optimization time.

Based on the optimized controllable process parameters of the deep drawing process, we are also able to optimize the geometry of the stringer. The idea is to

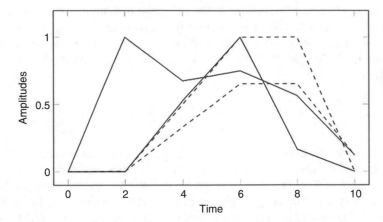

Fig. 5.27 Initial (*dashed*) and optimal (*solid*) control curve for the blank holder force (*red*) and the fluid pressure (*blue*)

Fig. 5.28 (**a**) Initial stringer height, (**b**) optimal stringer height

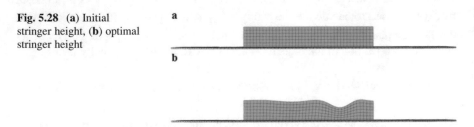

obtain a nonlinear stringer height, which is as high as possible, such that the forming process does not lead to buckling of the stringer. This is favorable because higher stringers lead to a higher stiffness, as mentioned in Sect. 3.2.4. The free stringer edge at the top of the stringer is modeled via a spline function with eight control points. We choose the degree of buckling as the objective function and apply the same approach as before with the new objective function to solve the optimization problem. Figure 5.28 shows the stringer height of the used stringer sheet. Figure 5.28a shows the initial stringer height of 18 mm and Fig. 5.28b the optimal nonlinear stringer height with values between 10.78 mm and 17.83 mm. The corresponding deep drawn stringer sheets are depicted in Fig. 5.29. A reduction of the objective function value by 95.97% within 80 iterations can be achieved.

Derivative-based Approach. Due to the high number of iterations and the time-consuming simulation, we are interested in a more efficient way to optimize the deep drawing process. Therefore, it is necessary to take information about the derivative into account. This leads to a derivative-based optimization approach which uses reduced order models, an efficient calculation of subgradients, and a bundle trust region method. To present the main idea of the optimization approach, a simplified mathematical model of the hydroforming process is considered and the focus lies on flat sheets without additional stringers. From a mathematical point of

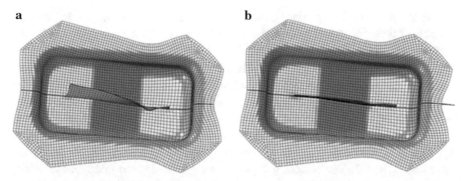

Fig. 5.29 Stringer sheet from above at the end of the forming process with (**a**) start stringer height and (**b**) optimal stringer height

view, the deep drawing process can be modeled as an elasto-plastic contact problem with friction and can be formulated as a quasi-variational inequality. To state the deep drawing process mathematically, Lagrange multipliers are introduced and a formulation consisting of a partial differential equation along with complementarity conditions corresponding to contact, plasticity, and friction is obtained. For further details of a rate-independent elasto-plastic problem with linear, isotropic, and kinematic hardening for linearized strain, we refer to Hager and Wohlmuth (2009) and Seitz et al. (2014). Due to the high drawing depth compared to the sheet thickness, the linearized strain does not fit exactly the material behavior during the deep drawing process. Therefore, it is necessary to use nonlinear elasticity to model the physical properties as explained in Bratzke (2015). This also has an impact on the plastic behavior of the material as, for example, the split of the deformation gradient into an elastic and a plastic part is multiplicative and not additive. The model for finite plasticity is currently being developed, hence this section deals with a nonlinear elasticity problem with frictionless contact. The notations and definitions are similar to Sect. 5.2.

Problem 5.5 (Nonlinear Elasticity Contact Problem without Friction)
Given the equilibrium and boundary conditions

$$
\begin{aligned}
-\mathrm{div}(P) &= f && \text{in } \Omega, \\
y &= 0 && \text{on } \Gamma_D, \\
Pn &= q && \text{on } \Gamma_N,
\end{aligned}
$$

find the displacement field $y : \Omega \to \mathbb{R}^d$ such that the contact conditions hold.

$$
\begin{aligned}
g(y) &\leq 0, \\
\lambda_n &\geq 0, \\
\lambda_n g(y) &= 0, \\
\lambda_T &= 0
\end{aligned}
$$

Here, the first Piola-Kirchhoff stress P relates the Cauchy stress to the reference configuration Ω. The second Piola-Kirchhoff stress S can be written as $S = F^{-1}P$, where the deformation gradient is defined as $F = I + \nabla y$. We adopt a St. Venant-Kirchhoff material model with constitutive law given by $S = \lambda \ tr(E)I + 2\mu E$ with Green-Lagrangian strain $E = \frac{1}{2}(\nabla y + \nabla y^T + \nabla y^T \nabla y)$. The gap function $g(y)$ is defined by $g(y) = \tau_n^C(y) - \bar{g}(y)$ as in Sect. 5.2, but here the gap between the elastic body and the rigid obstacle depends on the deformation. At this point we skip the variational formulation of the mentioned problem as well as the FE-discretization and a detailed approach for the semismooth reformulation and refer the reader to Bratzke (2015) and Popp et al. (2009) for further details. In the following we denote by $\vec{y}_h$ and $\vec{\lambda}_h$ the FE coefficient vectors of the discretized form, further details are given in Sect. 5.2. The resulting semismooth system for the nonlinear elasticity problem with contact can be written as follows:

Problem 5.6 (Discretized Nonlinear Elasticity Contact Problem without Friction)

Find the displacement vector $\vec{y}_h = \vec{y}_h(t_k)$ and the stress $\vec{\lambda}_h = \vec{\lambda}_h(t_k)$ such that for each time step t_k the nonlinear system

$$H_h^N\left(\vec{y}_h, \vec{\lambda}_h\right) := \begin{pmatrix} A_h^N(\vec{y}_h) + F_h^c\left(\vec{\lambda}_h, \vec{y}_h\right) - F_h(t_k) \\ C_{h,n}^c\left(\vec{y}_h, \vec{\lambda}_{h,n}\right) \\ C_{h,T}^c\left(\vec{\lambda}_{h,T}\right) \end{pmatrix} = 0 \qquad (5.10)$$

is satisfied with initial conditions $\vec{y}_0 = 0$.

The contact conditions can be written as

$$C_{h,n}^{c,i}(\vec{y}_h, \vec{\lambda}_{h,n}) := \vec{\lambda}_{h,n}^i - \max(0, \vec{\lambda}_{h,n}^{tr,i}), \quad C_{h,T}^{c,i}(\vec{\lambda}_{h,T}) := \vec{\lambda}_{h,T}^i,$$

where the superscripted i denotes the i-th entry of the corresponding coefficient vector. The trial value for the normal contact stress for each node $i \in N_c$ is defined by $\vec{\lambda}_{h,n}^{tr,i} := \vec{\lambda}_{h,n}^i + c_n g_h^i(\vec{y}_h)$, where $c_n > 0$ is fixed. The matrix $F_h^c\left(\vec{\lambda}_h, \vec{y}_h\right)$ represents the contact virtual work and $A_h^N(\vec{y}_h)$ is the discrete operator for the stiffness. Problem 5.6 can be solved by a semismooth Newton method as mentioned in Sect. 5.2.

In the context of deep drawing our aim is to minimize the volume Ω_z between the bottom of the sheet metal and a desired target shape. As the bottom of the sheet metal moves in time, the current position depends on the displacement y. The applied surface force is the fluid pressure force q_p depending on time $t \in [0, T]$ and control points $\vec{u}_h$. The objective function is given by

$$J_h(y_h, u_h) = \text{vol}(\Omega_z(y_h(T))) + \frac{\alpha}{2} \int_0^T q_p^2(u_h, t) \, dt,$$

where $\alpha > 0$. In order to apply the bundle trust region method (BTR), see Sect. 5.2 and Schramm and Zowe (1992), we introduce the reduced discretized optimal control problem

$$\min_{u_h \in \mathcal{U}_h} j_h(u_h) := J_h(u_h, y_h(u_h), \lambda_h(u_h)).$$

Here $u_h \mapsto (y_h(u_h), \lambda_h(u_h))$ denotes a solution mapping implicitly defined by Eq. (5.10). The reduced form results in a nonlinear, nonsmooth, and nonconvex problem. We assume the solvability of this problem and refer to Bratzke (2015) and Outrata et al. (1998) for further details about necessary assumptions. The required subgradients can be efficiently determined using the adjoint approach as shown in Sect. 5.2.4.

To speed up the solving procedure, the dimension of the system is reduced with the help of a reduced order model (ROM) which is a low-dimensional approximation of the large-scale model and has a similar input-output behavior as the original model. The solution space of system (Eq. (5.10)) can be approximated by a reduced space in a way that the physical properties of the system are captured. For the construction of a reduced basis (RB) for the approximation of the solution space of the displacements, proper orthogonal decomposition (POD) is applied. The key idea is to determine an orthonormal basis with the help of an optimization problem which minimizes the mean square error between the snapshots and their representation in the reduced space. The snapshots, which are solutions of the given system, are calculated by an FE simulation. This optimization problem can be handled by solving a symmetric eigenvalue problem (Volkwein 2013). An algorithmic description is given in Bratzke (2015) in Algorithm 5.3. This algorithm returns a reduced basis $\Theta_y := \left[\vartheta_1^y, \ldots, \vartheta_{l_y}^y \right]$ of the reduced space which approximates the snapshots best. The l_y basis functions are chosen such that the approximation of the system has a given quality. A reduced approximation of the displacement y^r can be represented as a linear combination of these basis functions with coefficient vector $\widehat{y}^r$.

A RB for the contact snapshot ensemble can be obtained by the Angle-Greedy algorithm proposed in Haasdonk et al. (2011). This algorithm returns the l_c indices which capture the largest volume, based on an angle criterion. We select the corresponding snapshots from the solution space of λ and normalize them, which leads to the reduced basis matrix $\Theta_c := \left[\vartheta_1^c, \ldots, \vartheta_{l_c}^c \right]$ for λ. The corresponding coefficient vector for the contact stress is denoted by $\widehat{\lambda}^r$. Due to the time-dependence of the normals and tangents for the contact boundary, additional matrices N_k and O_k for the current unit outer normals and tangents of the contact nodes are necessary. Furthermore, it has to take into account that the discrete operator A_h^N depends nonlinearly on the displacement vector. We use a discrete empirical interpolation method (DEIM) together with a sparse evaluation strategy, see Bratzke (2015). The ROM model related to Eq. (5.10) reads

$$\widehat{H}^N\left(\widehat{y}^r, \widehat{\lambda}^r\right) := \begin{pmatrix} A_h^N(\Theta_y \widehat{y}^r) + \Theta_y^T F_h^c(\Theta_c \lambda, \Theta_y \widehat{y}^r) - \Theta_y^T F_h^N(t_k) \\ (N\Theta_c)^T C_{h,n}^c\left(\Theta_y \widehat{y}^r, (N\Theta_c)\widehat{\lambda}^r{}_n\right) + (O\Theta_c)^T C_{h,T}^c\left((O\Theta_c)\widehat{\lambda}^r{}_T\right) \end{pmatrix}$$

Now the FE model can be replaced by the ROM model within the optimization process. Therefore, the reduced objective function has to be evaluated at the state y^r and λ^r instead of y_h and λ_h. The ROM-based reduced optimal control problem is then given by

$$\min_{u_h \in U_h} j^r(u_h) := J^r(u_h, y^r(u_h), \lambda^r(u_h)), \tag{5.11}$$

where the solution map is now defined through $\widehat{H}^N\left(\widehat{y}^r, \widehat{\lambda}^r\right)$. To solve this problem, the Bundle-ROM-based Trust Region method with Levenberg regularization from Bratzke (2015) is applied, which follows the TR-POD approach of Arian et al. (2000).

In the following a numerical result for solving the ROM-based optimal control problem for a simplified deep drawing process by optimizing the pressure control is presented. To solve the ROM-based optimal control problem for nonlinear elasticity with frictionless contact Algorithm 5.2 is applied.

Algorithm 5.2 (Bundle-ROM-based Trust Region algorithm with Levenberg regularization)

Input: Starting point $\vec{u}_h \in \mathbb{R}^m$, ratio of acceptance $\eta \in (0,1)$, and Levenberg regularization parameter ρ_L^0

Output: Solution of optimal control problem 5.11

For $k = 0, 1, 2, \ldots$ **do**

1. Compute time FE snapshot sets for $\vec{y}_{h,k}$, $\vec{\lambda}_{h,k}$ and the adjoint variable corresponding to $\vec{u}_{h,k}$.
2. Compute reduced basis Θ_y, Θ_c and build ROM.
3. Minimize

$$s_k = \arg \min j_k^r(\vec{u}_{h,k} + s_k) + \frac{\rho_L^k}{2}||s_k||_2^2$$

   using the BRT algorithm.

4. **STOP** if s_k satisfies a termination criterion.
5. Compute $j_{h,k}(\vec{u}_{h,k} + s_k)$ and set $\alpha = \frac{j_{h,k}(\vec{u}_{h,k}) - j_{h,k}(\vec{u}_{h,k} + s_k)}{j_k^r(\vec{u}_{h,k}) - j_k^r(\vec{u}_{h,k} + s_k)}$
6. **If** $\eta < \alpha$: Successful step, go to 1.

   Else: Non-successful step, go to 3.

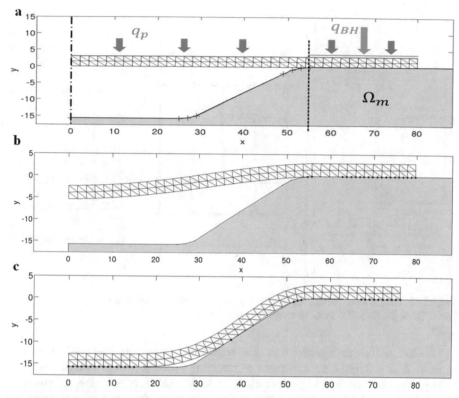

Fig. 5.30 (a) Initial configuration, deformed mesh at the end of the deep drawing process with (b) initial control and (c) optimal control

We consider a two-dimensional, symmetric problem illustrated in Fig. 5.30a, hence we take only half of the sheet and the die into account. The sheet metal has a total length of 160 mm and a thickness of 3 mm. The die is represented by Ω_m, where the red line describes the potential contact boundary.

The deep drawing height is 15.7 mm. The applied surface loads are the fluid pressure q_p and the blank holder force q_{BH} which acts only on the right end of the sheet metal. We choose an equidistant time grid for the interval [0, 3] with 11 steps. The optimization needs 10 Bundle-ROM-TR steps to reach a local minimum u_{opt} with a reduction of the objective function by 87.7%. Figure 5.30b illustrates the deformed mesh with the active contact nodes (blue) for the initial control and Fig. 5.30c the mesh for the optimal control.

5.3.2 Partitioning

Some profiles, as, for example, a solution from Sects. 5.2.1 or 5.2.2, are too complex to be produced out of just one piece of sheet metal. They can only be produced in separate pieces that are welded together. If different pieces are

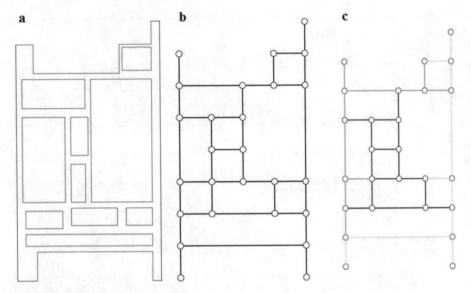

Fig. 5.31 (**a**) Example profile, (**b**) its derived graph G, and (**c**) partitioning into three isomorphic parts (*blue, green, yellow*) and one remaining part (*red*)

produced on the same production line, a retooling process is needed. The idea of this section is to reduce the number of retooling processes since a modification of the production line is very time consuming in general. The best case, in which no modifications of the production line are needed, arises from the separation of the sheet metal product into identical parts. The first part of this section deals with finding a partition into a given number of identical parts. Since this is not always possible, we allow a single remaining part that can be different from the other parts. The objective is to minimize the size of this remaining part.

Mathematically, given an integer K for the number of identical parts and a profile from which we derive a graph $G = (V, E)$ by defining a node $v \in V$ for each junction or bending of the sheet metal. The edges E of G correspond to the sheet metal joining those given vertices according to the profile, see Fig. 5.31a, b. The task is to find a partitioning $V_0, \ldots, V_K$ of the nodes V such that the induced graphs $G[V_i]$ are connected for every i in $\{0, \ldots, K\}$ and all except the first $G[V_0]$ are isomorphic, that is $G[V_i] \cong G[V_j]$ for $i, j \in \{1, \ldots, K\}$. The graph $G[V_0]$ induced by a set of nodes $V_0 \subseteq V$ is formed by the set of nodes V_0 and all edges from G that have both endpoints in V_0. Two graphs G and H are isomorphic, if there is an edge-preserving bijection $f : V(G) \rightarrow V(H)$ between the nodes of G and H, that is $\{u, v\} \in E(G) \Leftrightarrow \{f(u), f(v)\} \in E(H)$.

In our case this task will be applied to the line graph of G, that is, the edges will be partitioned instead of the nodes. Note that the only graphs for which an isomorphism of the line graph does not imply an isomorphism of the original graph are the claw $K_{1,3}$ and the triangle K_3. This is why it suffices to work with node partitions. In an extension of this model, also length and thickness information can be used. The partitioning problem alone is already NP-hard and this also holds for the problem stated here.

First we present the main ideas for an integer programming formulation for the partitioning problem described above and do not include the full model. Let $z_{v,w}$ be a binary variable that is 1 if node v is mapped to node w via an isomorphism and 0 otherwise. If v or w is a node from the remaining part V_0 then $z_{v,w}$ is set to be 0. The objective function is

$$\max \sum_{v \in V} \sum_{w \in V} z_{v,w},$$

which is equivalent to the minimization of V_0. Binary variables x_v^k for every $v \in V$ and $k \in \{0, \ldots, K\}$ are 1 if node v is in partition k and 0 otherwise. Further, $y_{v,w}$ for every $\{v, w\} \in E$ which is 1 if and only if v, w are in the same partition. Along with the obvious partitioning constraint

$$\sum_{k=0}^{K} x_v^k = 1 \quad \forall v \in V,$$

x_v^k and $y_{v,w}$ are coupled in an obvious way which will not be described here. For the connectedness we introduce flow variables and corresponding constraints.

Because the isomorphism can map one node to at most $K - 1$ other nodes and the isomorphic partitions contain a maximal number of $\lfloor |V|/K \rfloor$ nodes, we already know an upper bound of our objective function:

$$\sum_{v \in V} \sum_{w \in V} z_{v,w} \leq \left\lfloor \frac{|V|}{K} \right\rfloor K(K - 1).$$

This model itself contains symmetries which reduces solution speed. One type of symmetry can be restricted by the following inequality via limiting the order of the partitions:

$$\sum_{\substack{k \in \{1, \ldots, K\} \\ k \leq v}} x_v^k + x_v^0 \geq 1 \quad \forall v \in V,$$

where we assume the nodes to be numbered by $1, \ldots, |V|$. Connectivity can also be enforced directly on the x_v^k variables by the following separator inequalities:

$$x_v^k + x_w^k - \sum_{s \in S} x_s^k \leq 1 \quad \forall k \in \{1, \ldots, K\}, v, w \in V, S \text{ min } v\text{-}w\text{-separator}.$$

For two distinct nodes v and w in V with $\{v, w\} \notin E$, a subset $S \subseteq V \setminus \{v, w\}$ is called a v-w-(node)separator if and only if there is no path from v to w in the induced graph $G[V \setminus S]$. A separator S is *minimal* if S does not properly contain a v-w-separator.

Let $\mathcal{P}$ be the polytope given by the convex hull of the incidence vectors of the induced connected subgraphs. In general, it is NP-hard to optimize over this polytope (Johnson 1985). If G is a cycle or a complete bipartite graph, a complete description is known (Lüthen and Pfetsch 2016). Combined with the case of trees, which was discussed in Wang et al. (2015), the integer programming formulation can be further strengthened by additional inequalities.

We also developed an algorithm that directly solves the problem presented in the beginning, see Algorithm 5.3. Since most profiles in practical examples can be modeled as sparse graphs containing only slightly more edges than a tree, connectivity is a very strong constraint. Starting with a K-subset of the nodes, of which the single nodes are all isomorphic to another, we try to extend the isomorphism by adding nodes step by step to the different parts. Of course, this algorithm is only efficient if the number of neighbors is small, since otherwise the number of different possibilities grows exponentially. Practical examples could be solved in a few seconds. An example partitioning can be seen in Fig. 5.31c.

Algorithm 5.3 (Brute-force algorithm for finding an optimal partition)

Input: Graph $G = (V, E)$, $K \in \mathbb{N}$

Output: Partition $V_0, \ldots, V_K$ of the nodes, such that $G[V_i] \cong G[V_j]$ for $i, j \in \{1, \ldots, K\}$ and $G[V_i]$ is connected for $i = 0, \ldots, K$ and $|V_0|$ is as small as possible

For each K-subset $T := \{v_1, \ldots, v_K\}$ of V **do**

    $V_i := \{v_i\}, i = 1, \ldots, K, V_0 := V \setminus T$

    For each node $v \in V_0$ that is connected to V_1 **do**

        $V_1 := V_1 \cup \{v\}$ $V_0 := V_0 \setminus \{v\}$

        For $i = \{2, \ldots, K\}$ **do**

        For each node $w \in V_0$ such that $G[V_1] \cong G[V_i \cup \{w\}]$ **do**

            $V_i := V_i \cup \{w\}$

            $V_0 := V_0 \setminus \{w\}$

        If $G[V_0]$ is connected and $|V_0|$ is smaller than the best solution found

            save solution

        If $|V_1| = \left\lfloor \frac{|V|}{K} \right\rfloor$

            return optimal solution

    return best solution found

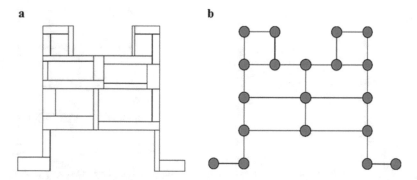

Fig. 5.32 (a) Example profile, not to scale, (b) associated graph G

5.3.3   Production Sequence

In this section we present an approach developed by Günther (2010) that computes
the best production sequence for a given geometry of a multi-chambered profile.
This means finding a sequence of production steps such that the given profile can be
manufactured on an integrated production line as presented in Chap. 3. The aim is
to find an optimal unrolling such that the difference between the geometry of the
produced profile and the desired profile is minimized and the technology chosen for
the intersection points is as good as possible. For example, possible technologies for
producing a three-way junction are welding three pieces together, bending one
piece and welding the third or splitting one piece (Chap. 3). Based on mechanical
analysis the different ways of creating junctions are rated to serve as input for the
objective function.

Here, we explain only one of the models discussed in Günther (2010). The main
idea is to create a graph $G = (V, E)$ that describes the given profile and solve a
Steiner tree problem on an extended graph with additional constraints. This can be
achieved by defining a node for each junction or bending of the sheet metal, see
Fig. 5.32 for an example profile and the derived graph. The directed Steiner tree
problem in a weighted directed Graph $D = (W, A)$ is defined by a subset of the nodes
$T \subseteq W$ called the *terminals*. The task is to find a minimal weighted directed tree in
D that contains all the terminal nodes T. An example of how to read off the
production sequence from a Steiner tree can be seen in Fig. 5.33b.

The edges E of G correspond to the sheet metal joining the given vertices
according to the profile as described in the previous section, see Fig. 5.31a, b. We
define two functions $l : E \rightarrow \mathbb{Q}$ and $t : E \rightarrow \mathbb{Q}$ to model the sheet metal length and
thickness. To accommodate the different possibilities for a junction, this graph
needs to be extended to the directed graph $D = (W, A)$, see Fig. 5.33c. Vertices of
G of degree d become star graphs on $d + 1$ nodes in D with arcs directed in both
ways. Bends are treated as two-way junctions. All undirected edges $\{u, v\}$ become
two directed arcs in W from one leaf of the star u to one leaf of the star v. As before,
the arcs are directed in both ways. The terminals for the Steiner tree problem are

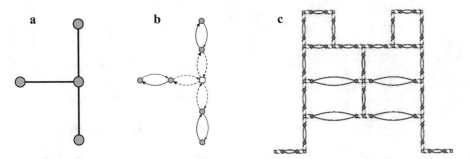

Fig. 5.33 (**a**) Directed graph D for the Steiner tree model, (**b**) exemplary three-way junction, (**c**) example profile. The *red* arcs in (**b**) display a possible construction: Bending on part from the *left downwards* and welding the third part from the *top*

defined as the leaves of the stars. Note that we can obtain a higher flexibility of the solutions by splitting edges in the graph G before deriving D, which allows to model a joining process in the interior of the edges.

The problem can now be modeled as a MIP, compare Sect. 5.2. One necessary constraint is to allow exactly one technology (as explained in the first paragraph) for every junction or bend by introducing binary technology variables. The full model will not be shown here, but it can be found in Günther (2010).

Furthermore, there are some technology-induced properties that have to be incorporated in the model. First, it has to be taken into account that the sheet metal has a maximal width which cannot be exceeded. Second, the length of every flange has to be bounded from above. To ensure the first condition, the model forbids paths P whose length $l(P)$ is larger than the maximal length and also contains the web (see Chap. 3 for a definition). The containment of the web is needed in order to ensure that splitted material is only measured once. Checking if path P contains the web can be accomplished by the technology variables. The flange condition can be treated analogously. Both conditions are dynamically handled by the solver in the form of cutting planes.

The model can be extended by including a condition that bounds the number of separate components from above. So far, this step does not regard any similarities between the components as was discussed in Sect. 5.3.2.

To reduce the running time of the MIP-solver, we first try to find a feasible starting solution by two different heuristics. The first one incorporates the more general algorithm of finding a minimum diameter spanning tree that contains a set of given edges. In this case, the diameter corresponds to the sheet metal width and hence it is used to compute an unrolling with small sheet metal width. Because the general problem of finding a minimum diameter spanning tree is NP-complete, we employ a heuristic (Günther 2010). The second heuristic computes an unrolling with small maximal flange length. As before, finding a minimum spanning tree with minimal flange length is NP-hard, such that a heuristic can be applied for fast solving (Günther 2010). Note that the heuristics are only used as a starting point of

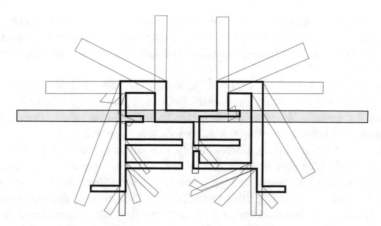

Fig. 5.34 Schematic flower pattern of the optimal unrolling for the example profile (not to scale). The *gray bar* represents the starting material, final geometry is *bold*

the algorithm such that an optimal solution is not necessary and faster heuristics suffice.

We also studied the polytope defined only by the variables that model the technology decision for a fixed vertex of arbitrary outdegree (the number of outgoing arcs). For this case a complete linear description can be given (Günther 2010).

Let $\mathcal{P}$ denote the projection of the full polytope onto the space of the technology variables. Define by $\mathcal{P}^+ := \mathcal{P} + \mathbb{R}_+^m$ the technology polytope as an upward monotonization (or dominant) of $\mathcal{P}$, since $\mathcal{P}$ is not full-dimensional. Because the components are connected, the facet-defining inequalities for $\mathcal{P}^+$ for connected subsets are of interest. We found inequalities for $\mathcal{P}^+$ that are satisfied by an exponential number of solutions. The separation problem of these inequalities can be solved by a variation on finding minimal cuts and can be found in Günther (2010).

Both polyhedral studies were implemented and lead to a decisive reduction of running time of the MIP-solver. The solution of the example profile from Fig. 5.32a is depicted in Fig. 5.34.

We also investigated algorithms that generate all possible unrollings ordered by its objective value and check for each one whether it satisfies the constraints. This procedure only works well if the constraints are easily satisfied. In most practical cases the procedure described above is faster.

5.4 Beneficial Effects of Formalization and Optimization

Product design is essentially about finding the best solution for a given technical problem. Not only customer concerns have to be considered. The design also has to satisfy demands regarding, for example, manufacturing, environmental or economic aspects. A concurrent consideration of product and process requirements is necessary to comprehensively address these aspects. The resulting complexity of the design task can be efficiently handled with the help of mathematical optimization methods.

The formalization of the design task is the basis for applying mathematical optimization methods within the manufacturing-integrated, algorithm-based product design approach. Based on product properties the entire solution space can be consistently described, enabling the mathematically supported product and process solution finding (Sect. 5.1.1). Unfortunately, product and process requirements are often not explicit or only vaguely formulated. With the help of process analyses, requirements can be comprehensively acquired and afterwards transformed and formalized into desired properties (Sect. 5.1.2). Derived from desired properties, the optimization problem comprising the objective function is formulated. Specialized algorithms, e.g., for profiles, surface structures, and manufacturing processes, such as the algorithm for partitioning profiles (Sects. 5.2 and 5.3), allow a fast and efficient solution finding.

Manufacturing technologies—being the link between the virtual product solution and its physical production—offer specific possibilities for the realization of innovative product solutions. Systematically exploiting this potential is one of the key elements of the manufacturing-integrated, algorithm-based product design approach. Processing (Sect. 4.2) and integrating the identified manufacturing-induced properties is essential for the approach: extending the solution space, manufacturing-initiated solutions such as linear flow split snap-fit elements can be provided. The manufacturing-initiated solutions not only offer benefits for the fulfillment of product requirements but also of process requirements (Sect. 5.1.3).

The consideration of process requirements, the application of mathematical optimization methods, and the early exploitation of manufacturing potential can reduce the amount of iterations during later development phases, leading to a more efficient solution finding process (Groche et al. 2012).

References

Andreassen E, Clausen A, Schevenels M, Lazarov BS, Sigmund O (2011) Efficient topology optimization in MATLAB using 88 lines of code. Struct Multidiscip Optim 43:1–16

Arian E, Fahl M, Sachs EW (2000) Trust-region proper orthogonal decomposition for flow control. Technical Report, Institute for Computer Applications in Science and Engineering

Bäcker F, Bratzke D, Groche P, Ulbrich S (2015) Time-varying process control for the stringer sheet forming by a deterministic derivative-free optimization approach. Int J Adv Manuf Technol 80(5):817–828

Bartels S, Carstensen C (2002) Each averaging technique yields reliable a posteriori error control in FEM on unstructured grids. Part II: higher order FEM. Math Comput 71(239):971–994

Birkhofer H (1980) Analyse und Synthese der Funktionen technischer Produkte. Dissertation, TU Braunschweig

Birkhofer H (2005) In fünf Minuten von der Aufgabe zur optimalen Lösung: Ein Beitrag zur Algorithmisierung der Frühen Phasen. In: Meerkamm H (ed) Proceedings of the 16th Symposium on Design for X, Neukirchen, Erlangen, 13–14 October 2005, p 47–58

Birkhofer H, Wäldele M (2005) Applied engineering design science: the missing link between design science and design in industry. In: Hosnedl S (ed) AEDS 2005 Workshop, Pilsen, 3–4 November 2005

Birkhofer H, Wäldele M (2008) Properties and characteristics and attributes and . . .—an approach on structuring the description of technical systems. In: Vanek V, Hosnedl S, Bartak J (eds) Proceedings of AEDS 2008 Workshop, Pilsen, p 19–34

Bornemann F, Erdmann B, Kornhuber R (1993) Adaptive multilevel methods in three space dimensions. Int J Numer Methods Eng 36(18):3187–3203

Braess D (1986) On the combination of multigrid method and conjugate gradients. In: Hackbusch W, Trottenberg U (eds) Lecture notes in mathematics, vol 1228. Springer, Heidelberg, pp 52–64

Braess D (2007) Finite elemente. Springer, Heidelberg

Bratzke D (2015) Optimal control of deep drawing processes based on reduced order models. Dissertation, TU Darmstadt

Brenner SC, Scott LR (2007) The mathematical theory of finite element methods. Springer, Heidelberg

Cartis C, Gould NIM, Toint PL (2011a) Adaptive cubic regularisation methods for unconstrained optimization. Part I: motivation, convergence and numerical results. Math Program 2 (127):245–295

Cartis C, Gould NIM, Toint PL (2011b) Adaptive cubic regularisation methods for unconstrained optimization. Part II: worst-case function- and derivative-evaluation complexity. Math Program 2(130):295–319

Curtis FE, Johnson TC, Robinson DP, Wächter A (2013) An inexact sequential quadratic optimization algorithm for large-scale nonlinear optimization. Technical report, Department of ISE, Lehigh University

Deutsches Institut für Normung (2012) Sachmerkmal-Listen—Teil 1: Begriffe und Grundsätze, DIN 4000-1:2012-09. Beuth, Berlin

Ehrlenspiel K (2009) Integrierte Produktenwicklung: Denkabläufe, Methodeneinsatz, Zusammenarbeit, 4. akt. Auflage. Hanser, München

Franke, HJ (1976) Untersuchungen zur Algorithmisierbarkeit des Konstruktionsprozesses. Dissertation, TU Braunschweig

Fügenschuh A, Hess W, Schewe L, Martin A, Ulbrich S (2008) Verfeinerte Modelle zur Topologie- und Geometrie-Optimierung von Blechprofilen mit Kammern. In: Groche P (ed) 2. Zwischenkolloquium des Sonderforschungsbereichs 666: Integrale Blechbauweise höherer Verzweigungsordnung—Entwicklung, Fertigung, Bewertung, Darmstadt, 12–13 November 2008

Göllner T (2014) Geometry optimization of branched sheet metal structures with a globalization strategy by adaptive cubic regularization. Dissertation, TU Darmstadt

Göllner T; Günther U, Hess W, Martin A, Ulbrich S (2010) Form- und Topologieoptimierung verzweigter Blechbauteile. In: Groche, P (ed) 3. Zwischenkolloquium des Sonderforschungsbereichs 666, Darmstadt, 29–30 September 2010

Göllner T, Günther U, Hess W, Pfetsch M, Ulbrich S (2012) Optimierung der Geometrie und Topologie flächiger verzweigter Blechbauteile und von Mehrkammerprofilen. In: Groche, P (ed) 4. Zwischenkolloquium des Sonderforschungsbereichs 666, Darmstadt, 14–15 November 2012

Gramlich S (2013) Vom fertigungsgerechten Konstruieren zum produktionsintegrierenden Entwickeln: Durchgängige Modelle und Methoden im Produktlebenszyklus. Dissertation, Technische Universität Darmstadt

Gramlich S, Birkhofer H, Bohn A (2011) Design process automation: a structured product description by properties and development of optimization algorithms. In: Culley SJ, Hicks

BJ, McAloone TC, Howard TJ, Clarkson PJ (eds) Proceedings of the 18th International Conference on Engineering Design (ICED 11), Lyngby, Copenhagen, 15–19 August 2011, p 299–309

Groche P, Schmitt W, Bohn A, Gramlich S, Ulbrich S, Günther U (2012) Integration of manufacturing-induced properties in product design. CIRP Ann 61(1):163–166

Günther U (2010) Integral sheet metal design by discrete optimization. Dissertation, TU Darmstadt

Günther U, Martin A (2006) Mixed integer models for branched sheet metal products. Proc Appl Math Mech 6(1):697–698

Haasdonk B, Salomon J, Wohlmuth B (2011) A reduced basis method for the simulation of american options. In: Cangiani A, Davidchack R, Georgoulis E, Gorban A, Levesley J, Tretyakov M (eds) Numerical mathematics and advanced applications 2011. Springer, Berlin

Hager C, Wohlmuth BI (2009) Nonlinear complementarity functions for plasticity problems with frictional contact. Comput Methods Appl Mech Eng 198:3411–3427

Haslinger J, Mäkinen RAE (2003) Introduction to shape optimization. SIAM, Philadelphia

Heidemann B (2001) Trennende Verknüpfung: Ein Prozessmodell als Quelle für Produktideen. Dissertation, Technische Universität Darmstadt

Heinkenschloss M, Vicente LN (2001) Analysis of inexact trust-region SQP algorithms. SIAM J Optim 12(2):283–302

Hess W (2010) Geometry optimization with PDE constraints and applications to the design of branched sheet metal products. Dissertation, TU Darmstadt

Hess W, Ulbrich S (2012) An inexact $\ell 1$ penalty SQP algorithm for PDE-constrained optimization with an application to shape optimization in linear elasticity. Optim Mehtod Softw 28 (5):943–968

Hinze M, Pinnau R, Ulbrich M, Ulbrich S (2009) Optimization with PDE constraints. Springer, Heidelberg

Hubka V (1973) Theorie der Maschinensysteme: Grundlagen einer wissenschaftlichen Konstruktionslehre. Springer, Berlin

Johnson D (1985) The NP-completeness column: an ongoing guide. J Algorithms 6(1):145–159

Johnson SG (2016) The NLOpt nonlinear optimization package. http://ab-initio.mit.edu/nlopt. Accessed 25 Jun 2015

Kikuchi N, Oden JT (1988) Contact problems in elasticity: a study of variational inequalities and finite element methods. SIAM, Philadelphia

Laursen TA (2002) Computational contact and impact mechanics. Springer, Heidelberg

Lindemann U (2009) Methodische Entwicklung technischer Produkte: Methoden flexibel und situationsgerecht anwenden, 3. korr. Auflage. Springer, Berlin

Lüthen H, Pfetsch ME (2016) On the connected subgraph polytope. Manuscript, TU Darmstadt

Mattmann I, Roos M, Gramlich S (2014) Transformation und Integration von Marktanforderungen und fertigungstechnologischen Erkenntnissen in die Produktentwicklung. In: Groche P (ed) Tagungsband 5. Zwischenkolloquium SFB 666, Mörfelden-Walldorf, 19–20 November 2014, pp 5–14

Mattmann I, Gramlich S, Kloberdanz H (2015a) The malicious labyrinth of requirements: three types of requirements for a systematic determination of product properties. In: Weber C, Husung S, Cascini G, Cantamessa M, Marjanovic D, Rotini F (eds) Proceedings of the 20th International Conference on Engineering Design (ICED 15), Milan, 27–30 July 2015, p 31–40

Mattmann I, Gramlich S, Kloberdanz H (2015b) The inscrutable jungle of quality criteria: how to formulate requirements for a successful product development. In: Shpitalni M, Fischer A, Molcho G (eds) Procedia CIRP, vol 36, CIRP 25th Design Conference Innovative Product Creation, p 153–158

Mattmann I, Gramlich S, Kloberdanz H (2016a) Mapping requirements to product properties: the mapping model. In: Marjanovic D, Storga M, Pavkovic N, Bojcetic N, Skec S (eds) Proceedings of the DESIGN 2016 14th International Design Conference, Dubrovnik, p 33–44

Mattmann I, Gramlich S, Kloberdanz H (2016b) Getting requirements fit for purpose—improvement of requirement quality for requirement standardization. In: Wang L, Kjellberg R (eds) Procedia CIRP, vol 50, 26th CIRP Design Conference, p 466–471

Nemhauser GL, Wolsey LA (1988) Integer and combinatorial optimization. Wiley-Interscience, New York

Nocedal J, Wright SJ (2006) Numerical optimization. Springer, Heidelberg

Outrata J, Kocvara M, Zowe J (1998) Nonsmooth approach to optimization problems with equilibrium constraints, vol. 28 of Nonconvex Optimization and Its Applications. Kluwer Academic Publishers, Heidelberg

Pahl G, Beitz W, Feldhusen J, Grote KH (2007) Engineering design: a systematic approach, 3rd edn. Springer, London

Piegl L, Tiller W (1997) The NURBS book. Springer, Heidelberg

Popp A, Gee MW, Wall WA (2009) A finite deformation mortar contact formulation using a primal-dual active set strategy. Int J Numer Methods Eng 79(11):1354–1391

Röder J (2014) Entwicklung einer clusterbasierten Methodik zur Anforderungserfassung auf Basis eines Modellraums zur Kategorisierung von Anforderungen. Dissertation, TU Darmstadt

Röder B, Birkhofer H, Bohn A (2011) Clustering customer dreams: an approach for more efficient requirement acquisition. In: Culley SJ, Hicks BJ, McAloone TC, Howard TJ, Dong A (eds) Proceedings of the 18th International Conference on Engineering Design (ICED 11), Lyngby, Copenhagen, 15–19 August 2011, p 11–20

Röder B, Gamlich S, Birkhofer H (2012) Von der abstrakten Anforderung zur formalisierten Entwicklungsaufgabe: Algorithmenbasierte Entwicklung am Beispiel komplexer, spaltprofilierter Blechbaugruppen. In: Groche P (ed) Tagungsband 4. Zwischenkolloquium SFB 666, Darmstadt, 14–15 November 2012, p 5–14

Roos M, Horn B, Gramlich S, Ulbrich S, Kloberdanz H (2016) Manufacturing integrated algorithm-based product design: case study of a snap-fit fastening. In: Wang L, Kjellberg R (eds) Procedia CIRP, vol 50, 26th CIRP Design Conference, p 123–128

Roy R, Hinduja S, Teti R (2008) Recent advances in engineering design optimisation: challenges and future trends. CIRP Ann 57(2):697–715

Schramm H, Zowe J (1992) A version of the bundle idea for minimizing a nonsmooth function: conceptual idea, convergence analysis, numerical results. SIAM J Optim 2(1):121–152

Seitz A, Popp A, Wall WA (2014) A Semismooth Newton method for orthotropic plasticity and frictional contact at finite strains. Comput Methods Appl Mech Eng 285:228–254

Suh NP (1998) Axiomatic design theory for systems. Res Eng Des 10(4):189–209

Suh NP (2001) Axiomatic design: advances and applications. MIT-Pappalardo series in mechanical engineering, Oxford University Press

Tekkaya AE, Allwood JM, Bariani PF, Bruschi S, Cao J, Gramlich S, Groche P, Hirt G, Ishikawa T, Löbbe C, Lueg-Althoff J, Merklein M, Misiolek WZ, Pietrzyk M, Shivpuri R, Yanagimoto J (2015) Metal forming beyond shaping: predicting and setting product properties. CIRP Ann 64(2):629–653

Ulbrich M (2002) Semismooth Newton methods for operator equations in function spaces. SIAM J Optim 13(3):805–841

Ulbrich M, Ulbrich S, Bratzke D (2016) A multigrid semismooth Newton method for semilinear contact problems. Technical report, TU Darmstadt, TU München

Verein Deutscher Ingenieure (1993) Methodik zum Entwickeln und Konstruieren technischer Systeme und Produkte, Richtlinie VDI 2221. VDI, Düsseldorf

Volkwein S (2013) Proper orthogonal decomposition: theory and reduced-order modelling. Lecture Notes

Wagner C, Roos M, Gramlich S, Kloberdanz H (2016) Process integrated design guidelines: systematically linking manufacturing processes to product design. In: Marjanovic D, Storga M, Pavkovic N, Bojcetic N, Skec S (eds) Proceedings of the DESIGN 2016 14th International Design Conference, Dubrovnik, p 739–748

Wäldele M (2012) Erarbeitung einer Theorie der Eigenschaften technischer Produkte: Ein Beitrag
    für die konventionelle und algorithmenbasierte Produktentwicklung. Dissertation, TU
    Darmstadt
Wang Y, Buchanan A, Butenko S (2015) On imposing connectivity constraints in integer pro-
    grams. Optimization Online
Wohlmuth BI (2000) A mortar finite element method using dual spaces for the lagrange multiplier.
    SIAM J Num Analysis 38(3):989–1012
Wriggers P, Laursen TA (2006) Computational contact mechanics, 2nd edn. Springer, Heidelberg
Yserentant H (1993) Old and new convergence proofs for multigrid methods. Acta Numerica
    2:285–326
Ziems JC, Ulbrich S (2011) Adaptive multilevel inexact SQP methods for PDE-constrained
    optimization. SIAM J Optim 21(1):1–40

Chapter 6
Computer-Integrated Engineering and Design

T. Weber Martins, S. Abedini, L. Ahmels, K. Albrecht, R. Anderl,
E. Bruder, P. Groche, H. Kaufmann, P. Mahajan, T. Melz, M. Özel,
H. Pouriayevali, J. Reising, S. Schäfer, Y. Tijani, A. Tomasella,
and B.-X. Xu

Virtual product development aims at the use of information modeling techniques and computer-aided (CAx-) tools during the product development process, to represent the real product digitally as an integrated product model (Anderl and

T. Weber Martins (✉) • K. Albrecht • R. Anderl
Department of Computer Integrated Design (DiK), Technische Universität Darmstadt,
Darmstadt, Germany
e-mail: Weber@dik.tu-darmstadt.de

S. Abedini • J. Reising • S. Schäfer
Institute of Construction Design and Building Construction (KGBauko), Technische
Universität Darmstadt, Darmstadt, Germany

L. Ahmels • E. Bruder
Physical Metallurgy (PhM), Technische Universität Darmstadt, Darmstadt, Germany

P. Groche • P. Mahajan • M. Özel
Institute for Production Engineering and Forming Machines (PtU), Technische Universität
Darmstadt, Darmstadt, Germany

H. Kaufmann
Fraunhofer Institute for Structural Durability and System Reliability LBF (LBF),
Fraunhofer-Gesellschaft zur Förderung der angewandten Forschung e.V., Munich, Germany

T. Melz
Research group System Reliability, Adaptive Structures, and Machine Acoustics (SAM),
Technische Universität Darmstadt, Darmstadt, Germany

Fraunhofer Institute for Structural Durability and System Reliability LBF (LBF),
Fraunhofer-Gesellschaft zur Förderung der angewandten Forschung e.V., Munich, Germany

H. Pouriayevali • B.-X. Xu
Mechanics of Functional Materials (MfM), Technische Universität Darmstadt, Darmstadt,
Germany

Y. Tijani • A. Tomasella
Research group System Reliability, Adaptive Structures, and Machine Acoustics (SAM),
Technische Universität Darmstadt, Darmstadt, Germany

© Springer International Publishing AG 2017
P. Groche et al. (eds.), *Manufacturing Integrated Design*,
DOI 10.1007/978-3-319-52377-4_6

Trippner 2000). Thereby, data related to the product as well as product properties are generated and stored as result of the product development process (e.g., product planning, conceptual design) (Pahl et al. 2007; VDI 2221 1993). Within virtual product development CAx process chains have been established. They comprise the concatenating of the applied tools and technologies within the steps of the virtual product development process enabling the consistent use of product data (Anderl and Trippner 2000). The computer-aided design (CAD) technology aims at the integration of computer systems to support engineers during the design process such as design conceptualization, design, and documentation. It provides the geometry of the design and its properties (e.g., mass properties, tolerances) which is abstracted to be used in computer-aided engineering (CAE) systems (e.g., finite element method (FEM)) for design analysis, evaluation, and optimization. The computer-aided process planning (CAPP) technology provides tools to support process planning, Numerical Control (NC) programming, and quality control (Hehenberger 2011; Lee 1998; Vajna 2009). The advantages are continuous processing and refinement of the product model, minimizing the modeling efforts regarding time as well as costs, and avoiding error sources. In addition, all relevant data and information related to the product can be provided for subsequent processing (Anderl and Trippner 2000). CAx technologies have been widely established within the product development processes in industry. They have been further developed in the last years; however, efforts to integrate and to automate them are still a topic of research. Especially, with the introduction of innovative manufacturing technologies such as linear flow and bend splitting new methods and tools for the virtual product development process are required. These technologies enable the production of a new range of sheet metal products with characteristic properties (e.g., Y-profile geometry, material properties) that are not addressed in state-of-the-art methods and tools.

In this context, a computer-integrated engineering and design approach for innovative sheet metal products and production technology is introduced in this chapter. This approach is briefly described in Fig. 6.1 using the structured analysis and design technique (SADT) (Ross 1977). The main functional components are CAD modeling, simulation of forming processes, simulation of local properties, and simulation of fatigue strength. This approach enables that input data such as design requirements, constraints, and results of the mathematical optimization are stepwise and continuously refined through functional components of the CAx process chain. An integrated information model is the backbone of this process chain since it represents the fundamental structure including all data as well as their semantics and relationships. A core model realizes a basis for adding further partial models to represent an integrated product and manufacturing process model for innovative sheet metal products (Sect. 6.1.1).

In addition to these outcomes, the integration of the simultaneous engineering (SE) approach enhances the product development process of innovative sheet metal products by parallelizing their steps (Sect. 6.1.2). It enables that subsequent tasks and process steps start once sufficient information are available. Data and information exchange between the steps of the CAx process chain are improved.

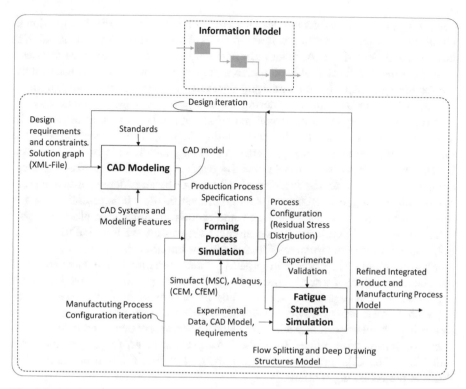

Fig. 6.1 CAx process chain for innovative sheet metal products and production technology

The product geometry is realized within the CAD modeling steps based on the data provided by the integrated information model. However, new design modeling approaches need to be explored in order to capacitate state-of-the-art CAD systems to create 3D CAD models of bifurcated sheet metal products. The direct modeling approach introduces design features for generation of feature-based, parametric 3D CAD models of such products (Sect. 6.2.1). In addition, the algorithmic modeling approach enables the automatic generation of CAD models to support the algorithm-3D-based product development (Sect. 6.2.2). The panelization approach realizes a geometrical simplification of freeform surfaces to enable a feasible production of complex products (Sect. 6.2.3).

3D CAD models provide fundamental geometric information to generate the finite element model (FE model) needed in the next steps of the CAx process chain. FEM simulation is a very useful method to obtain information about the product as well as the process without having to carry out expensive experiments (Anderl and Binde 2014; Schäfer 2006). However, there are still some challenges (e.g., computing time or quality of FE model) to be solved in order to obtain all relevant information of innovative sheet metal products as well as their manufacturing processes. Especially the 3D simulation of multistep rolling process requires higher computing power. To overcome this challenge, a more efficient FEM simulation to obtain the necessary process configuration as well as residual stress distribution is introduced in Sect. 6.3.1. Another important output of FEM simulations for metal forming is the distribution of local material properties, which is highly relevant for the load-bearing capacity and structural durability. In case of linear flow splitting, this aspect is fundamental considering the substantial changes in local material properties which represent important manufacturing-induced properties (Chap. 4). Yet, the severe plastic strains that occur during the process are a challenge for the prediction of local mechanical properties via FEM simulations, which is reviewed in Sect. 6.3.2. These outcomes (especially residual stress distributions and local strength changes within the working piece) need to be considered for an accurate simulation of fatigue strength and service life estimation. It is essential that the manufacturing process history is taken into account in order to exploit the light-weight potential. Therefore, fatigue strength simulation for flow splitting (Sect. 6.4.1) and deep drawn structures (Sect. 6.4.2) are introduced.

Such an integrated approach anticipates the determination of relevant properties and behavior about integral bifurcated sheet metal products as well as their corresponding manufacturing process. Thus, product development process including embodiment design and detail design is being accelerated. Subsequent processes in particular specific production planning and subsequent production benefit from the information gained from the CAx process chain (e.g., cost and error reduction). The CAx process chain as introduced in Fig. 6.1 will be explored in detail in the following sections.

6.1   Integrated Information Model

Information models are used during integrated, virtual product development to integrate information concerning the whole product life cycle. The innovative technologies of linear flow splitting and bend splitting require new methods and processes for development, evaluation, and manufacturing of the bifurcated sheet metal profiles. Therefore, the algorithm-based product development process was introduced. The complex requirements and new approaches have to be supported in a continuous way to successfully realize the products. All available information and process chains have to be integrated into this process. Therefore, an integrated information model provides the suitable basis. It enables phase-specific use of provided information and enhances collaboration and information exchange.

Models are abstractions of reality. They simplify complex correlations (e.g., details, functionalities) and are often used to emphasize specific scopes of the original object, system, or any desired part of the world (Stachowiak 1973; Hesse et al. 1994). In particular, information models are used in software engineering to represent concepts, relationships, rules, or structures (Chen 1976). During product creation they provide the backbone of computer-integrated processes.

We developed an integrated information model in which all data and information available during the product creation process is represented. It enables seamless integration of product information from all phases of the product life cycle (Anderl and Trippner 2000). The integrated information model supports the developed approaches for successful development of bifurcated profiles. The integrated representation of linear flow split and bend split parts enables continuous enhancement of the integrated information model. The integrated information model illustrated in Fig. 6.1. Section 6.1.1 provides an overview of the integrated information model as well as the enhancement of the core model to represent bifurcated sheet metal parts and their multifunctionalities. The core model, which represents the linear flow split profile, enables continuous advancement of the information model when new information is available. Furthermore, Sect. 6.1.2 introduces an insight into the simultaneous engineering approach applied to the algorithm-based product development. The integrated information model supports this approach by providing necessary information to the respective domain.

6.1.1   Core Model

The integrated information model was developed to support the product creation process of bifurcated profiles. We developed an integrated information model to support the product creation process of bifurcated profiles. It contains a *CoreModel* which serves as a basis for the whole information model. It represents the cross section of linear flow split parts (Groche et al. 2005; Wu et al. 2008b). Any changes and advancements during the product creation process are always connected to the product structure. To represent this information, the partial models are associated with the core model (Groche et al. 2005; Anderl et al. 2006). In Fig. 6.2, the *CoreModel* as well as the assigned partial models are depicted. The partial models are classified into four groups. The first group represents the product development phase with an *OptimizationModel* and a *RequirementsModel*. Subsequently, the

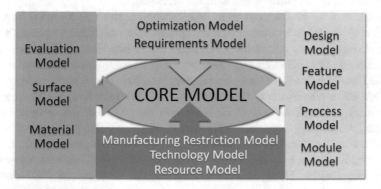

Fig. 6.2 Structure of integrated information model

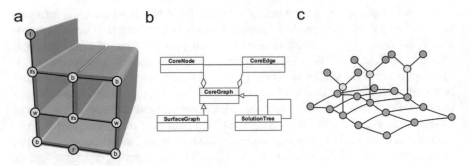

Fig. 6.3 (**a**) Multichambered profile with assigned *TechnologyNodes*, (**b**) *CoreGraph*, (**c**) freeform structure adapted from Weitzmann et al. (2012)

virtual product development is represented by a *DesignModel*, a *FeatureModel*, a *ProcessModel*   (Rollmann   2012),   and   a   *ModuleModel*.   The *ManufacturingRestrictionModel* (Albrecht et al. 2016), *TechnologyModel, and ResourceModel* (Weitzmann and Anderl 2012) represent the manufacturing part. For the product evaluation, the *EvaluationModel*, the S*urfaceModel* (Weitzmann and Anderl 2013), and the *MaterialModel* (Weitzmann 2014) are used.

To represent the information of the integrated information model, Unified Modeling Language (UML) class diagrams are used (Pilone 2003). The *CoreModel* class in combination with partial models allows representation of the life cycle of bifurcated profiles. New research outcomes are integrated via extension of existing partial models or addition of new partial models to the *CoreModel*. In Fig. 6.3, the *CoreModel* contains the class *CoreGraph* comprising the two classes *SolutionTree* and *SurfaceGraph*. *SolutionTree* contains a graph-based structure representing linear flow split parts, which was developed based on the results of the mathematical optimization introduced in Chap. 5. The investigation of freeform sheet metal parts leads to an enhancement of the *SolutionTree* through the *SurfaceGraph* (Fig. 6.3). The *SurfaceGraph* represents the freeform surface of bend split parts. This combination enables the representation of linear bifurcated profiles as well as freeform sheet metal parts (Weitzmann et al. 2012).

We defined the product structure of bifurcated profiles as a rooted tree representation. A rooted tree consists of a designated root which has no parent node. A root node has at least one child node while a leaf node has no children (Diestel 2000). Hence, Fig. 6.3a shows a sheet metal profile based on *SolutionTree*. The *SolutionTree* contains *CoreNodes* and *CoreEdges*. *CoreEdges* are associated with a *TailNode* and a *HeadNode* of the class *CoreNode*. The *CoreNodes* can either exist as a *RootNode l*, *TechnologyNode,* or *LeafNode*. A *RootNode r* may have an additional technology associated. *TechnologyNodes* represent the manufacturing technology used to produce this part of the sheet metal structure. In Fig. 6.3a, the used technologies are *b*—bending, *lfs*—linear flow splitting, and *w*—welding (Wu et al. 2008b). Using the tree structure, the profiles are fully described (Weitzmann and Anderl 2013). Furthermore, in addition to the *SurfaceGraph*, the tree structure is combined with freeform sheet metal surfaces. A tree structure based on a freeform surface represents the linear bend splitting process. Therefore, the initial root node is substituted by the *lbs*—linear bend splitting technology node (Weitzmann et al. 2012).

The structure of the *CoreModel* is represented in an eXtensible Markup Language (XML) data exchange format to allow a system-independent data exchange for all domains. The structure of the XML format file was developed taking into account the outcomes of mathematical optimization.

An example for adding partial models to the integrated information model is provided by the partial model *ModuleModel*. It represents multifunctional modules (Fig. 6.4). Multifunctionalities indicate the usage of manufacturing-induced properties and add specific functions to a sheet metal structure. The class *Module* is a specialization of *CoreGraph*. It can be used to describe every given sheet metal part. The subclass *MultifunctionalModule* represents a Module which has specific functions assigned. The Module is connected to Boundary Representation (B-Rep) data structures (Stroud 2006) of 3D CAD models to be able to represent the functionalities. This enables the graph structure to be processed in a 3D CAD system, see Sect. 6.2.1 for examples. Therefore, a *Module* consists of *GeometricElements* and *TopologicElements*. For the representation of functionalities, the *TopologicElements* are used. A *TopologicElement* can be a *Vertex*, *Edge*,

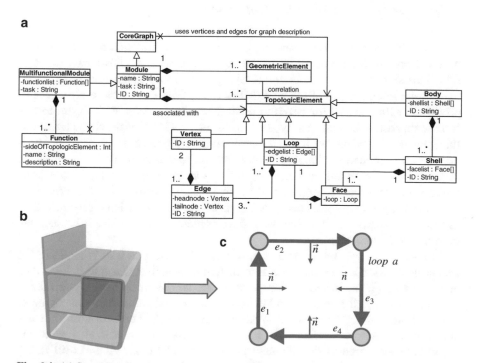

Fig. 6.4 (**a**) Structure of multifunctional module model, (**b**) 3-chamber model with water supply chamber, (**c**) representation of example chamber

Loop, Face, Shell, or *Body* (Anderl and Trippner 2000). Based on this representation, the Euler–Poincaré formula regarding STEP conformity is evaluated. A *MultifunctionalModule* consists of different *Functions*. Each *Function* is assigned to a *TopologicElement*.

An example is depicted in Fig. 6.4 at the bottom. The *MultifunctionalModule* is a 3-chamber profile. The marked chamber illustrates the exemplary task of water supply for a 2D representation. To define this function, the loop *a* is determined. Loop *a* consists of the four edges limiting the chamber. Every edge of this loop is now linked to the water-task information on the side of the normal *n*. An extension to 3D representation is necessary when profiles are modified in longitudinal way. For example, the addition of holes to join a sheet metal part to another structure is supported by the different *TopologicElements* as described before.

The *CoreModel* is the essential part of the integrated information model. Due to associating the partial models with the *CoreModel,* it is possible to describe the different product creation steps. The enhancement of the *CoreModel* by *ModuleModel* enables applying multiple functions to bifurcated profiles. During the next CAx steps multifunctionalities can be considered for simulation of the product or for manufacturing planning.

6.1.2 Products and Production Processes

As mentioned in the previous section, the integrated information model is used to describe products and production processes. Based on this knowledge, we integrated the simultaneous engineering (SE) approach into the algorithm-based product development process (Rollmann 2012). SE is used during the early stages of product development to improve the product creation process with regard to less time-to-market, cost, and quality (Eversheim et al. 1997). Main characteristic of this approach is the integration and parallelization of processes for product development and production planning. In comparison to classic product development, sequential processes are applied (Fig. 6.5 upper process). This leads to a large number of iterations emerging out of conflicts concerning the fulfillment of manufacturing requirements and process steps, because there is no cross communication. Furthermore, the following development step only begins when the current step is completed.

In contrast to sequential processes, during SE while one task is realized, the following task starts once sufficient information is available. Changes in input data are considered later. Based on this practice, the communication and information exchange between the domains and therefore the understanding between engineers involved increases. Errors and disadvantages are identified and corrected in early stages of product development. In Fig. 6.5, the SE approach is applied to the algorithm-based product development process. In Schüle and Anderl (2013), the approach was also adapted for specific needs of the product creation process of freeform sheet metal.

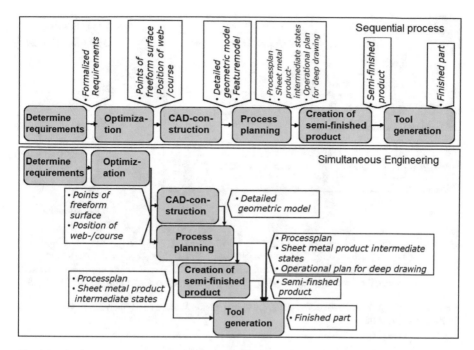

Fig. 6.5 Simultaneous engineering during algorithm-based product development process adapted from (Rollmann 2012)

The simultaneous engineering approach is supported by the integration of product development and manufacturing process planning introduced in Rollmann et al. (2011). Therefore, we developed an approach for sequencing processes and operations using feature-based precedence graphs. The definition of features in the context of geometric modeling and CAD models is explained in Sect. 6.2.1. The convenient side effect of the usage of features is the ability to integrate CAD and CAPP (computer-aided process planning) simultaneously, while following the guidelines of simultaneous engineering (Rollmann et al. 2011).

In this context, features are high-level entities in virtual product development. In Fig. 6.6, the process from an advanced feature model for bifurcated sheet metal parts to the processing of features for manufacturing process planning based on Rollmann et al. (2011) is depicted. The basis of feature precedence graphs is the core model introduced in Sect. 6.1.1. Due to the knowledge of available technology nodes, the manufacturing process steps are derived. A feature, e.g., for linear flow splitting, is extracted to a production process model which consists of the necessary steps needed to conduct linear flow splitting processes. In Fig. 6.6, the features are linear flow splitting, which is applied first, followed by roll forming to achieve the T-profile. Based on this step, the operational model is deducted. Every step is defined to manufacture the needed Y-profile via linear flow splitting and after that a number of necessary roll forming processes until the T-profile is manufactured. In the example of Fig. 6.6, the linear flow splitting process consists

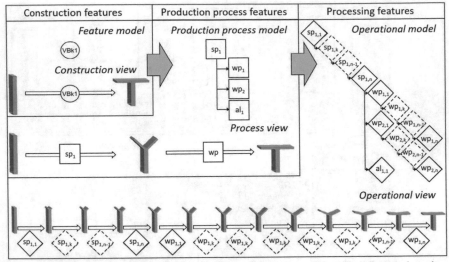

VBk *bifurcation of sheet metal edge;* **sp** *flow splitting;* **wp** *roll forming;* **al** *cut to length*

Fig. 6.6 Starting from construction features to processing features and operational view adapted from Rollmann (2012)

of four steps until the final flange length is formed. The following roll forming *wp* needs eight steps to form the T-profile part. This process is defined inside the operational model, and the creation steps are depicted as an operational view.

Using this framework, knowledge about products and processes supported by the integrated information model of bifurcated sheet metal parts is derived. The knowledge about the process sequence is used to derive the final products' profile pattern. The latter is combined with the achievements of Sect. 5.1 and used to declare the different production processes at the nodes. In Fig. 6.7, the process of deriving the profile pattern is shown. The given feature model depicts the process of creating a 3-chamber profile with flow splitting and bending. Starting with root node *w* there are two bending operations *b*1 and *b*2 applied. They are followed by two more bending operations and the linear flow splitting on both sheet metal edges. The result is similar to the chamber profile of Fig. 6.3a. However, to produce this profile the process has to be reversed. This is illustrated inside the process plan of Fig. 6.7 bottom. At first the linear flow splitting of the edges is processed (Step 1). This is followed by a sequence of bending operations. The first is used to create an I-profile out of the Y-profile (Step 2). Consequently, the following bending operations are used to create the final profile (Step 3–10). The profile pattern in Fig. 6.7 shows the different operations and their sequence. Based on this approach the macro process is described. Every manufacturing step consists of a number of partial steps to reach the resulting profile geometry. Therefore, e.g., *b*1 needs two steps for the final result. Depending on the roll positions and the installation space it is possible to perform forming processes simultaneously. The linear flow splitting is conducted on either side of the sheet metal as well as the bending operations.

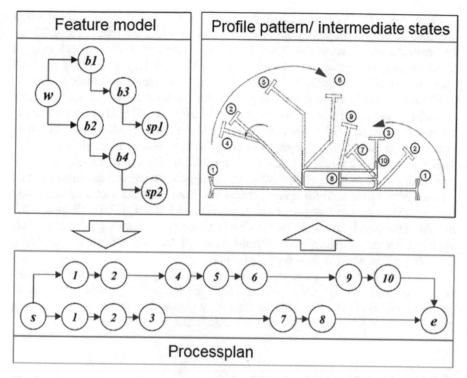

Fig. 6.7 From feature model to process plan and profile pattern adapted from Rollmann (2012)

The SE approach we introduced in this section enables faster development of bifurcated sheet metal parts. Due to usage of feature precedence graphs and process planning, profile patterns are derived just from the 3D CAD model.

This section provided an insight into the structure and application of an integrated information model. Further applications during application of CAx process chains are discussed in subsequent sections.

6.2   CAD Modeling for Bifurcated Sheet Metal Parts

Computer-aided design (CAD) systems enable the generation of product data based on dimensions, constraints, and parameters in order to support the product development and product creation processes. CAD models provide a fundamental element for information and data storage, representation and visualization obtained during all steps of the product development process. Therefore, CAD models containing single part as well as assembly information support its up- and downstream processes such as simulation (FEM, etc.), planning (CAPP), and manufacturing (computer-aided manufacturing (CAM)) (Anderl and Trippner 2000; Anderl and Mendgen 1996; Lee 1999; Picard 2015).

State-of-the-art CAD systems do not provide native functionalities for efficient and user-friendly modeling of bifurcated sheet metal structures (Anderl et al. 2008). For example, the characteristic Y-profile of linear flow split parts implies topologic changes in the data structure. Therefore, new CAD modeling approaches to represent and visualize integral, bifurcated sheet metal products and their geometric manufacturing-induced properties (e.g., geometric properties such as flange length as well as sheet metal thickness (Gramlich et al. 2015) in 3D CAD systems need to be explored. Section 6.2.1 describes design features for linear flow and bend split operations in 3D CAD systems. Section 6.2.2 introduces modeling approaches which automatically generate 3D CAD models of bifurcated profiles from the results of the mathematical optimization to support the algorithm-based product development process. Both approaches profit from products and manufacturing data provided by the integrated information model (Sect. 6.2.1). Section 6.2.3 describes the panelization of freeform surfaces for architectural applications. This approach realizes a geometrical simplification of freeform surfaces to enable a feasible production of such complex products.

The 3D CAD models generated with these modeling approaches are an important contribution to the CAx process chain for integral bifurcated sheet metal products (Fig. 6.1). They provide the necessary geometric information for forming process (Sect. 6.3.1) and fatigue strength simulation (Sect. 6.4) steps.

6.2.1 Direct Modeling Approach

State-of-the-art CAD systems do not provide modeling features to represent and visualize geometries of integral, bifurcated sheet metal products and their properties in the virtual product creation process (Groche et al. 2005; Wu et al. 2008a). In particular, they cannot handle the specific properties of linear flow splitting and bend splitting, such as variable sheet metal thickness (Schüle et al. 2009). Therefore, new methods and approaches to enable the modeling of CAD models for integral, bifurcated sheet metal products have been developed. Based on the definition of design features and feature-based modeling as the aggregation of product properties (e.g., constraints, dimensions, material, and manufacturing information) to geometric elements (VDI 2209 2009), the direct modeling approach has been established. The direct modeling approach uses existing modeling functionalities and software architectures available in state-of-the-art CAD systems to build integral, bifurcated sheet metal product models. Relevant requirements necessary to accomplish stable and accurate CAD design features are: assumption of volume constancy regarding forming process properties (Doege and Behrens 2010), verification of solids validity regarding topology and geometry (constructive solid geometry (CSG)/boundary representation (B-REP)—Euler–Poincaré), processing of sheet metal operations (e.g., bending) within the product model, and integration of technological information from information model (Sect. 6.1.1) in geometric elements.

Table 6.1 Euler operators (Rollmann 2012; Stroud 2006)

Euler operator	Description	Basis vector
mef	Make edge and face	$a_1^T = (0, 1, 1, 0, 0, 0)$
mev	Make edge and vertex	$a_2^T = (1, 1, 0, 0, 0, 0)$
mvfs	Make vertex, face, and shell	$a_3^T = (1, 0, 1, 0, 1, 0)$
kemr	Kill edge make ring	$a_4^T = (0, -1, 0, 0, 0, 1)$
kfmrh	Kill face make ring and hole	$a_5^T = (0, 0, -1, 1, 0, 1)$

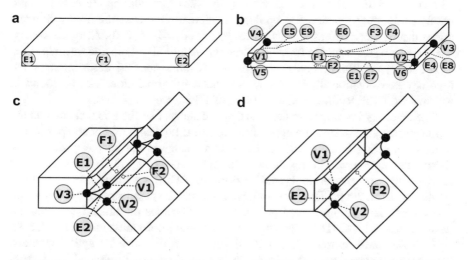

Fig. 6.8 Topology manipulation using Euler operators of a flow split operation (Rollmann 2012; Rollmann et al. 2009)

Topology-geometry investigations of linear flow split as well as linear bend split operations are necessary in order to get an insight into the implications of the data structure. Based on their outcome, modeling functions can be developed and implemented in the geometric kernel of CAD systems (e.g., Parasolid, ACIS).

Beginning with a simple block Euler operators (Table 6.1) are consequently applied to represent linear flow split and bend split operations. The fundamental linear-independent Euler operators have been defined as basis vector using Euler–Poincaré's coefficients $a_i^T = (v, e, f, h, s, r)$ (v for the number of vertices, e for the number of edges, f for the number of faces, h for the number of holes, s for the number of shells, and r for the number of rings) and combined with each other to obtain the topology structure of integral bifurcated sheet metal profiles (Rollmann 2012; Stroud 2006). Figure 6.8 shows the topology manipulation with Euler operators to obtain Y-profile from a simple block representing a single flow split operation. In the first sub-operation (Fig. 6.8a), the solid body is branched out to represent the inner face of the flanges, which is defined by 5 new vertices, 9 new edges, and 3 new faces. Thereby, $8 \times mev$ as well as $3 \times mef$ are applied (Fig. 6.8b).

Consecutively each flange side is bent. To represent a bend operation on a single side, the bending radius must be considered. Therefore, 8 new vertices, 16 new edges, and 8 new faces must be defined by the application of $8 \times mev$ as well as $8 \times mef$ (Fig. 6.8c). In order to represent the splitting center correctly, the vertices on the bending radius at the inner flange faces are connected with the definition of 2 new edges. Additionally, the remaining edges and faces inside of the solid body are removed. In this sub-operation $3 \times kef$ and $2 \times kev$ are applied. Thus, a valid topology of an integral bifurcated sheet metal profile needs in total $1 \times mvfs$, $23 \times mev$, and $19 \times mef$ to be described (Fig. 6.8d). Considering the initial simple block, a modeling function for topology manipulation to describe flow split operations can be described as a basis vector $p_{FSp} = (19,22,0,0,0,0)$. Also, the consistency check of the resulting topology structure using Euler–Poincaré formula $v - e + f = 2(s - r) + h$ (Stroud 2006) is fulfilled. A more detailed description of the topology investigation of linear flow and bend split operations can be found in Rollmann (2012), Rollmann et al. (2009), and Schüle et al. (2009).

The topology investigation described above shows that there is a change on the topology structure during the application of linear flow split and bend split operations as modeling functions. It implies also a change in the geometry due to the B-Rep data structure. The resulting Euler operators can be realized as modeling function in any geometric kernel (Parasolid, ACIS) to model such products (Rollmann et al. 2009). The results above resemble the graph structure to represent sheet metal parts as introduced in Sect. 6.1.1. It enables the linkage between the design features for CAD systems and the integrated information model (Sect. 6.1).

User-Defined-Features (UDFs) for linear flow split and bend split operations have been developed to enable a modular design of integral bifurcated sheet metal profiles. A UDF is a set of geometric elements using available modeling functionalities in the CAD system in order to extend it with a new customized feature (VDI 2209 2009). Thereby, each modeling step is defined using parameters, constraints, and mathematical relationships according to the respective manufacturing process to achieve the target geometry. Mathematical relationships are defined to describe the geometry of linear flow splitting and bend splitting with a y-formed shape in order to ensure volume constancy (Rollmann 2012; Schüle and Anderl 2012; Schüle et al. 2009).

$A_0 = y_0 s_f$	l_f: flange length in mm
$A_0 = A_f + A_1 + A_2 + A_3 + A_4$	y: splitting depth in mm
$A_1 = \frac{R^2 \tan \alpha - \widehat{\alpha}}{2}$	s_f: flange thickness in mm
$A_2 = \frac{s_f^2}{2} \tan \alpha$	s_0: sheet metal thickness in mm
	s_f/s_0: ratio
$A_3 = h^2 \left(\tan \frac{\gamma}{2} - \frac{\widehat{\gamma}}{2} \right)$	R: splitting roller radius in mm
	α: splitting roller flank angle in degrees
$A_4 = \frac{s_f^2}{2} \tan \alpha$	h: support roller radius
$l_f = A_f s_f$	γ: auxiliary angle
	A_f: total flange area
	A_i: area of flange section i
	$[\,\widehat{}\,]$: angle in radians

Fig. 6.9 Geometric and mathematical description of 2D flow split profile (Rollmann 2012; Schüle et al. 2009)

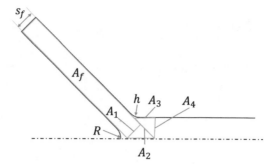

Fig. 6.10 Geometric and mathematic description of 2D bend split profile (Schüle et al. 2010a; Schüle et al. 2010b; Schüle et al. 2011)

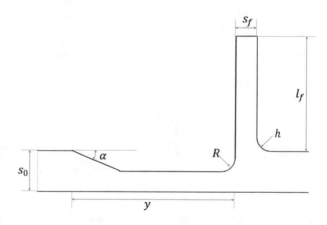

The UDFs for linear flow splitting and bend splitting are basically composed of an initial block, datum planes as reference objects, parametrized sketches, and extrude features. In general, the approach is to subtract a part of the initial block volume using simple geometric elements, e.g., 2D sketches with sweep operations (extrude), and to add the same amount of volume using a sketch of linear flow and bend splitting specific geometry as described above (Figs. 6.9 and 6.10). In this way, the required assumption of volume constancy can be fulfilled. All relevant geometric elements and operations are parametrized so that users can generate 3D CAD models of integral bifurcated sheet metal profiles considering relevant manufacturing information and restrictions. Modeling applications within CAD systems can be implemented using the UDFs to enable the iterative design of integral bifurcated sheet metal profiles. Figure 6.11 illustrates the dialog boxes of the design features for linear flow and bend splitting implemented in Siemens NX using NX Open API. Thereby, the user merely needs to select the faces, the reference objects, and the parameters. The UDFs are automatically modeled and placed on 3D geometry.

The design features for linear flow and bend splitting are extended to generate 3D CAD models of freeform bifurcated sheet metal products. Examples for freeform modeling functionalities in CAD systems are Bezier, B-Splines, and especially nonuniform rational B-Splines (NURBS). However, during the

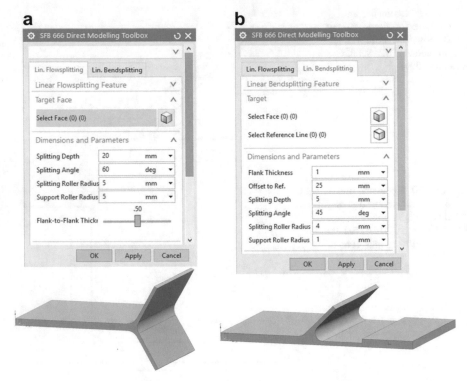

Fig. 6.11 Design features for (**a**) linear flow splitting and (**b**) bend splitting in 3D CAD system

modification of freeform surfaces, the parameters of linear flow and bend splitting are not automatically adjusted and need to be adapted manually afterwards. The reason for this is due to the different data structure used to represent solid bodies and freeform surfaces in CAD systems. Freeform surface functionalities are defined by control points or control polygons based on local coordinate systems with dimensionless coordinates. Thereby, geometry with its dimensions is not determining as in the case of conventional solid body modeling methods. Freeform surface and linear flow as well as bend splitting based on solid body modeling are not completely linked to each other. In this context, a modeling method needs to combine them to enable the generation and manipulation of parameterized freeform bifurcated sheet metal product model. The approach is to realize the bidirectional association allowing the mutual manipulation between freeform features and solid body modeling features. For that, a local coordinate system and area of effect are defined on the freeform surface (Fig. 6.12a).

The local coordinate system is set on the corner of the freeform surface at the nearest position related to the global coordinate system. It is composed by the dimensionless coordinates u and v, which are tangential to the freeform surface and have a value between 0 and 1. In addition, the dimensionless w coordinate is set on the normal direction on the neutral fiber of the sheet metal to describe thickness

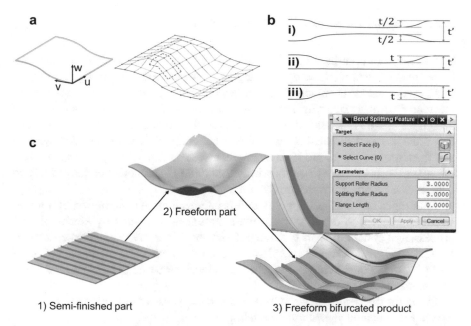

Fig. 6.12 Freeform modeling approach for freeform bifurcated sheet metal products (Schüle et al. 2010a; Schüle et al. 2011)

variations as well as vertical deformations. The area of effect is a geometric area on the freeform surface including the corresponding control points to enable the local manipulation of the feature. The thickness variation is defined as an absolute value. The gradient of the sheet metal thickness variation is described by the transition degree (Fig. 6.12b). Following this approach, the design features of freeform bifurcated sheet metal can be carried out in CAD systems (Fig. 6.12c) (Schüle et al. 2011).

Technology features connect the technology properties contained in the information model with the geometric elements of feature-based 3D CAD models (Schüle 2014). The concept makes use of the B-Rep data structure of 3D CAD models to link semantics into the corresponding nodes and edges of the product data. They are visualized in the 3D CAD models using Product and Manufacturing Information (PMI) or surface properties models as introduced in Weitzmann (2014). The design features for linear flow and bend splitting as presented above enable the generation of feature-based, parametric 3D CAD models of integral bifurcated sheet metal products. They are essential contributions to the CAx process chain as introduced in Fig. 6.1 since they provide products geometric information for the documentation as well as further evaluation using numerical methods. Such methods based on FEM simulation are discussed in Sects. 6.3 and 6.4.

6.2.2 Algorithmic Modeling Approach

Within the algorithm-based product creation approach mathematical optimization tools are applied to find the optimal topology or geometry of a design. Based on these results, 3D CAD models are generated for further processing (e.g., adaption of the model to manufacturing limits and refinement of the model by inclusion of design features). However, this step requires further time and effort, since the input data from the mathematical optimization is not compatible with 3D CAD systems and therefore need to be modeled manually by users. Thus, the algorithmic modeling approach investigates methods to generate 3D CAD models automatically to enhance the algorithm-based product creation process.

Challenges of the algorithmic modeling approach are: (1) definition of a representation concept to describe mathematic optimization results by means of data processing, (2) definition of the integral and consistent interface between mathematical optimization results and CAD modeling, (3) embodiment and adaption of CAD model by considering manufacturing-induced properties, and (4) visualization of optimized product model data.

Based on the requirements introduced above, a systematic approach has been developed to enable algorithmic modeling:

1. Elementary processing of topologic elements
2. Derivation and processing of geometric elements
3. Identification of semantics and allocation of semantics to the corresponding topologic-geometric elements
4. Processing of topologic-geometric elements containing semantics with features
5. Fully integrated product model of integral bifurcated sheet metal products

Characteristics of bifurcated sheet metal products enabling such an algorithm-based approach are: common part design pattern, common product properties, functional design element granularity, and parametric functional design elements (Anderl et al. 2008).

The results of the topology and geometry optimization are represented as a solution graph, which is defined as a set of ordered edges and nodes $G = (V, E)$ (Hazewinkel 2001). A data model based on the eXtensible Markup Language (XML) file format enables the representation of optimized integral bifurcated sheet metal products as solution graphs (Weitzmann et al. 2012). XML file format is a neutral data format realizing the data exchange between heterogeneous systems (CAD systems, MATLAB, or MS Office) which is supported by a wide number of application programming interfaces (API) (Harold and Means 2004). Therefore, it enables the continuous connection and communication between optimization data, information model, and product model data as CAD models. Figure 6.13 shows the structure of the data model. Thereby, XML elements describe nodes and edges with an id. The x- and y-coordinate of each node is defined. An edge connects two nodes based on their ids.

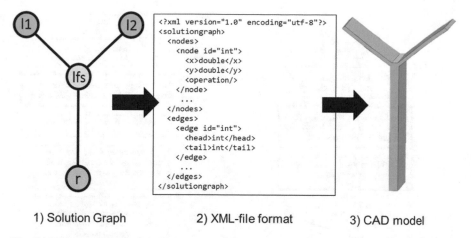

```
<?xml version="1.0" encoding="utf-8"?>
<solutiongraph>
  <nodes>
    <node id="int">
      <x>double</x>
      <y>double</y>
      <operation/>
    </node>
    ...
  </nodes>
  <edges>
    <edge id="int">
      <head>int</head>
      <tail>int</tail>
    </edge>
    ...
  </edges>
</solutiongraph>
```

1) Solution Graph   2) XML-file format   3) CAD model

Fig. 6.13 Data model based on XML format file to represent optimized integral bifurcated sheet metal products as solutions trees (Weitzmann et al. 2012)

During the conception of the integrated information model (Sect. 6.1.1) and data model, the B-Rep data structure is considered and resembled. It facilitates therefore the processing of the XML file to CAD models. An algorithm and an interface taking into account this data model enable the automated creation of CAD models from XML format files. Figure 6.14 explains the algorithm.

The first step is to load the XML file containing the optimized data. Furthermore, methods suitable to handle XML documents and Document Object Model (DOM) are applied to read the XML file and split its content into lists or arrays for nodes and edges. Next steps are to identify topological elements, get their data (ids and x, y-coordinate), and transform the information into the according geometric elements. For loops, points representing head and tail nodes are generated based on the x,y-coordinate data. Afterwards, a line representing an edge connects the points. After running through all edge elements on the list, a closed two-dimensional geometry is generated within the CAD system. In the end, using the sweep representation approach, the geometry is extruded in z-direction. The final result is the 3D CAD model of an integral bifurcated sheet metal profile which represents the best solution by mathematical means. The algorithm described in Fig. 6.14 is implemented to realize an interface in Siemens NX. The algorithm is implemented in C# as well as a Java application using NX Open API with a Graphical User Interface (GUI) to facilitate the use of the application (Weber Martins et al. 2015). Figure 6.15 shows the application of the implemented algorithm model feature within the Siemens NX CAD system. On top an example for an integral bifurcated sheet metal profile and on bottom a freeform sheet metal part are automatically generated based on their input XML file. Both examples show that even more complex geometries are automatically generated as CAD models with only one step. However, further refinement steps to include design features (e.g., bending radius, holes) to the optimized geometry are carried out manually.

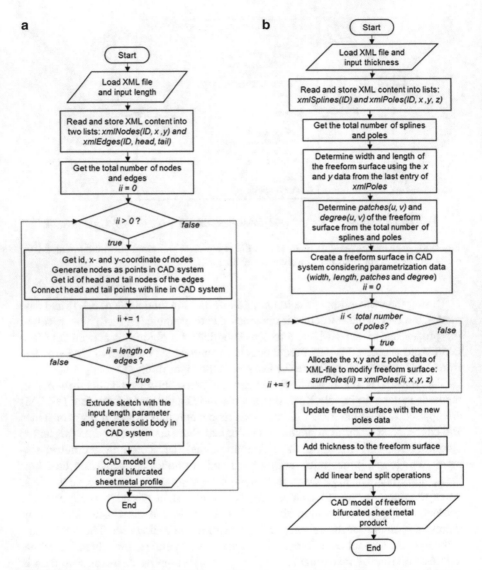

Fig. 6.14 Algorithm for reading XML format files, processing topologic and geometric elements, and generating CAD models automatically—(**a**) linear, (**b**) freeform surfaces (Weber Martins et al. 2015)

State-of-the-art web-based technologies (e.g., Hypertext Markup Language (HTML), Representational State Transfer (REST)) provide a promising framework for development of innovative tools supporting the algorithm-based product creation process. A web-based application yields to facilitate the information exchange of heterogeneous data files within the interdisciplinary steps of the product creation process. It also enables engineers, mathematicians, and process planner to design a whole new product or adapt the geometry design by given input data without the

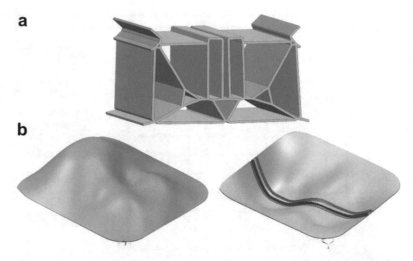

Fig. 6.15 Application of the algorithm modeling feature to generate automatically 3D CAD models: (**a**) integral bifurcated sheet metal profile, (**b**) integral bifurcated sheet metal part

need of any CAD system. The web-based application implementation is based on technologies such as HTML 5, Cascading Style Sheets (CSS), and JavaScript. Its main functions are to visualize, to generate, and to manipulate design geometries within the web browser in PC or Tablet-PC. Additionally, an interface has been implemented based on a web server and remote access between the web browser and CAD system in order to generate CAD models (Weber Martins et al. 2016b). Figure 6.16 shows a Y-profile created with the web-based application.

Both algorithmic modeling approaches create 3D geometries based on mathematical optimization input data. However, in order to fulfill a fully integrated product model, specific manufacturing-induced properties need to be added to the CAD models. An approach to solve this is to extend the data model by considering engineering and technological data as non-geometric data to the corresponding geometric element. The objective of this approach is to include engineering requirements such as manufacturing restrictions into the 3D CAD models and to reach higher level completeness (Weber Martins et al. 2016a).

In conclusion, the algorithmic modeling approach implements a gapless and continuous interface between mathematic optimization and CAD modeling. The developed data model using XML file format enables data and information exchange between all steps. The potential of the use of XML as neutral file format between heterogeneous systems has been shown with the native CAD and web-based application. Benefits as well as potential for future developments for both approaches are identified. In comparison with conventional modeling approaches in CAD (e.g., feature-based), the algorithmic modeling approach enables an efficient modeling method of CAD models. Thereby, less iterations and modeling time are necessary to accomplish a CAD model with higher level of completeness. Once the CAD models are generated, they deliver geometric

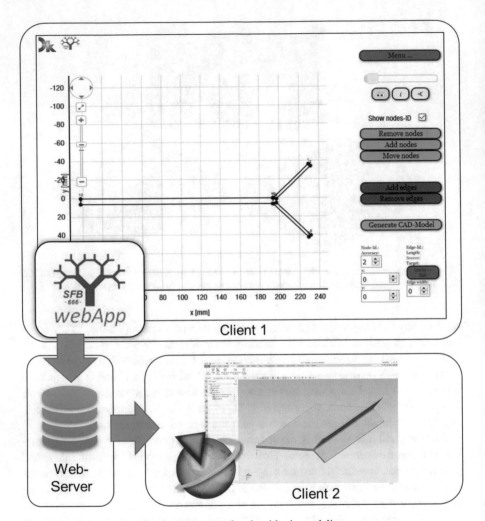

Fig. 6.16 Web-based application to support the algorithmic modeling

information for further processing (e.g., additive manufacturing, FEM analysis (Sects. 6.3 and 6.4)). Further investigations to integrate design features (e.g., bending radius, linear flow, and bend split operations) into the current algorithm for a more detailed CAD model are very promising in order to avoid further refinement steps of the model.

6.2.3 Panelization of Freeform Building Facades

Since NURBS (Non-Uniform Rational Basis Spline) surfaces already mentioned in Sect. 6.2.1 are available for modeling in many CAD applications, they are also

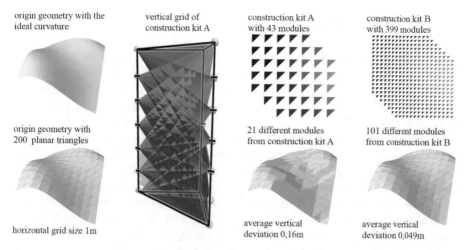

origin geometry with the
ideal curvature

vertical grid of
construction kit A

construction kit A
with 43 modules

construction kit B
with 399 modules

origin geometry with
200 planar triangles

21 different modules
from construction kit A

101 different modules
from construction kit B

horizontal grid size 1m

average vertical
deviation 0,16m

average vertical
deviation 0,049m

Fig. 6.17 Penalization of a freeform surface with two different construction kits (Schäfer et al. 2012)

increasingly used in architecture design processes. The Guangzhou Opera House designed by Zaha Hadid is an example of this architectural design language. However, it also illustrates that the construction of such architecture still represents a major challenge so far, which is only realized in individual prestige projects where the building costs typically are no major concern (Menges and Castle 2006). The serial production is a solution to lower the costs of such buildings. To use the advantages of the serial production within the generation of freeform geometries, solutions for the division of freeform geometry in modular systems have to be found.

With the Software Autodesk Maya (2012) in combination with the scripting language Python a tool was developed, which examines the possibilities of a modular system using planar triangles. The first step to achieve a limited number of different triangles is the parallel projection of a horizontal grid of triangles into a freeform surface. As a result, the modules are congruent to each other in the top view though their dimensions are still different. Moving the vertexes of the projected triangles vertically into a grid, a limited number of possible modules arises (Fig. 6.17 construction kit A). This number depends on the defined dimension of the vertical grid. The amount of modules of the construction kit results from the square of the used vertical grid divisions.

It is shown that a bigger construction kit with more modules allows a better approximation of the initial geometry (Fig. 6.17) (Schäfer et al. 2012). The geometric simplification by tessellation is a significant factor allowing a better feasibility of freeform architecture. Further information about the panelization of freeform geometry can be found in Sect. 7.2.1.

6.3 Process Simulation

Finite element analysis (FEA) is a practical application of the finite element method (FEM) which is a powerful and useful method for analysis of forming processes. In a development phase of a process, FEA can be used to design a tool system by estimating evolving forces and the final profile geometry before performing time- and cost-intensive experiments. In addition to estimation of forces, FEA can be utilized to predict stresses and strains evolved in the profiles. However, besides these benefits there are some challenges in implementing FEA for forming processes. The computational cost involved in analysis of forming processes is generally high. This is also applicable for a numerical analysis of linear flow splitting.

Linear flow splitting is a multistage continuous process which induces high deformations in the work piece (Chap. 3). For an FEA of linear flow splitting, two main challenges arise. The workpiece undergoes a high deformation at the splitting center for each stage. This makes it necessary to develop a remeshing algorithm which reduces the distortion of finite elements during the simulation of the forming stages. Furthermore, the computational cost is directly proportional to the number of finite elements considered in the analysis. Therefore, the length of the profile considered for the simulations is reduced. However, the transient effects in the front and back end of the simulated profile reduce the stationary-state zone in the sheet. In reality, the profile length is much longer and the transient zone in the front and back end has no importance which is often cut away to evaluate the stationary-state zone only. To reduce the computational time without reducing the accuracy, a number of methods are developed by researchers to optimize FE simulations. However, only few consider the transient effects (Rullmann et al. 2012a ; Kim et al. 2005; Hailing et al. 2006; Komori and Koumura 2000).

6.3.1 Linear Flow Splitting Simulation of the Macro Geometry

The main challenge in the conventional finite element analysis of linear flow splitting process is the transient effect. The stepwise movement of the splitting roll results in formation of flanges. The elongation in the splitting center results in variation in longitudinal strains in the front and back end of the sheet compared to the stationary-state part in the middle (regarding the longitudinal direction) of the sheet. Figure 6.18 displays the longitudinal elongation for the conventional method compared to a real part in stage 5. Furthermore, the elongation affects the duration of the simulation and its accuracy because of the large local deformations. The elongation affects the accuracy further as the profile deforms in subsequent stages.

The FE model of the linear flow splitting process is generated with the software package Simufact Forming SFM 9.0. All rolls are modeled as rigid bodies.

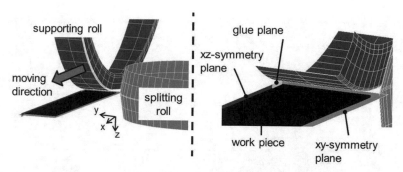

Fig. 6.18 Longitudinal strain in FE compared to the experiment (Rullmann et al. 2012a)

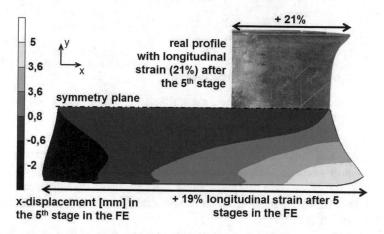

Fig. 6.19 FE model of linear flow splitting (Rullmann et al. 2012a)

The boundary conditions to model the symmetry are implemented with program-specific symmetry planes as shown in Fig. 6.19.

Due to symmetry of the final profile, a quarter of the model is simulated. Therefore, two symmetry planes (*xz, xy*) are defined. The sheet metal is fixed in all directions through the glue plane at the beginning of the part. In contrast to the real process, the rolls move along the work piece. In the simulation the sequence of motion of the rolls are repeated while the infeed of the rolls is increased.

The high deformation is induced by the process at the splitting center. This results in considerable deformation of finite elements which also lead to mesh convergence problems. To eliminate this problem, a new method is developed which is described in the following section.

Cut-Expand-Method. The Cut-Expand-Method (CEM) is developed to overcome the challenges regarding the transient effects in the forming and to handle the high distortion of the elements in the FE mesh. The required flange length determines the necessary number of forming steps. Each step of the Cut-Expand-Method is shown

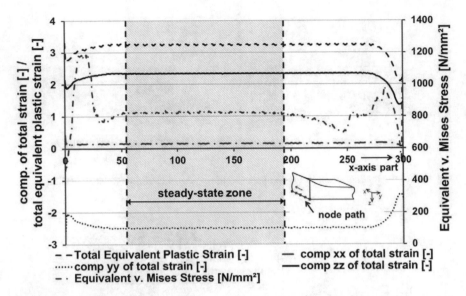

Fig. 6.20 Components of total strain and steady-state zone along the sheet (Rullmann et al. 2012a)

in Fig. 6.21. First the conventional FE simulation of the first forming step is conducted and checked whether the simulation reaches the steady-state zone. With CEM, the results are analyzed along a longitudinal user-defined path. For example, Fig. 6.20 shows the strain and stress distributions at a longitudinal node path in the splitting center.

It is observed that the steady-state zone is reached after 55 mm of the sheet. As the front and back end is in the transient zone, the values are not constant. However, the front and back end has no importance in real parts as they are cut away. Therefore, these regions are not considered further in the simulations. The idea of CEM is to extrude the steady-state zone to form a new profile without the transient zones. This new profile is used for simulations of further stages.

Following this idea, the steps of the CEM are shown in Fig. 6.21. Once a steady-state status is reached, the FEA is stopped. A cut plane is set in the steady-state zone afterwards and the element information of the elements in this plane is stored. In the next step, the results will be transferred from 3D into a 2D grid. Due to the high deformation of the elements in the splitting center, a new mesh has to be generated with a 2D mesh generator. As the mesh in 2D can be structured more easily, this step allows generation of a finer mesh region in the splitting center, where the highest strains evolve. Due to the much smaller plastic strains in the web, a coarser mesh is used which leads to savings in simulation time.

In the next step of CEM, the information from the old mesh is mapped onto the new mesh. Finally, the resulting 2D results are extruded to a 3D mesh.

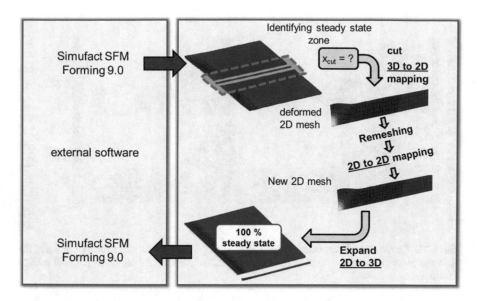

Fig. 6.21 The procedure of the Cut-Expand-Method (CEM) (Rullmann et al. 2012b)

The 2D to 2D mapping from the old mesh to a new mesh can be performed by three methods. In the first method *"Inter- and Extrapolation mapping,"* the values are transferred from the old grid into the new grid by using the shape function. In the second method *"Inverse Distance Weighted,"* a patch is created around every new integration point and the values in the new grid are calculated according to the distance between the new and the old mesh. The third method *"Super convergent patch recovery (SPR)"* computes a continuous function of stress within a patch. This method accounts for minimal mapping error of 12.5% compared to 14% and 33% for the first two methods. Therefore, SPR mapping algorithm is used in the CEM method (Rullmann et al. 2012a). Figure 6.22 shows the comparison of the cross sections of the steady-state region of the sheet and the same section after applying the Cut-Expand-Method.

For validation of the FE results there are already split profiles produced in different projects. One possibility of comparing geometrical values is the flange length of the profile. Because of exploitation of the symmetrical conditions of the process, there is only one flange length in the FE simulations. Contrarily, the flange lengths do not evolve perfectly symmetrical on both sides of the web. Therefore, two flange lengths are measured in experiments and compared with the result of FE simulations. The error bars of the experimental data are also depicted in the diagram (Fig. 6.23).

It can be shown that the flange length is continuously increasing in the numerical simulations over the forming steps. This relation is also observed in the experiments between step 2 and 8. In last three stages differences between upper and lower flange occur. The reason for this can be attributed to the compliance of the stands. Contrary to the experiments, in the FE simulations, the stands are modeled as rigid

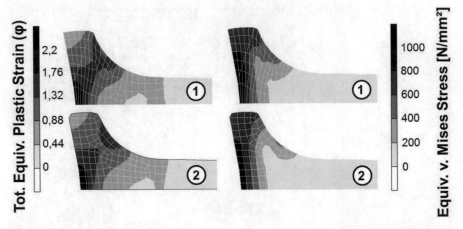

Fig. 6.22 Plastic strain distribution before and after mapping the results; stationary section of the 3D work piece (end stage 2) (marked with 1); generated cross section by CEM (marked with 2) (Rullmann et al. 2012a)

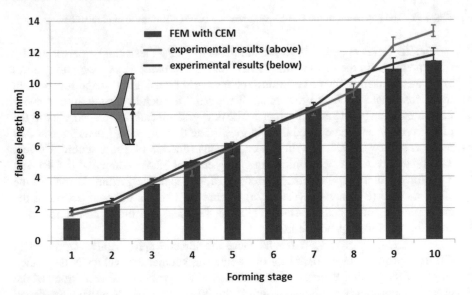

Fig. 6.23 Verification of FE results using the flange length (Rullmann et al. 2012b)

bodies without any compliance. The compliance of the stands increases with increasing flange length and exerted forces. Therefore, deviations appear in the numerical results for stand 8–10.

Time Reduction with CEM. CEM uses the steady-state region in the sheet and creates a new workpiece out of this information. Consequently, it is possible to stop the simulation once the steady-state region is evolving. Therefore, for simulations it

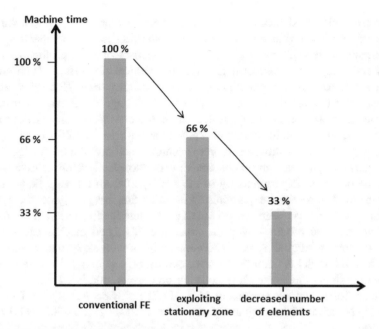

Fig. 6.24 Time reduction by CEM (Rullmann et al. 2012b)

is sufficient to use a sheet length of 300 mm. Regarding the von Mises stresses, after a simulation time of 66%, the steady-state region evolves. Therefore, the numerical model is not simulated until the end. The conventional simulation is carried out without a remeshing rule. Therefore, the total number of finite elements remains constant. Whereas, the simulations with CEM consider twice the number of elements in the fine mesh region (near splitting center). By considering this fact, the actual simulation time is reduced by 50%. Therefore, the overall simulation time is reduced by 66% (Fig. 6.24).

Conclusion. For multistage rolling processes, like linear flow splitting, time-intensive and cost-intensive experiments can be reduced by the application of the FEM. However, computational cost of simulations itself is reduced by application of "Cut-Expand-Method (CEM)" without reduction in the quality of the results. With the help of the CEM, the simulation of the linear flow splitting becomes practicable and more efficient.

6.3.2   Simulation of Local Properties

It has been detailed in this chapter that FEM simulations are powerful tools providing valuable information about a manufacturing technology that then can be included into the product development process. Previous subchapters have numerically addressed process forces and profile geometry in the simulation of bifurcated profiles produced by linear flow splitting and linear bend splitting. Furthermore, investigation of graded local properties of such profiles as well as steady state regarding microstructure and mechanical properties which emerges after a certain total splitting depth or number of splitting steps (Bohn et al. 2008) is of great interest as these unique properties (Chap. 4) induced during the manufacturing process can be exploited within the product development process.

However, simulation of linear flow splitting as a formation of the band edge into two flanges while the splitting roll drives continuously into the band edge (Müller et al. 2007) implies a continuous increase in the plastic strain instead of the saturation observed experimentally in the area apart from the very flange tip and the web. In this section, a particular FE simulation by justifying the loading-unloading scenario has been carried out to investigate the formation of the forming zone and the steady state as well as expansion of graded mechanical properties which are compared qualitatively and quantitatively with experimental results.

Since a 3D model is numerically very expensive, especially for a large number of splitting steps, the linear flow splitting process is simulated via a simplified two-dimensional plain strain model in the FEM software ABAQUS. The strain-hardening behavior of the simulated 6 mm sheet of DD11 grade mild steel (Chap. 4) is defined in ABAQUS via a compression test data which is extrapolated by a Hollomon equation with a prefactor of 580 MPa and an exponent of 0.145. The setup consists of a rigid splitting roll with a tip radius of 5 mm and a splitting angle of 120° as well as two rigid supporting rolls with a corner radius of 5 mm and faces being parallel to the sides of the splitting roll (Fig. 6.26). In 2D simulations, these parallel faces are implemented to prevent bending of the flanges, which in 3D is prevented by the constraint exerted via the profile geometry. During the loading, the splitting roll remains fixed, the material is pushed towards the splitting roll, and simultaneously the two supporting rolls move upwards. All three rolls are retracted parallel to the infeed axis during unloading. A total number of 42 splitting steps with an incremental splitting depth of 1 mm are modeled to facilitate the observation of a steady state which emerges only after a certain number of splitting steps.

Figure 6.25a shows the calculated equivalent plastic strain in one half of the linear flow split profile. A red line in a constant distance X below the top surface illustrates the path along which the equivalent plastic strain (PEEQ) is captured and represented in Fig. 6.25b. A gradient of equivalent plastic strain in the direction of top to bottom surfaces of the flanges is easily observed in Fig. 6.25b. In addition, apart from the very flange tip and the web area, the equivalent plastic strain levels at an approximately constant value for a given distance below the top surface. It can be concluded that an increase in the length of flanges does not induce further plastic deformation in the flanges area, and thereby the formation of a steady state is confirmed. The simulation result qualitatively resembles the hardness measurements and microstructural observations presented and discussed in Sect. 4.1. The area with greatest magnitude of equivalent plastic strain, e.g., strain of 4 in Fig. 6.25b, may represent the ultrafine-grained zone (Lin et al. 2012; Azushima

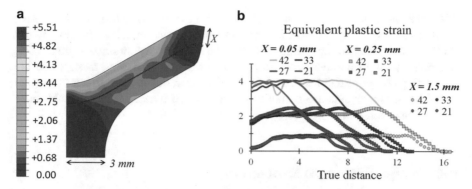

Fig. 6.25 (**a**) Distribution of equivalent plastic strain (PEEQ) after 27th splitting step, (**b**) PEEQ parallel to the flange surface in a distances of $X = 0.05$, 0.25, and 1.5 mm below the top surface, numbers denote the number of steps

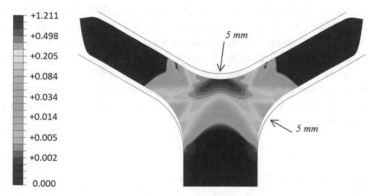

Fig. 6.26 Change in the equivalent plastic strain during the deformation in the 29th splitting step using a logarithmic scale

et al. 2008) which is accompanied with a maximum magnitude in the hardness measurements (Sect. 4.1).

Figure 6.26 shows the change in the equivalent plastic strain occurring in one splitting step. While material flows from the web to the flanges, an increase in the plastic strain occurs primarily at the splitting center and directly below which defines the size of the forming zone. Here, it can be concluded that plastic deformation is kept unchanged in the flanges and it confirms the observation of constant magnitudes of equivalent plastic strain while the flange length increases. In order to quantitatively compare the simulated material properties to measured ones, the simulated equivalent plastic strain is averaged in the mid length of the flanges over a thickness of 0.8 mm from the split surface towards the bottom surface which represents the geometry of the tensile specimen detailed in Sect. 4.1.1. An average equivalent plastic strain of 2.1 is calculated and corresponds to a yield strength of 650 MPa obtained from the Hollomon equation. This value is approximately

100 MPa lower than the yield strength measured in the tensile tests shown in Sect. 4.1.1, i.e., the Hollomon hardening equation fitted to the uniaxial compression test underestimates the strengthening behavior in the severely strained regions. This discrepancy could be addressed using more advanced hardening models that are based on physical principles (Estrin 1998). However, due to the number of internal variables, those models are numerically much more intricate which would make simulations even more expensive, especially in 3D.

6.4 Fatigue Strength Simulation

Numerical fatigue strength analysis is important in view of the requirement to reduce time and costs for product development. A reliable computational approach also enables the evaluation of large components. The methodology for numerical fatigue strength assessment of components manufactured by linear flow splitting is described in Sect. 6.4.1. Related examples are also provided. Furthermore, the extension of this approach is demonstrated for fatigue strength assessment of deep drawn structures in Sect. 6.4.2. As a result of the inhomogeneous material properties in both forming processes, a local strain approach based on elastic–plastic stresses and strains is applied. With the presented approach a reliable design can be reached taking into consideration the positive effects of the forming process on the cyclic material behavior.

6.4.1 Linear Flow Split Structures

A durable design in the development phase considering the influences of the manufacturing process on the material behavior is essential in order to avoid component fatigue failure during service. To utilize the lightweight potential of linear flow split structures, local properties of the flanges have to be incorporated in the fatigue assessment. The characteristic microstructure of the linear flow split parts is shown in Sect. 4.1, with a strong work hardening in the flange and ultrafine-grained microstructure (UFG) on the flange top. As indicated in Sect. 4.1, this has a significant effect on the fatigue properties. Due to this influence on the cyclic material properties, an assumption of a homogeneous material behavior cannot be used for a reliable fatigue analysis. The inhomogeneous material properties which develop during the forming process should be taken into account during the simulation of fatigue strength in linear flow split structures (Müller et al. 2007).

The local strain concept is presented in this subsection for the evaluation of fatigue strength of linear flow split parts. For this approach, the strain–life curve and the material parameters to obtain the cyclic stabilized stress–strain curve are required. This information can be obtained from experimental investigations of linear flow split parts and serves as the major input for the numerical fatigue

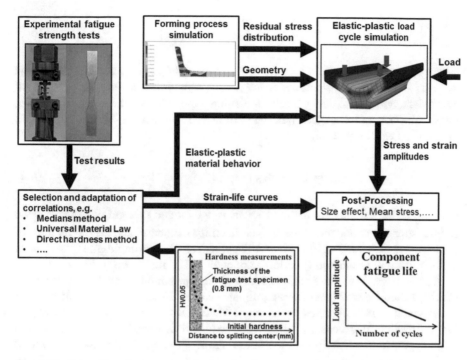

Fig. 6.27 Connections between experiments, forming process simulation, and fatigue strength simulation of linear flow split structures

strength assessment. The correlation between hardness and cyclic material properties would be utilized. By using the Medians method (Meggiolaro and Castro 2004), the correlation can be mathematically fitted to experimental results of fatigue tests on smooth, nonhomogeneous specimens of the investigated profiles. Thereby the characteristic manufacturing-induced gradient in the microstructural properties would be considered in the numerical analysis. In addition, the experimental fatigue test results shall be used to validate the numerical analysis. Furthermore, the residual stress distribution obtained either from the forming process simulation or potentially also from experimental measurements is also required for the numerical fatigue assessment. The information from the forming process simulation is provided in the form of an ASCII file that contains the node number, node coordinates, and associated node averaged stress components.

The residual stress can then be integrated into the elastic–plastic finite element model of the linear flow split parts. The connections between the experiments, process simulation, and numerical fatigue analysis are shown in Fig. 6.27. For complex component analysis, an ideal geometry from the CAD modeling can also be used for the subsequent finite element simulation as indicated in Sect. 6.2.

For the fatigue strength simulation, elastic–plastic finite element calculations are required. The cyclic stress–strain curves derived from the Ramberg–Osgood relationship (Ramberg and Osgood 1943) in Eq. (6.1) are required,

$$\varepsilon_{a,t} = \varepsilon_{a,e} + \varepsilon_{a,p} = \frac{\sigma_a}{E} + \left(\frac{\sigma_a}{K'}\right)^{\frac{1}{n'}} \tag{6.1}$$

where $\varepsilon_{a,t}$ is the total strain amplitude with its elastic part $\varepsilon_{a,e}$ and its plastic part $\varepsilon_{a,p}$, σ_a is the stress amplitude, E the Young's modulus, K' the cyclic strength coefficient, and n' the cyclic strain hardening exponent. For the strain–life curve, the Coffin–Manson–Basquin equation (Coffin 1954; Manson 1965; Basquin 1910) in Eq. (6.2) would be used,

$$\varepsilon_{a,t} = \varepsilon_{a,e} + \varepsilon_{a,p} = \frac{\sigma_f'}{E}(2N_i)^b + \varepsilon_f'(2N_i)^c \tag{6.2}$$

where ε_f' is the fatigue ductility coefficient, σ_f' the fatigue strength coefficient, b the fatigue strength exponent, c the fatigue ductility exponent, and N_i the number of cycles to crack initiation. The strain–life curves resulting from fatigue experiments and the estimation method (Landersheim et al. 2012) are shown in Fig. 6.28.

Although the Medians Method gives a good approximation of the experimental results, the slopes are higher for the estimated curves. Therefore, the estimated local stress–strain curves obtained by Medians Method would be adapted to the experimental results. In this case, a better fit can be obtained. The adapted estimations are more appropriate to interpolate the stress–strain curve within the linear flow split parts.

The hardness-based parameters required for both Eqs. (6.1) and (6.2) are presented in Landersheim et al. (2012) and Karin et al. (2013). The damage

Fig. 6.28 Approximation of experimental results (HC480LA) in as-received material state and from flange specimens of linear flow split parts (Landersheim et al. 2012)

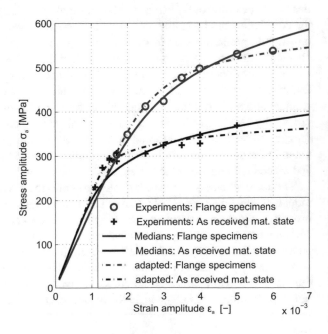

parameters for the fatigue analysis can be determined according to Bergmann P_B (Bergmann 1983) or Smith, Watson, and Topper P_{SWT} (Smith et al. 1970),

$$P_B = \sqrt{\left(\sigma_a + \left(\left(\frac{0.5\sigma_{UTS}}{1600MPa} + 1\right)^2 - 1\right)\sigma_m\right)\varepsilon_a E} \qquad (6.3)$$

$$P_{SWT} = \sqrt{(\sigma_a + \sigma_m)\varepsilon_a E} \qquad (6.4)$$

where σ_m is the mean stress and σ_{UTS} the ultimate tensile stress. The maximum endurable damage parameter would be defined by P_X–life curve which can be derived by

$$P_X(N_i) = \sqrt{\sigma_f'^2(2N_i)^{2b} + \sigma_f'\varepsilon_f'E(2N_i)^{b+c}} \qquad (6.5)$$

where P_X is the used damage parameter (P_B or P_{SWT}).

The application of this methodology facilitates the fatigue assessment of various linear flow split parts and component-like structures. In addition, the fatigue analysis of a linear guide is presented. In all cases, elastic–plastic FE models with the derived material property distribution are implemented.

The result of the numerical fatigue strength evaluation of a 4-point-bending test of a component-like linear flow split structure made from HC480LA is presented in Fig. 6.29 (Landersheim et al. 2012). For this example, the damage parameter

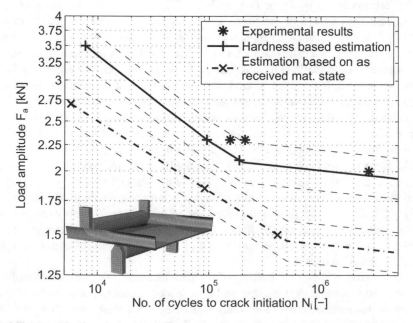

Fig. 6.29 Numerically estimated load–life curves of component-like structure and validation with experimental results (Landersheim et al. 2012)

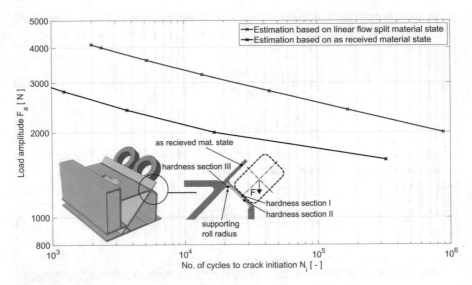

Fig. 6.30 Numerically estimated load–life curves of a linear guide (Tijani et al. 2015)

P_B according to Bergmann was used. The results of the numerical investigations show good correlations with experimental results.

Another application example for this methodology is the numerical fatigue strength assessment of a linear guide made of high strength low alloy steel HC340LA produced by linear flow splitting (Tijani et al. 2015). The assessment considered the carriage of the linear guide. By using the finite element program ABAQUS 6.13, the local stresses and strains in the linear guide are calculated. Based on the hardness distribution, an individual stabilized cyclic stress strain curve is assigned to each hardening state. The hardness values were measured at a linear flow split semi-finished part and are transferred to the profile geometry. In order to reduce the computation time, the symmetry in longitudinal direction of the profile is considered. For the carriage region of the linear guide, a force $F = 4$ kN is introduced via rolls on a frictionless surface-to-surface contact to the sliding surface. The component is loaded with a pulsating load, i.e., load ratio $R_F = 0$ and the local mean stresses are considered. The results of the investigation are shown in Fig. 6.30. The calculated numbers of cycles of the linear flow split part are higher than the values of the as-received profile. At a force amplitude of $F_a = 2$ kN, the fatigue life of the linear flow split guide increased by about 50-folds in comparison to the estimation based on the as-received material state. This example demonstrates the potential for lightweight design of linear flow split structures for cyclically loaded components.

The previously described investigations are carried out using constant amplitude loading. An extension of this numerical fatigue assessment to linear flow split parts subjected to variable amplitude loads is an important step from an application perspective. This has been attempted by Tomasella et al. to evaluate fatigue specimens made from HC340LA steel (Tomasella et al. 2013). However, in this regard, there is still a challenge to establish a satisfactory correlation between the numerically estimated strain–life curves and experimental investigations. Nevertheless, for constant amplitude loading, the developed approach allows an estimation of the fatigue strength of linear flow split parts. An adequate consideration of the inhomogeneous material properties leads to significant improvements in fatigue assessments.

6.4.2   Deep Drawn Structures

As shown in the previous chapters, the consideration of the forming history is of fundamental importance for a reliable estimation of fatigue life as well as for the exploitation of the lightweight potential.

By analogy with linear flow split components, the material properties in deep drawn structures are not homogeneous, but change depending on the different local plastic deformation. For a more reliable fatigue assessment of deep drawn components, these microstructural changes should be taken into account. Moreover, the effect of residual stresses should not be disregarded and a method for their consideration in the fatigue life estimation is required. An extension of the methodology used for linear flow split components consists here in the use of information about the total equivalent plastic strain distribution for estimating the material properties, allowing the possibility of simulating the component without the need of hardness measurements. In cooperation with industry we developed a method for the fatigue life estimation of cold worked components by considering the local hardening and residual stresses estimated by a forming process simulation. The procedure was applied as example to the planet carrier shown in Fig. 6.31.

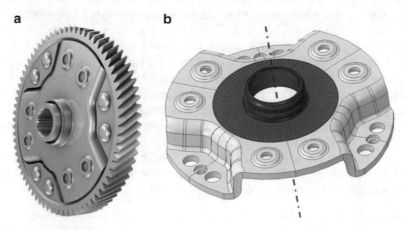

a   **b**

Fig. 6.31 Cold worked planet carrier

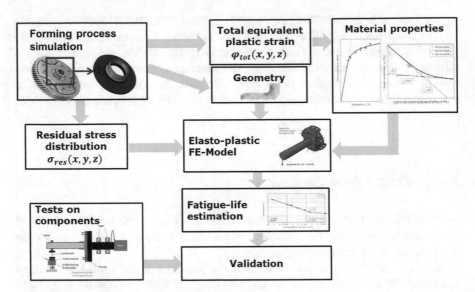

Fig. 6.32 Schematic procedure for fatigue life estimations

The attention of the analysis was concentrated on the radius of the flange of the planet carrier, normally loaded under rotating bending during service. The flange was manufactured through deep drawing, and the plastic deformation was not constant through its thickness.

The method for the fatigue life estimation is shown schematically in Fig. 6.32. The process simulation of deep drawing aimed to deliver information about the geometry and the total equivalent plastic strain distribution. Moreover, a reliable estimation of the residual stresses distribution was required. Using these results, an elasto-plastic FE model simulating rotating bending, i.e., with a bending moment applied to the flange during the rotation of the component, was used to determine the local stress and strain distribution and the fatigue damage, i.e., the fatigue life.

The local work hardening was taken into consideration by subdividing the elements into classes dependent on the estimated total equivalent plastic strain. The whole component was then obtained by associating a different stress–strain relation to each class. In the model, the stabilized cyclic stress–strain curves shown in Fig. 6.33 were used. The stress and strain distribution in the component under rotating bending was determined by the simulation and the fatigue life of the elements of each class was then estimated for each class according to the strain–life curve associated to the class of elements.

We carried out the fatigue life estimation considering constant amplitudes and the damage parameter of Smith–Watson–Topper (Smith et al. 1970) extended to the analysis of multiaxial stress–strain states (Socie 1987) (Eq. (6.6)).

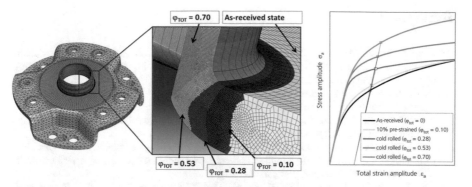

Fig. 6.33 Division of the cold worked flange of the planet carrier into classes and related cyclic stress–strain curves

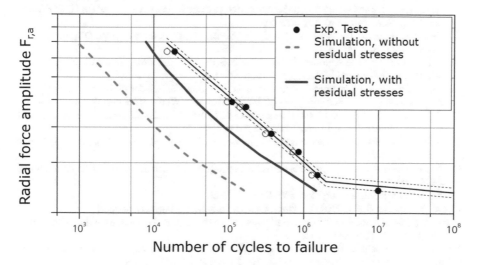

Fig. 6.34 Estimation of fatigue life of the cold worked flange under rotating bending with and without considering residual stresses and comparison with experimental results

$$P_{\text{SWT}} = \sqrt{\sigma_{n1,\max} \cdot \frac{\Delta\varepsilon_1}{2} E} = \sqrt{\sigma_f'^2 (2N_A)^{2b} + E \cdot \sigma_f' \cdot \varepsilon_f' (2N_A)^{b+c}} \qquad (6.6)$$

Figure 6.34 shows the results of fatigue life estimations when taking into consideration the local hardening with and without residual stresses (Waterkotte et al. 2015). The results show for this example that the consideration of the total equivalent plastic strain alone does not allow a reliable estimation of fatigue life, i.e., a factor 20 between estimated and experimentally determined fatigue lives. An improvement is obtained when also residual stresses are taken into account, reducing the error to a factor of 2.

References

Albrecht K, Weber Martins T, Anderl R (2016) Fertigungsrestriktionsmodell zur Unterstützung des algorithmisierten PEP fertigungsgerechter Blechprodukte. In: Stelzer R (ed) Proceedings of Entwerfen Entwickeln Erleben, Dresden, 2016

Anderl R, Binde P (2014) Simulations with NX—Kinematics, FEM, CFD, EM and Data Management. With Numeroous examples of NX 9, Hanser, Munich

Anderl R, Mendgen R (1996) Modelling with constraints: theoretical foundation and application. Comput Aided Des 28(3):155–168

Anderl R, Trippner D (eds) (2000) STEP Standard for the exchange of product model data. Eine Einführung in die Entwicklung, Implementierung und industrielle Nutzung der Normenreihe ISO 10303 (STEP). Teubner, Stuttgart, Leipzig

Anderl R, Wu Z, Rollmann T, Kormann M (2006) Integrationsmodell für integrale Blechbauweisen höherer Verzweigungsordnung. CAD-CAM Report 10(200):24–27

Anderl R, Rollmann T, Wu Z, Chahadi Y (2008) Algorithm-based product development—refined concept and example applications. In: ASME 2008 International Design Engineering Technical Conferences and Computers and Information in Engineering Conference IDETC/CIE 2008, Brooklyn, New York, 3–6 August 2008

Azushima A, Kopp R, Korhonen A et al (2008) Severe plastic deformation (SPD) processes for metals. CIRP Ann Manuf Technol 57(2):716–735

Basquin OH (1910) The exponential law of endurance tests. In: Proceedings of the American Society for Testing and Materials, vol 10, p 625–630

Bergmann JW (1983) Zur Betriebsfestigkeit gekerbter Bauteile auf der Grundlage der örtlichen Beanspruchungen. Dissertation, Institut für Stahlbau und Werkstoffmechanik, TU Darmstadt

Bohn T, Bruder E, Müller C (2008) Formation of ultrafine-grained microstructure in HSLA steel profiles by linear flow splitting. J Mater Sci 43:7307–7312

Chen P (1976) The entity-relationship model—toward a unified view of data. ACM Trans Database Syst 1(1):9–36

Coffin LF (1954) A study of the effect of cyclic thermal stresses on a ductile metal. Transactions of ASME 76:931–950

Diestel R (2000) Graph theory, 2nd edn. Springer, New York

Doege E, Behrens B-A (2010) Handbuch Umformtechnik. Springer, Heidelberg

Estrin Y (1998) Dislocation theory based constitutive modelling: foundations and applications. J Mater Process Technol 80-81:33–39

Eversheim W, Bochtler W, Grässler R et al (1997) Simultaneous engineering approach to an integrated design and process planning. Eur J Oper Res 100(2):327–337

Gramlich S, Roos M, Ahmels L, Kaune V, Müller C, Bauer O, Karin I, Tomasella A, Melz T (2015) Ein wissensbasierter fertigungsintegrierender Produktentwicklungsansatz. Paper presented at the Stuttgarter Symposium für Produktentwicklung 2015—Entwicklung smarter Produkte für die Zukunft Stuttgart, Stuttgart

Groche P, Wu Z, Anderl R, Ulbrich SU, Martin A, Günther U, Greif G, Rollmann T (2005) An algorithm based approach for integral sheet metal design with higher order Bifurcations. Paper presented at the PACE Annual Forum 2005, Darmstadt

Hailing Y, Xianghua L, Xianming Z, Kusaba Y (2006) FEM analysis for V-H rolling process by updating geometric method. J Mater Process Technol 180:323–327

Harold ER, Means WS (2004) XML in a Nutshell, 3rd edn. O'Reilly Media, Sebastopol

Hazewinkel M (2001) Graph theory encyclopedia of mathematics. Springer, New York

Hehenberger P (2011) Computerunterstützte Fertigung: Eine kompakte Einführung. Springer, Berlin

Hesse W et al (1994) Terminologie in der Softwaretechnik—Ein Begriffssystem für die Analyse und Modellierung von Anwendungssystemen. Teil 1: Begriffssystematik und Grundbegriffe. Informatik Spektrum 17:39–47

Karin I, Tomasella A, Landersheim V et al (2013) Application of the local strain approach on a rolling point contact model. Int J Fatigue 47:351–360

Kim S, Lee H, Min J, Im Y (2005) Steady-state finite element simulation of bar rolling processes based on rigid—viscoplastic approach. Int J Numer Methods Eng 63:1583–1603

Komori K, Koumura K (2000) Simulation of deformation and temperature in multi-pass H-shape rolling. J Mater Process Technol 105:24–31

Landersheim V, Jöckel M, el Dsoki C et al (2012) Fatigue strength evaluation of linear flow split profile sections based on hardness distribution. Int J Fatigue 39:61–67

Lee K (1999) Principles of CAD/CAM/CAE Systems. Addison Wesley, Reading

Lin J, Balint D, Pietrzyk M (2012) Microstructure evolution in metal forming processes. Woodhead Publishing Limited, Cambridge

Manson SS (1965) Fatigue: a complex subject—some simple approximation. Exp Mech 5:193–226

Meggiolaro MA, Castro JTP (2004) Statistical evaluation of strain-life fatigue crack initiation predictions. Int J Fatigue 26:463–476

Menges A, Castle H (eds) (2006) Techniques and technologies in morphogenetic design. Wiley, London

Müller C, Bohn T, Bruder E et al (2007) Severe plastic deformation by linear flow splitting. Mat Sci Eng Tech 38:842–854

Pahl G, Beitz W, Feldhusen J, Grote K-H (2007) Engineering design—a systematic approach. Springer, London

Picard A (2015) Integriertes Werkstückinformationsmodell zur Ausprägung werkstückindividueller Fertigungszustände. Dissertation, TU Darmstadt

Pilone D (2003) UML: pocket reference. O Reilly Media, Beijing

Ramberg W, Osgood WR (1943) Description of stress–strain curves by three parameters. Technical note No. 902 National Advisory Committee for Aeronautics, Washington DC

Rollmann T (2012) Simultaneous Engineering von integralen Blechbauweisen höherer Verzweigungsordnung. Ein Beitrag zur Integration von Konstruktion und Produktionsprozessplanung. Dissertation, TU Darmstadt

Rollmann T, Schüle A, Anderl R, Chahadi Y (2009) Three-Dimensional Kernel Development with parasolid for integral sheet metal design with higher order bifurcations. In: ASME 2009 International Design Engineering Technical Conferences and Computers and Information in Engineering Conference, San Diego, California, 30 August–2 September 2009

Rollmann T, Schüle A, Weitzmann O, Anderl R (2011) Feature Precedence Graphs as an Approach for the forming operations planning of integral sheet metal parts. In: Proceedings of International Symposium on Assembly and Manufacturing ISAM 2011, Tampere

Ross DT (1977) Structured analysis (SA): a language for communicating ideas. IEEE Trans Softw Eng SE-3(1):16–34

Rullmann F, Abrass A, Kruse M, Groche P (2012a) The Cut-Expand-Method for the FE-simulation of steady-state rolling processes. Steel research international Special Edition:31–34

Rullmann F, Bauer O, Landersheim V, Groche P, Hanselka H, Tijani Y (2012b) Numerische durchgängige Prozesskettenbewertung mittels FEM. In: 4. Zwischenkolloquium des Sonderforschungsbereichs 666: Integrale Blechbauweise höherer Verzweigungsordnung— Entwicklung, Fertigung, Bewertung, Darmstadt, 14–15 November 2012

Schäfer M (2006) Computational engineering—introduction to numerical methods. Springer, Berlin

Schäfer S, Reising J, Abedini S, Bäcker F (2012) Definition of a Tool Library for the Approximation of Freeform Surfaces. In: Zhang L et al. (eds) ICAMMS 2012. International Conference on Advanced Material and Manufacturing Science 2012, Beijing, December 2012

Schüle A (2014) Ein technologiebasiertes Informationsmodell zur Unterstützung des Produktentstehungsprozesses verzweigter Blechbauteile. Dissertation, TU Darmstadt

Schüle A, Anderl R (2012) Implementierung von Modellierungsfunktionen für Verzweigungen auf mehrsinnig gekrümmten Oberflächen. In: Groche P (ed) 4. Zwischenkolloquium des Sonderforschungsbereichs 666: Integrale Blechbauweise höherer Verzweigungsordnung—Entwicklung, Fertigung, Bewertung, Darmstadt, 14–15 November 2012

Schüle A, Anderl R (2013) A simultaneous engineering approach for free form bifurcated sheet metal products. In: Proceedings of CIRP Design Conference, Bochum

Schüle A, Rollmann T, Anderl R (2009) Realization of design features for linear flow splitting in NX6. In: World Academy of Science Engineering and technology, vol 58, p 496–501

Schüle A, Rollmann T, Anderl R (2010a) Ein rechnerunterstützter Produktentwicklungsprozess für flächige Blechbauteile mit Verzweigungen. In: 8. Gemeinsame Kolloquium Konstruktionstechnik. docupoint Verlag, Barleben, p 97–102

Schüle A, Rollmann T, Anderl R (2010b) Rechnergestützte Modellierung flächiger verzweigter Blechbauteile. In: Groche P (ed) 3. Zwischenkolloquium des Sonderforschungsbereichs 666: Integrale Blechbauweise höherer Verzweigungsordnung—Entwicklung, Fertigung, Bewertung, Darmstadt, 29–30 September 2010, pp 33–39

Schüle A, Weitzmann O, Anderl R (2011) Feature-based Modeling of Bifurcated Sheet Metal Products. Proceedings of the 12th Asia Pacific Industrial Engineering and Management Systems Conference (APIEMS 2011), Beijing, China

Smith KN, Watson P, Topper TH (1970) A stress-strain function for the fatigue of metals. J Mater 5:767–778

Socie D (1987) Multiaxial fatigue damage models. J Eng Mater Technol 109:293–298

Stachowiak H (1973) Allgemeine modelltheorie. Springer, New York

Stroud I (2006) Boundary representation modelling techniques. Springer, London

Tijani Y, Bauer O, Kaufmann H et al. (2015) An enhanced numerical fatigue assessment of linear flow split profile produced by severe plastic deformation. In: Zoch HW, Lübben T. (eds) Proceedings of the 5th International Conference on Distortion Engineering, IDE 2015 Bremen, p 95–104

Tomasella A, Bauer O, Landersheim V et al. (2013) An experimental and numerical fatigue assessment of ultrafine-grained microstructures produced by severe plastic deformation under constant and variable amplitude loading. In: Seventh International Conference on Low Cycle Fatigue, Aachen, 9–11 September 2013

Vajna S (2009) CAx für Ingenieure: Eine praxisbezogene Einführung (2., völlig neu bearb. Aufl ed.). Springer, Berlin

Verein Deutscher Ingenieure (1993) Methodik zum Entwickeln und Konstruieren technischer Systeme und Produkte, Richtlinie VDI 2221, Düsseldorf

Verein Deutscher Ingenieure (2009) 3-D-Produktmodellierung, Technische und organisatorische Voraussetzungen. Verfahren, Werkzeuge und Anwendungen. Wirtschaftlicher Einsatz in der Praxis. Richtlinie 2209, Beuth, Düsseldorf, p 167

Waterkotte R, Tomasella A, Hundertmark A, Zeng X, Weninger J, Melz T (2015) Bewertung der Schwingfestigkeit von kaltumgeformten Planetenträgerblechen eines Stirnraddifferentials unter Berücksichtigung des lokalen Verfestigungszustandes. Paper presented at the 42. Tagung des DVM-Arbeitskreises Betriebsfestigkeit, 7–8 June 2015, Dresden

Weber Martins T, Albrecht K, Anderl R (2015) Automated import of XML-files containing optimized geometric data to 3D–CAD-models of non-linear integral bifurcated sheet metal parts. Paper presented at the ASME 2015—International Design Engineering Technical Conferences and Computers and Information in Engineering Conference IDETC/CIE 2015, Boston, MA, USA

Weber Martins T, Albrecht K, Anderl R (2016a) An extended data model for production-orinetated integral bifurcated sheet metal products. Paper presented at the Tools and Methods of Competitive Engineering, Aix-en-Provence, France

Weber Martins T, Albrecht K, Anderl R (2016b) Web-based Application for Algotihm Based Product Development Process of Integral Bifurcated Sheet Metal Parts. Paper presented at the

ASME 2016—International Design Engineering Technical Conferences and Computers and Information in Engineering Conference IDETC/CIE 2016, Charlotte, NC, USA

Weitzmann O (2014) Informationsmodell zur Repräsentation Technischer Oberflächen zur Unterstützung der virtuellen Produktentwicklung. Dissertation, TU Darmstadt

Weitzmann O, Anderl R (2012) Integration von Betriebsmittel- und Fertigungsrestriktionen in das integrierte Informationsmodell für Blechbauteile mit verzweigten Strukturen. In: Groche P (ed) Tagungsband 4. Zwischenkolloquium SFB 666, Darmstadt, 14–15 November 2012, pp 33–40

Weitzmann O, Anderl R (2013) A model for the representation of surface roughness and form deviation in geometry models. In: Proceedings of ASME DETC/CIE, 4–7 August 2013

Weitzmann O, Schüle A, Rollmann T et al. (2012) An object-oriented information model for the representation of free form sheet metal parts in integral style. In: Tools and Methods of Competitive Engineering, Karlsruhe

Wu Z, Anderl R, Kormann M, Rollmann T (2008a) Algorithm-based product development—concepts and example application. In: Proceedings of the International Conference on Comprehensive Product Realization (ICCPR) 2008, Beijing, China

Wu Z, Rollmann T, Anderl R (2008b) Information modeling and representation of sheet metal parts with higher order bifurcations. In: Proceedings of the 9th International Conference on Technology of Plasticity, Gyeongju, Korea, 2008.

Chapter 7
New Challenges: Technology Integrated Market-Pull

S. Abedini, P. Groche, V. Monnerjahn, M. Neuwirth, M. Özel, J. Reising, S. Schäfer, and A. Zimmermann

Newly developed products usually do not only face the market requirements. With the manufacturing induced properties it is possible to satisfy the constantly rising market requirements, e.g., in the mechanical engineering market or the building industry. Besides the original functions, the new products often have to meet other requirements such as lightweight construction, multifunctionality, reliable joints, or aesthetical demands.

The new manufacturing processes described in Sect. 7.1 enable the production of variable and curved objects. With the new manufacturing method of flexible flow splitting (Sect. 7.1.2), it is possible to produce sheet metal panels with integrally produced bifurcations. As shown in Sect. 3.2.2, a bifurcated sheet metal panel has a much higher load capacity than a simple sheet metal without bifurcations and thus offers a higher potential for lightweight constructions.

The new manufacturing techniques also enable the development of new product families. The bifurcation itself can be used as a joining for various functions, e.g., additional multifunctional elements and cable ducts (Sect. 7.2.3), or the bifurcation itself includes the functions, e.g., as joints with flow split flanges (Sect. 7.2.2) (Fig. 7.1).

In addition to the rising market requirements on product properties, the digitization of architectural design in CAD results in a new and more expressive free-form architectural language (Sect. 6.2). With this manufacturing process, the panelization of the new free-form architecture in modular systems is becoming

S. Abedini (✉) • J. Reising • S. Schäfer • A. Zimmermann
Institute of Construction Design and Building Construction (KGBauko), Technische Universität Darmstadt, Darmstadt, Germany
e-mail: abedini@kgbauko.tu-darmstadt.de

P. Groche • V. Monnerjahn • M. Neuwirth • M. Özel
Institute for Production Engineering and Forming Machines (PtU), Technische Universität Darmstadt, Darmstadt, Germany

© Springer International Publishing AG 2017
P. Groche et al. (eds.), *Manufacturing Integrated Design*,
DOI 10.1007/978-3-319-52377-4_7

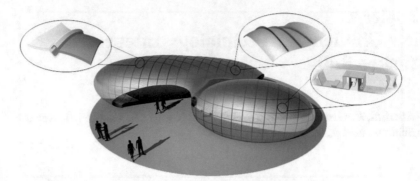

Fig. 7.1 Integration of multifunctionality, lightweight construction, and the joining technique in a free-form architecture

increasingly important. This kind of a technical challenge can be met by applying various sets of scripting tools and the approximation of the original geometry (Sect. 7.2.1).

While designing and approximating new geometries of free-form architecture, it is also important to identify aesthetic requirements. Section 7.7.3 shows some approaches for aesthetical requirements in architectural drafts.

7.1 New Processes

7.1.1 Integrated Bending

For a cost-efficient production of thin-walled profile geometries out of sheet metals, the continuous roll forming process with high quantities and low costs has been established. A large number of products, e.g., for the building, automotive, or aircraft industry, require curved profiles. A processing of straight into curved profiles can be realized either by a separate bending process or by bending during linear flow splitting (Groche et al. 2016).

In Groche et al. (2016), bending methods for a continuous vertical bending of linear flow split profiles are evaluated regarding the requirements of process integration and adjustment of the tool system. Subsequently, four bending strategies are developed and evaluated with numerical simulations, which are generated and solved by the software Marc Mentat with an implicit solver. In the three-roll bending method, three bending rolls are arranged after the linear flow splitting process. With this method, high bending forces which can lead to buckling and a high springback behavior are necessary. The purpose of partial local rolling is to thin a profile locally, which results in local elongation of the thinned zone and finally to a bending operation in transverse direction of the profile. An issue is the high force required for the thinning, which usually leads to deflections of the tool system and requires a very stiff system structure. The bending process after a straightening station is also evaluated negative regarding the disadvantages of the three-roll bending method. The selected and realized bending method for curved linear flow split profiles is presented schematically in Fig. 7.2. In this bending method, a bending roll is arranged after a linear flow splitting station, which can be moved either transversally or longitudinally, depending on the required bending radius.

This new tool concept uses the plastification in the web of the split profile for a superposition of bending stresses during the linear flow splitting process. In Fig. 7.3, the forming zone and the distribution of longitudinal strains during the integrated bending process are shown. The linear flow splitting process induces high compressive stresses in the forming zone, which result in a plastification of the entire web of the profile and lead to a symmetrical strain distribution in path

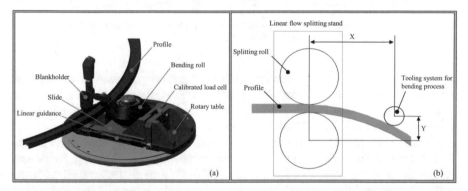

Fig. 7.2 (a) Tool system for integrated bending, (b) roll arrangement in a bending process (Mahajan et al. 2016)

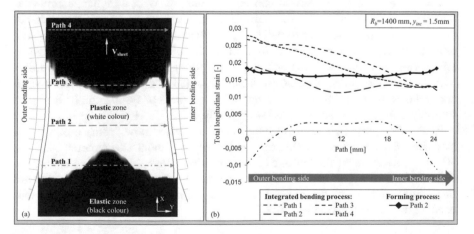

Fig. 7.3 (a) Forming zone in the split profile, (b) distribution of strains (Groche et al. 2016)

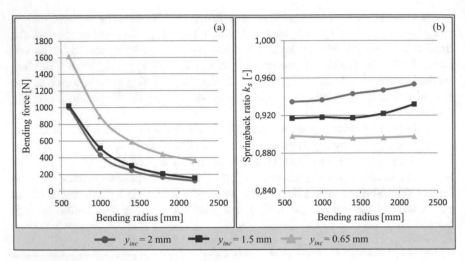

Fig. 7.4 (**a**) Effect of stress superposition on the bending force, (**b**) effect of stress superposition on the springback ratio (Mahajan et al. 2016)

2 (forming process). On the contrary, the superposition of stresses through the integrated bending process leads to an asymmetrical distribution of stains. The inner bending side is compressed and the outer bending side is stretched, which can be seen in path 2 (integrated bending process).

Groche et al. (2016) determined the location of the bending operation entirely within the forming zone, which has a positive effect on the bending forces and tolerances of the profile. In Mahajan et al. (2016), a numerical model with the software package Marc Mentat 2012 is developed for the integrated bending of split profiles. The bending forces and the springback behavior are successfully verified, whereby the stress superposition is determined as a decisive factor. In Fig. 7.4, the effect of the stress superposition can be seen clearly. With a higher incremental infeed y_{inc} of the splitting roll, the bending forces are decreased significantly due to the increased stress superposition. The same effect can be seen for the springback ratio k_s, which is defined as the ratio of the radius before springback to the radius after springback. However, the bending forces for curved linear flow split profiles with the described stress superposition can be valued as significantly lower than in conventional three-roll bending machines. Hence, the tool system can be designed for lower forces and realized considerably lighter.

With the presented tool system, curved linear flow split profiles are manufactured continuously. For the production, an initial sheet metal width of 50 mm, a sheet metal thickness of 2 mm, a microalloyed fine-grain steel with a yield strength of approx. 865 MPa, the sheet metal velocity of 3 m/min, and seven linear flow spitting stands are used. The manufactured profiles with the different bending radii are shown in Fig. 7.5.

Fig. 7.5 Curved linear flow split profiles with different bending radii

7.1.2 Flexible Flow Splitting

Transition to Flexible Processes. Since modern industrial production shows a demand for flexible manufacturing processes (Lorenzer 2010), the forming processes of linear flow splitting and linear bend splitting, presented in Sect. 3.1, are intended to gain flexibility as well. Terkaj et al. (2009) regard flexibility in production processes and describe different levels of common comprehension and appreciation by the value companies assign to manufacturing flexibility. The definition and quantification of flexibility is concluded as a key for the comparison and rating with other demands. Flexibility is referred to internal and external changes in Chryssolouris (1996), and different types of flexibility are classified. For the flexibilization of flow splitting processes, especially machine flexibility, process flexibility, and product flexibility apply.

With the advancements of flexible flow splitting and flexible bend splitting, the forming processes are enriched with the geometrical flexibilities of producing parts with varying cross sections through nonlinear bifurcation lines (Kummle et al. 2012).

Requirements and Boundary Conditions. For the flexible forming processes, a precut sheet is used. Since the roller tool cannot form or process a sharp corner or a jump discontinuity in the progression of the bifurcation line, there are some geometrical restrictions on the design of the nonlinear bifurcation. The bifurcation line, which is located at the band edge in case of flexible flow splitting or at the bending edge in case of flexible bend splitting, has to meet the following constraints (Neuwirth et al. 2016):

- Planar: the forming zone can only be located in the plane of the sheet metal
- Continuous: the bifurcation line has to be continuous
- Continuously differentiable: the progression of the bifurcation line has to be continuously differentiable

Regarding these basic conditions, the smooth bifurcation line can consist of straight segments, which are parallel or with positive or negative inclination to the sheet motion, concave and convex radii, and transition points as shown in Fig. 7.6.

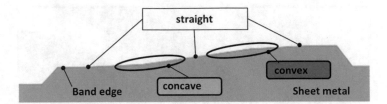

Fig. 7.6 Nonlinear bifurcation line consisting of a sequence of defined sections

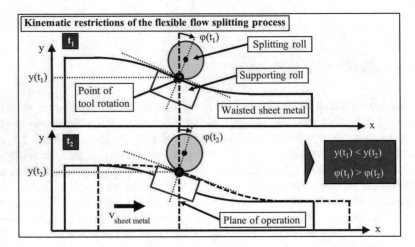

Fig. 7.7 Degrees of freedom of the tool system, necessary to follow the progression of the nonlinear splitting line (Neuwirth et al. 2014b)

For the flexible flow and bend splitting, the same process principles apply as for the linear versions of the forming processes. For each processing side there has to be one splitting roll and there are an upper and a lower supporting roll with the opportunity to adjust the roller gaps (Groche and Schmitt 2010). In the middle of the web, there is no support by roller tools and a free space that allows transition of the tool systems. The web has to provide sufficient stiffness to bear the process loads and to prevent instability phenomena.

To assure proper support and interaction of the roller tools with the sheet metal, the connecting line between the axles of the splitting roll and the supporting rolls has to be perpendicular to the nonlinear bifurcation line at any time (Schmitt et al. 2012). For that reason, two additional degrees of freedom are necessary to allow the tool system to follow the nonlinear progression of the forming zone's location (Kummle et al. 2012). This relation is presented in Fig. 7.7.

Figure 7.8 presents the tool system carrying the roller tools that is provided with the additional degrees of freedom for translation and rotation. The support and application of the process force into the forming zone is also shown and is analog to the process of linear flow splitting.

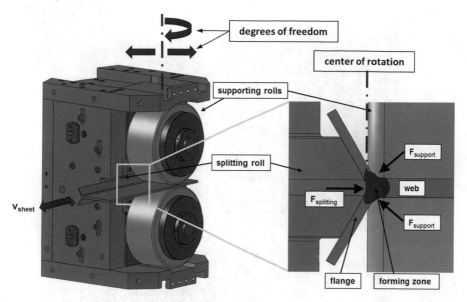

Fig. 7.8 Additional degrees of freedom to the tool system for the flexible flow splitting (Neuwirth et al. 2016)

Realization of the Tool System. For the accurate production of flexible flow split parts, the gaps between the roller tools have to be adjustable. The splitting roll can be positioned continuously in height direction and prestressed by the application of a wedge mechanism. We chose this design to ensure a precise positioning while bearing high process forces and bending moments. The upper and lower supporting rolls can be adjusted in height direction by spacers and positioned continuously in lateral direction using wedge mechanisms (Groche and Schmitt 2010). Figure 7.9 shows the elements to adjust the roller gaps.

The tool system for flexible bend splitting additionally features four pairs of symmetric double-sided wedge mechanisms for the continuous adjustment in height direction and the defined prestressing of the supporting rolls.

Schmitt et al. (2012) present the estimated loads and bending moments to which the tool system is designed on. Groche and Schmitt (2010) examine the available construction space, motion spaces for components of the tools and the sheet. Additionally, the following influencing factors on the process window are identified: angle of bifurcation, angle of pre-bent flange, diameter of the roller tools, pivot angle of the tool system, incremental splitting depth, translational range, and the location and relation of tool center point and point of rotation. The interdependencies of these parameters influence the tool design and the available process window regarding maximum pivot angle, minimum radius, minimum and maximum bifurcation angle, and maximum flange length.

Regarding the identified interdependencies and influencing factors, we designed and constructed the presented tool system and we especially designed and defined

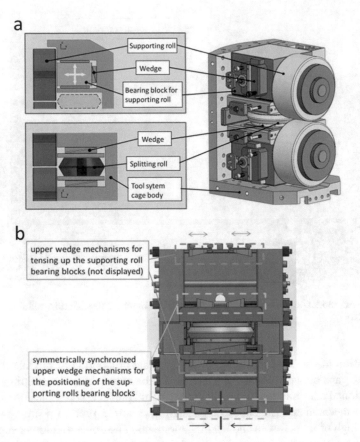

Fig. 7.9 Devices to adjust the position of roller tools inside the tool system (**a**) for the flexible flow splitting and (**b**) for flexible bend splitting

the tool center point to be congruent with the center of rotation (Fig. 7.8) (Groche and Schmitt 2010). This is the reason why there is no transversal degree of freedom to the splitting roll in the above-presented adjustment devices.

The additional degrees of freedom on the tool system are provided by a parallel geometry realized in the kinematic system of each side of the stand (Kummle et al. 2012). The tool system is pivot-mounted in a frame, which can be translated in lateral direction, as it is shown in Fig. 7.10. The parallel geometry shown in Fig. 7.10 provides translation and rotation to the tool center point by simultaneously actuating both drives. The main process force direction is directly supported by the stand, and the motion direction is perpendicular to that. With a high transmission ratio of the parallel geometry, we realize precise positioning of the tool even at high and nonsteady loads (Groche and Schmitt 2010).

Specific Characteristics. For flexible flow splitting and flexible bend splitting, the forming operation takes place along a nonlinear bifurcation line, and tailored sheets

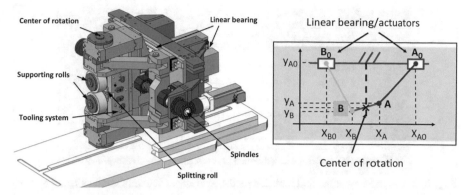

Fig. 7.10 Components of tool system and kinematic chain for the single-sided flexible flow splitting (Neuwirth et al. 2014b)

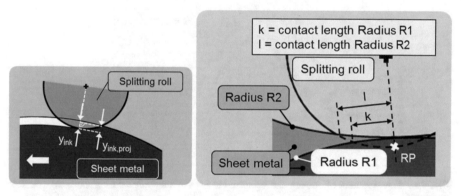

Fig. 7.11 Dependency of the tool intrusion on the band edge contour at constant splitting depth (Schmitt et al. 2012)

are utilized. Schmitt et al. (2012) show that the flexible processes are subject to various nonsteady process states.

Whereas the linear flow splitting process can be characterized by the adjustment of the roller gaps and the incremental splitting depth, the flexible process, however, is not sufficiently described by these values. As described in Schmitt et al. (2012), the tool intrusion into the material varies along the bifurcation line, even when the incremental splitting depth is kept constant. Reasons can be found in the geometrical interaction of the splitting roll with concave, straight, and convex sections of the bifurcation line. Figure 7.11 introduces the projected splitting depth that depends on the sheet metal radius. The same relation is valid for the contact length (Kummle et al. 2012). In convex sections (R_1) the projected splitting depth and the contact length are decreased while they are increased in concave sections (R_2), compared to a straight section (Schmitt et al. 2012).

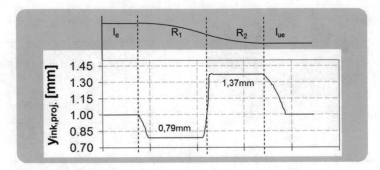

Fig. 7.12 Dependency of the tool intrusion on the band edge contour at constant splitting depth (Schmitt et al. 2012)

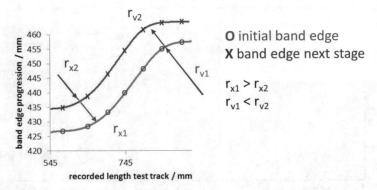

Fig. 7.13 Development of convex and concave radii at the band edge during multistage flexible flow splitting

Figure 7.12 presents an exemplary progression of the projected splitting depth according to a nonlinear bifurcation line, represented through a double bend. The above-described relation is visible.

Equipollent to an increased or decreased projected splitting depth or contact length, the material volume displaced by the splitting roll is increased or decreased. This results in varying process forces (Neuwirth et al. 2014b) and varying material strains (Özel et al. 2014). This behavior is a specific characteristic of the flexible process and represents challenges for the processing.

Groche and Schmitt (2010) describe that there is a nonsteady condition regarding the development of the bifurcation line as well, next to the nonsteady process factors tool intrusion, process force, and material strain. Figure 7.13 illustrates this dependency. Since the splitting depth is defined perpendicularly at the bifurcation line, the bifurcation line is not kept constant or shifted parallel from one step to the next one. Instead, the new bifurcation line is projected perpendicularly at the recent one resulting in an increase of concave and a decrease of convex radii in every step.

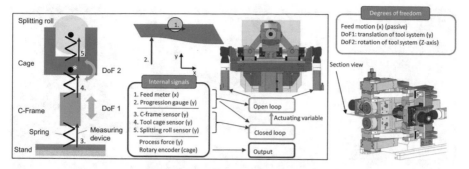

Fig. 7.14 Signals and process control in the process of flexible flow splitting (Neuwirth et al. 2014a)

For that reason, a new bifurcation line curvature is generated after each stage (Groche and Schmitt 2010).

Strategy for Process Guiding and Signal Management. To meet the challenges of an accurate forming process that is subject to varying process forces and changes in the development of the nonlinear tool path during a multistage process, strategies for the process control have been implemented.

Figure 7.14 presents an analogous model for the tool system and the kinematic chain for the one-sided flexible flow splitting process as well as the internal measuring devices 1–5. The analogous model displays that the splitting roll is pivoted in the tool cage, which also bears the supporting rolls and is pivot-mounted (DoF 1) in a frame. The frame can be translated (DoF 2) relatively to the stationary stand (Groche and Schmitt 2010). Elastic behavior can occur in every bearing or contact zone. Therefore, the measuring devices 3–5 are installed, which measure these deflections. The deviation of the tool referred to its ideal position can be calculated by the serial concatenation of the single values (Erbar and Schmitt 2012).

Within the tool system, but prior to the forming zone, there are a feed meter (1) and tactile gauges (2) recording the progression of the sheet metal band edges. The tool path is generated simultaneously from the actual sheet geometry, which is recorded by these measuring devices. By that, no default geometric cam is necessary, and parts which require changes in the development of the nonlinear tool path during a multistage process (Fig. 7.13) can be processed (Neuwirth et al. 2014b).

Additional to that, changes in part geometry can easily be implemented just by the use of another initial sheet geometry but without changing of any tools, adjustment, or programming of process control.

To compensate elastic deflections of the tool system under varying process forces, the internal signals from measuring devices 3–5 can be fed into a closed loop control to adjust tool positioning and obliterate deviations in the tool position (Neuwirth et al. 2014b).

The presented strategy for process control and the tool system enable the production of parts with nonlinear bifurcations in integral style while the process is characterized by nonsteady process factors. Due to the self-adjusting design of

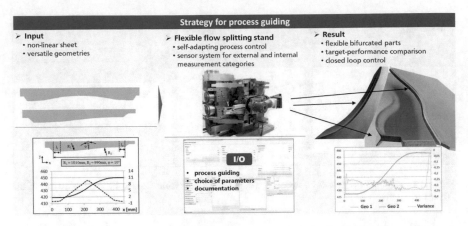

Fig. 7.15 Flexibility in process conduct by the process strategy for the flexible flow splitting

the process strategy, we can realize a change in part geometry by solely changing initial sheet geometry (Input in Fig. 7.15). The control of the flexible flow splitting stand automatically generates the intended tool path from the actual part geometry and the appropriate part is produced (result).

Experimental Results. We carried out experimental investigations with sheet metals out of a *DD11* grade mild steel and of an aluminum alloy *AlMg3*. The sheet thicknesses are 6 mm and geometrical varieties regard combinations of investigated target values for pivot angles from 8.5° to 12.5° and for concave and convex radii from 750 to 1250 mm and the sequence of defined sections (Neuwirth et al. 2014c). In single- and double-sided flexible flow splitting experiments the quantities to be measured are the process force and elastic deflection of tool components to characterize the process. For the characterization of produced parts geometrical features are recorded.

Fig. 7.16a presents a produced flexible flow split part with radii of 1000 mm and a max. pivot angle of 10°. In Fig. 7.16b, the band edge progression is shown after stages 1, 5, and 10. In Fig. 7.16c, the progression of the achieved total splitting depths after stages 5 and 10 are presented and compared to their ideal values (dotted lines). In the target-performance comparison, a significant deviation is visible.

The deviations from the target values result from elastic deflections of tool components due to the varying process forces. A closed loop control can compensate these deviations as described above. The process force and the deflections of the tooling-cage ($dy2$) and of the frame ($dy1$) are presented in Fig. 7.17a. In Fig. 7.17b, the superimposed deflection of $dy1$ and $dy2$ is presented for the case of an open loop and for the case of a closed loop control. As presented in Neuwirth et al. (2014b), the closed loop control is capable of reducing deviations in tool position to a level close to zero, even under varying process force.

Numerical Simulation of the Flexible Flow Splitting Process. Changes and optimizations in real forming processes are time and cost intensive. Due to the setup of the tools and complex geometries, process parameters cannot be measured

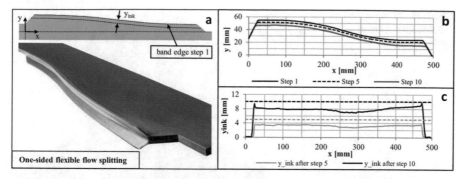

Fig. 7.16 (**a**) Manufactured flexible flow split part, (**b, c**) geometric characteristics (Neuwirth et al. 2014b)

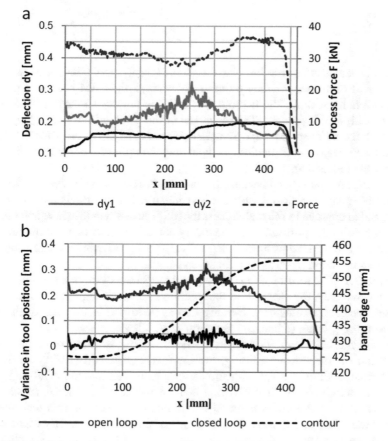

Fig. 7.17 (**a**) Influencing values on process, (**b**) variance depending on process control strategy (Neuwirth et al. 2014b)

Fig. 7.18 FE-model of
flexible flow splitting
process

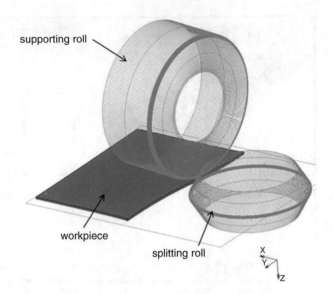

accurately and state variables like the stress tensor cannot be measured at all.
Therefore, to improve forming processes nowadays, the FEM became a powerful
tool. This holds also for the flexible flow splitting process so that an FE-Model is
built to analyze this process. The FE-model of the flexible flow splitting process is
built with the software package *Simufact sfFormingSFM 9.0*. All rolls are defined
as perfectly rigid solid bodies, without any compliance behavior (Fig. 7.18)
(Neuwirth et al. 2016).

A quarter model is used because the process is assumed to be symmetrical. The
applied boundary conditions ensure this symmetry. The process is modelled quasi-
static, and in contrast to the real process the rolls move along the sheet and the work
piece is held in the position using boundary conditions. During the simulation, the
work piece is deformed in several steps. Due to post-processing the large computed
data, the simulation is carried out in several steps.

In Fig. 7.19, the results of the total equivalent plastic strain of a forming step are
shown. Obviously, the elements at the splitting ground are highly deformed, which
requires remeshing after each forming step as described in the case of the linear
flow splitting process (Chap. 6) (Neuwirth et al. 2016).

As the numerical simulation of the flexible flow splitting is similar to the
simulation of linear flow splitting, the Cut-Expand-Method (CEM) can be also
used for the flexible flow splitting simulations. However, due to the different cross
section in longitudinal direction, this method has to be extended as shown in
Fig. 7.20. After a 2D cut is made in the middle of the profile, this deformed 2D
mesh is remeshed and expanded in longitudinal direction, to exactly represent the
geometry in longitudinal direction. Contrary to the CEM, the state variable values
are mapped from the original 3D–Mesh to the remeshed 3D–Mesh. This
whole process is repeated according to the forming steps that are modelled
(Neuwirth et al. 2016).

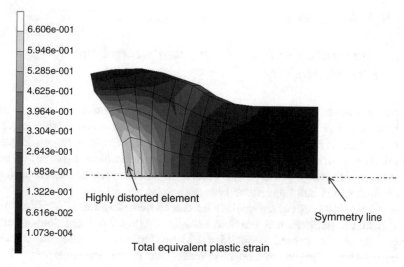

Fig. 7.19 Deformed mesh in the splitting ground (Neuwirth et al. 2016)

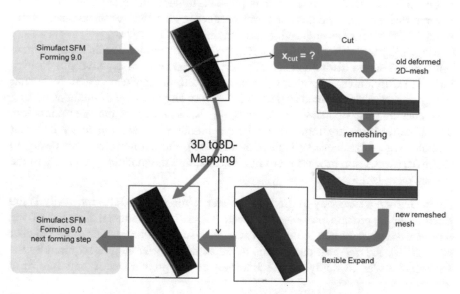

Fig. 7.20 Extended Cut-Expand-Method for flexible flow splitting process (Neuwirth et al. 2016)

7.2 New Products of Sheet Metal Panels

7.2.1 Panelization of Free-Form Architecture Using Flexible Modules

In order to transfer the advantages of industrial series production to free-form architecture, the principle of the geometry's division into transportable and producible modules is important. In doing so, one important factor is the consideration of production restrictions such as possible and useful radii of curvature already during the initial design phase of the project. To achieve this, an examination and classification of the building geometry is necessary.

The classification of curved geometries can be done differently (Pottman et al. 2007). A first classification is possible between single-curved and double-curved shapes. The double-curved geometries can be further classified into standard geometries (for example spherical, ellipsoid, or torus) and free-form shapes (Fig. 7.21).

Even the division of standard geometries results in a lot of different modules (Abedini et al. 2012). However, it is possible to summarize them into a limited amount of identical modules for serial production, but this set of modules offers much less flexibility as required in most architectural projects.

Regarding free-form geometries, it is possible to classify certain areas of them only. Hence, a first distinction can be made between areas of negative and positive Gaussian Curvatures. Applying these curvatures to the modular system for planar triangles, that was presented in Sect. 6.2.2, a very large amount of modules occurs. The final number of elements depends on the accuracy of the approximation. Therefore, a modulary logic is not longer possible because too many different modules would be needed (Schäfer et al. 2012). Furthermore, it is necessary to make the production process more flexible adapting the module's geometry to the chosen manufacturing technology.

Deep Drawing. One way to produce stringer reinforced modules integrally is the deep drawing process presented in Sect. 3.2.4. By incremental forming of the stringers, single parts can be modified according to particular specifications (Bäcker et al. 2010). A large cost-part of this manufacturing technology is required by the molds. In order to apply this technology to the production of any free-form

Fig. 7.21 Different surfaces: *plane, single curved, double curved* with positive and negative Gausian curvature, free-form geometry (*left* to *right*)

geometry, a special mold is needed for every single module, because the curvature varies across the whole surface.

Consequently, it is necessary to limit the number of different molds in order to use this production technique more efficiently. Therefore, the curvature of the geometry should be approximated with the curvature of a limited set of mold geometries. There are several options to approximate a given geometry in order to be applied to an available mold set. By rotating or moving the final outline of the module along the mold surface, different curvatures can be generated with the same mold (Figs. 7.22 and 7.23). Therefore, it is helpful to define an individualized outline of the modules. This could be done CNC controlled and thus automated.

A disadvantage of this approach is the approximation's usual deviation from the planned original geometry. This may result in kinked angles and breaks between the modules.

Flexible Flow Splitting. Choosing a different manufacturing technology, a new set of conditions is created. To utilize the benefits of the fast production speed of the roll forming technology, downstream manufacturing steps should be minimized.

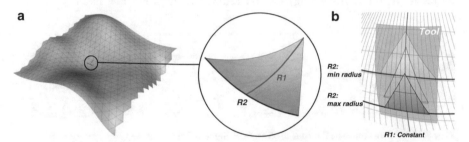

Fig. 7.22 (**a**) A free-form surface and the two approximated radii on one panel, (**b**) the secondary radius changes across the surface of the mold from *low* to *high* (Schäfer et al. 2012)

Fig. 7.23 Generation of different curvatures by rotating the module´s outline within the mold surface

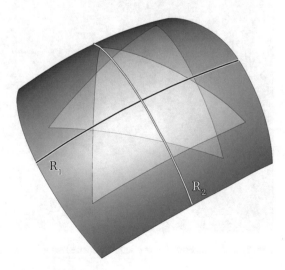

Therefore, it would be more efficient to use the strips produced in flow production directly as building parts. Since buildings are usually unique and their geometry varies widely, the manufacturing processes need to be highly flexible. The flexible flow splitting process as described in Sect. 7.2.1 provides this flexibility.

The potentials that result from flexible roll forming machines can be seen already in the cladding of complex free-form architecture with curved metal panels. Examples for this technology are the "Messehalle 3" in Frankfurt (Krase and Fischer 2002), or the roof of the Southern Cross Railway Station in Melbourne (Grimshaw and Hurd 2011), both designed by Grimshaw Architects, GB. To construct these roofs, a very complex substructure of beams, panelling, insulation, and finally covering is still required. This structure type leads to high demands for the dimensional tolerances that must be respected by the successive building companies.

By adding many profile elements, wide span shell structures can be produced as depicted in Figs. 7.24, 7.25, and 7.26. Splitting the sheet, edge branches can be produced, which can be used to connect the individual profiles with screws. The improved material properties resulting from the forming processes (UFG microstructure, material hardening) can be utilized to improve the stability of these

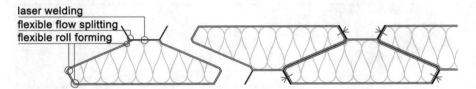

Fig. 7.24 Section through the roof profiles (Reising and Schäfer 2015)

Fig. 7.25 Perspective view of a hall constructed with curved profiles (Reising and Schäfer 2015)

Fig. 7.26 (**a**) A sequence of cross sections is showing the variation of the profile, (**b**) roof construction with curved profiles (Reising and Schäfer 2015)

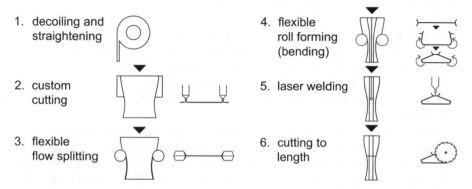

Fig. 7.27 Manufacturing steps of the curved profiles (Reising and Schäfer 2015)

critical points. Later the screws can be covered with an additional lid. The individual profiles can be filled with foam to achieve a thermal insulation effect and extra rigidity. Initial FEM simulations show that with a steel sheet thickness of 3 mm and a shell thickness of 150 mm, a span of more than 30 m can be obtained.

The sheet metal is decoiled and straightened. Then, the sheet is cut to obtain the desired flexible cross section. Subsequently, the edges are split by a flexible flow splitting process. With flexible roll forming the defined profile cross sections can be formed during the next steps. Now the profile gets closed by laser welding. Finally, the profile can be cut by a movable saw without interrupting the ongoing production process. These flat profiles then are bent and twisted during a subsequent process (Fig. 7.27).

For the example shown in Figs. 7.25 and 7.26, very large curvature radii are required. The current test facility (Sect. 7.1.2) is designed for much smaller radii.

By means of the proposed manufacturing technology, it would be possible to produce shell structures more efficiently and economically. The initial costs for developing and building the necessary machines are certainly not yet justified for just a single building. However, this method can be used to manufacture wide spanning structures more efficiently at lower cost.

In opposition to the traditional planning and building process of architecture, the planner needs to be aware of the industrial production processes and their limitations. Additionally, it is getting more necessary to plan very precisely in order to utilize the full potential of these technologies. This precision can be achieved through the use of CAD systems and customized planning tools.

Through geometric simplification and the use of new industrial production methods, complex free-form shell structures could be realized economically in the future.

7.2.2 Joints with Flow Split Flanges

New manufacturing processes (Sect. 7.1) allow a gain of prefabricated components. Already Wachsmann (1959) described the importance of the joining technique for the assembly of prefabricated components. Therefore the joining techniques require very low tolerances, a high load-bearing capacity, and offer an easy assembling on the site. Moreover, the joining technique also should satisfy aesthetical requirements as well.

With the continuous cold forming process (Sect. 3.1) bifurcated structures made of sheet metal can be produced integrally. Within this process, the material is subject to very strong pressure and is being deformed continuously. One visual advantage of this process among others is the missing machining traces on the final product (Herbert and Schäfer 2008). Another advantage of this processing step is a hardening of the material in the deformed zones, which results in an ultrafine grained microstructure (Sect. 4.1). The special characteristic of this zone is a good ductility and a high tensile strength (Schäfer et al. 2013). To use these optimized material characteristics, conceptual joining technique solutions were developed involving the linear flow splitting process.

In Abedini and Schäfer (2016b), a qualitative assessment of joining techniques with flow split flanges was described, where a set of conceptional joining techniques were compared with conventional ones (Fig. 7.28).

First, the assessment categories for a qualitative assessment of new joining techniques have to be identified. The most important categories are their mechanical strength and, due to maintenance matters of the whole component, their reversibility. Another important assessment criterion is the optimal use of

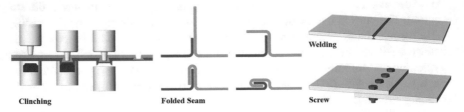

Fig. 7.28 Selection of conventional joining techniques (Abedini and Schäfer 2016b)

process-related properties such as the increased strength in the flanges. Furthermore, the aspects of the imperviousness, the assembling complexity, the aesthetics, and the functionality integration into the joining techniques were assessed. Second, the range of assessment for the aforementioned categories has to be identified depending on the use of the component. Third, all developed conceptional joining techniques are divided into main connection categories (Fig. 7.29). The main categories are snap joints, plug-in joints, and bayonet joints. Finally, some representatives of the main connection categories (green point in Fig. 7.29) were selected for the qualitative assessment.

Comparing the conceptional with the conventional joining techniques, Abedini and Schäfer (2016b) described a comparable usability of both kinds of joining techniques. Moreover, joining techniques using integral bifurcations have improved local material properties (Schäfer et al. 2013).

After the qualitative assessment the favorite joining technique (Fig. 7.30) of each connection categories can be considered in detail.

In Sect. 9.2.2 a snap joint for moveable hangers was considered in the algorithm based process chain. Furthermore, for the transmission of the conceptual solution to real applications, a plug-in connection with an additional coupling element was manufactured as a real demonstrator with flow split raw material (Fig. 7.31). This joining technique offers good mechanical strength and reversibility. Moreover, the positive properties of the UFG structure along the contact area are used.

The integral production of joining techniques along the component offers various advantages. All manufacturing steps can be implemented continuously into the process chain and at the production line-end a ready-to-assemble product is obtained. Moreover, the integrally formed joining techniques can positively contribute to the total load-bearing capacity of a construction. The material properties of the UFG microstructure also increase the load-bearing capacity.

These aspects expand the current product range and enable new joining techniques that were not realizable previously.

7.2.3 Multifunctional Building Modules

The technology-integrated market pull in building industry contributes to even more complex components. This includes among others the requirement of multifunctional and lightweight design focusing on resource-saving and sustainable construction as well as seeking for synergies. Another important element in the construction industry are prefabricated components with a very high processing quality.

The linear flow- and bend splitting process (Sect. 3.1) enables the production within a continuous cold forming process with ribs-stiffened sheet metal panels. Groche et al. (2007) showed already that buckling stress compared to the surface structure mass with and without bifurcations is many times higher. This characteristic has also been used in the demonstrator on Sect. 3.3.2. Thereby all new

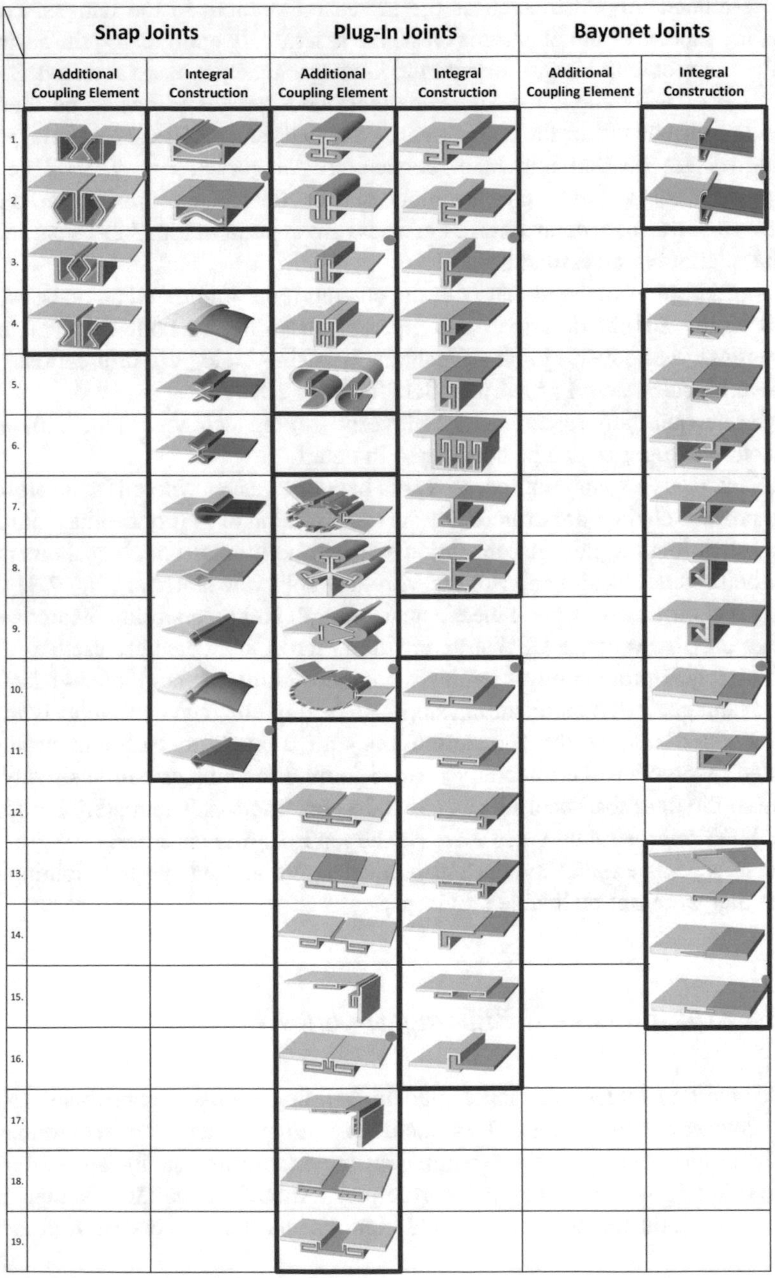

Fig. 7.29 Main connection categories and the final representatives (*green points*)

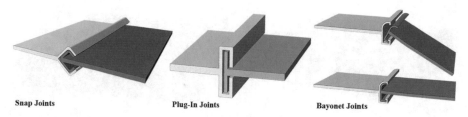

Snap Joints   Plug-In Joints   Bayonet Joints

Fig. 7.30 Favorite joining techniques after a qualitative assessment (Abedini and Schäfer 2016b)

Fig. 7.31 Demonstrator for a plug-in joint with an additional coupling element using linear flow split profiles (Schäfer et al. 2013)

developed components are both lightweight construction and industrially prefabricated.

Another important aspect for the surface quality of building modules is their appearance. With the linear bend splitting process it is possible to produce bifurcated sheet metal panels without any traces on their visible surfaces (Schäfer et al. 2013).

Another example of a lightweight design application in building industry are standard sandwich panels. Sandwich panels can incorporate multiple functions such as weather protection, thermal insulation, high stiffness properties as well as visual protection (Abedini and Schäfer 2016a). Furthermore, they are of economic interest, because they can be manufactured serially. Metal cover sheets made by linear bend splitting technology have an aesthetic advantage against conventionally manufactured metal cover sheets that are typically used for sandwich panels. Conventional strategies to increase the rigidity of thin metal sheets are usually limited to roll forming and profiling processes (Fig. 7.32b). However, this impairs the aesthetical appearance; hence sandwich panels are mainly used for industrial and commercial buildings (Abedini and Schäfer 2016a). On the other hand, metal cover sheets with integrally formed stiffening ribs obtain a visible surface, which remain traceless, clean, and smooth. Thus, they remain suitable for sophisticated building envelopes (Fig. 7.32).

Applying the hydroforming process (Sect. 3.2.4), deep drawing a stringer sheet with bifurcated undestroyed ribs onto a curved surface structure is possible. Figure 7.33 shows a selection of possible curved sandwich panels. In Abedini and Schäfer (2012) further examples of sandwich panels using the linear bend splitting process are presented.

The components complexity in building industry is increasing regarding the quality of the production and in terms of functionality. A multifunctional

Fig. 7.32 (**a**) Sandwich panel with bifurcations, (**b**) conventional sandwich panel

Fig. 7.33 Examples of curved sandwich panels

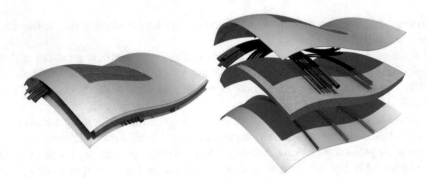

Fig. 7.34 Curved building component with the integration of various functions

component may cause space saving and allows a better utilization of the material and thus generates weight saving. Figure 7.34 shows an example of a multifunctional building component with the integration of photovoltaic, weather protection, bearing structure, thermal insulation, and other technical equipments. Moreover, industrial prefabrication contributes to smaller tolerances between several elements. These aspects facilitate the assembly and thus also decrease the related assembly cost.

The development of multifunctional modules is highly relevant for the building industry. The potential for material saving through synergy effects in multifunctional modules is very large. Necessary compartments for the multifunctional components can be manufactured integrally within the cold forming process of linear flow- and bend splitting (Sect. 3.2).

Figure 7.35 shows an application example in which in addition to the joining technique also various function chambers are incorporated integrally. By

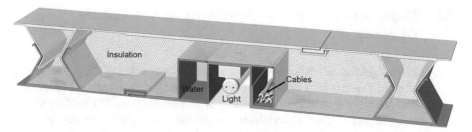

Fig. 7.35 Multifunctional building component with integrally produced chambers for technical equipment

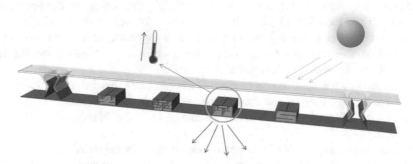

Fig. 7.36 Building component activated as a solar module

optimizing the component's cross section, the necessary chambers can be positioned perfectly according to a further increase of the load-bearing capacity (Sect. 5.2). This contributes that technical equipment like insulation, installation shafts for lighting, cable or media guides, and others can be already integrated in the finished component.

Both the avoidance of material doublings and the hardening through the UFG microstructure (Sect. 4.1) contribute to material saving. The possible integral production from the coil to the final product reduces the total manufacturing costs. Moreover, the use of reversible joining techniques also leads to a reduction of the maintenance and dismantling costs. The limitation to use just one material improves its recyclability and the building dismantling-caused waste.

Figure 7.36 shows an example with a component used as a functional solar module. The increasing complexity of components requires a change of the assembly process. Also, it should be ensured that in case of a function failure, the entire component must not be replaced completely.

In Sect. 9.2.2, the facade-cleaning demonstrator exemplarily shows the integration of various functional elements into the linear motion system. Furthermore, the manufacturing process within the production line of the linear motion system is shown in this section. For further information about the facade-cleaning system, see Sect. 9.2.2.

7.3   Implementation of Our Approach on Aesthetical Demands

It is an overall goal to achieve an algorithm-based process chain (Chap. 2). Hereby the number of the required iterations during the product development process is reduced and hence a maximum of efficiency is generated (Sect. 2.5). In this context it is useful to detect market requirements in the early phases of the product development phase applying them onto the product features (Sect. 5.1). Up to now, the evaluation of product characteristics in terms of identification of product requirements usually refers to the functional context exclusively. In this section additionally the aesthetic context will be taken into consideration.

Aesthetic evaluation standards vary indisputably depending on the perceptual subject and are therefore highly subjective. The existence of universal ideals for aesthetic values and the possibility of a respective approach can hardly be detected. Nevertheless, there are object-specific features, which are—at least among certain groups of people—rated as aesthetically appealing.

Furthermore, there are some object-specific features that initiate transcultural positive reactions. One of the most famous example is the "Golden Ratio" and variations thereof such as the "Golden Rectangle" and the "Golden Spiral" (Fechner 1876).

The aesthetic judgement of an object is determined by the individual experience of the contemplator and is also subject to social and cultural influences. The functional approach to design is generally well received and often results in the "form follows function-concept" associated with architecture and industrial design. For example, tension-oriented forms, such as hyperbolic paraboloids, which are typically used in membrane structures, are often considered as "beautiful." Not least, a perception usually is influenced by the spatial and temporal environment of the relevant objects and the contemplator. The totality of these relationships is shown in Fig. 7.37.

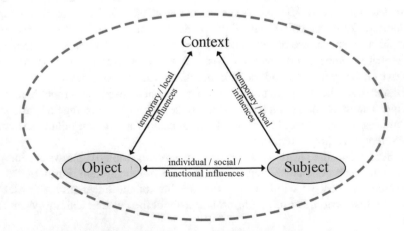

Fig. 7.37 Aesthetic interaction between object (=perception subject) and subject (=recipient)

Structures can remain in complete harmony with their environment, or set new trends, as Heraclitus of Ephesus already stated (Diels and Kranz 1956): *"The unlike is joined together, and from differences results the most beautiful harmony [...]."* In this regard, two design concepts were defined (Zimmermann et al. 2013):

(a) The (harmonic) "embedment in the spatial context"
(b) The (harmonic) "embedment in the dynamic context"

The former approach (a) refers to the dependency of a structure on the static component of its vicinity. For example, in parks either organic and natural shapes or nonnatural cubatures can be used for buildings. In the latter design concept (b) the dynamic building appearance is the primary goal. This serves to integrate a building aesthetically into the "traffic flow" of its vicinity. The increasing urbanization and the global development of transport networks results in an important relevance, especially for this kind of a design concept.

Therefore—based on known laws of perception—simple design principles can be formulated. A compliance with these design principles offers better results in dynamic structures. Such principles are, for example, quantitative "limit slenderness" and "critical angle," which principally should be obtained. A subsequent combination of all design principles provides a kind of mathematical functions, which allow the creation of dynamic building forms and joint patterns (Fig. 7.38). These functions include variables, which enable to adapt the building geometries with functional conditions. Hence, practice-relevant building geometries can be generated and relevant requirements for dimensions, angles, and curves of sheet metal elements are derived.

In a subsequent study approved deviations are defined for components, as to establish a direct link between general concepts of aesthetics and the specific process chain. On the basis of Schmitt et al. (2011) and a visual inspection of work pieces, a list of typical component faults was made (Table 7.1, left column).

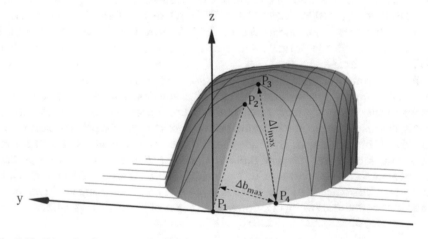

Fig. 7.38 Example of an automatically generated dynamic structure

Table 7.1 Assignment of product characteristics of the product facade cleaner (Sect. 9.2.1)

Product Characteristics	Product zones			
	Web	Flange inside	Flange outside	Profile cross-section
~~Asymmetry of the flange width~~				X
~~Asymmetry of the flange thickness~~				X
Variable flange width		X	X	
~~Thickening of the web~~				X
~~Thickening of the flange~~				X
Buckling of the web	X			X
Buckling of the flange		X	X	X
Waviness of the flange inside		X		X
~~Scratches of the flange outside~~			X	X
Scratches of the web	X			

The effective suppression of these component faults has been resolved technically (Groche et al. 2005; Groche and Vucic 2006), but is associated with a larger production effort in many cases. Therefore, it is appropriate to only accept allowable tolerances as small as necessary for the product and from the customer's perspective.

To determine the permissible deviations, a matrix was developed and implemented into the already quantitative recommendations for specific products. The client may override these values. Table 7.1 displays an example of such a matrix: All typical components can be fully described by means of the product zones "web", "flange inside", "flange outside", and "profile cross section" (Table 7.1, columns).

For the product "facade cleaner" (Sect. 9.2.1) for example, only the "web" and the "flange inside" are relevant in terms of aesthetics. Therefore, the other two product zones have not been taken into account. Based on the specified crosses within the matrix, all product features relevant for the specific product can be read directly (Table 7.1, lines). A corresponding table offers recommendations for the transformation of the relevant product characteristics on quantitative product requirements or —as shown in this example —to permissible deviations. Thus, product- or customer-specific requirements can be considered already in an initial stage of the product development process in terms of aesthetic stipulations.

References

Abedini S, Schäfer S (2012) Die Eigenschaftsverbesserung von Bauprodukten durch neue Fertigungstechnologien des SFB 666. In: Tagungsband 4. Zwischenkolloquium SFB 666, Darmstadt, 14–15 November 2012, p 131–136

Abedini S, Schäfer S (2016a) Sandwichelemente—Konventionelle Lösungen im Vergleich zu innovativen Weiterentwicklungen aus der Forschung. In: 11th Conference on Advanced Building Skins, Bern, Switzerland

Abedini S, Schäfer S (2016b) Qualitative assessment of joining techniques with flow split flanges. In: Anderson MST, Anderson PCO (eds) ACE 2016. Conference Proceedings of the 4th Annual International Conference on Architecture and Civil Engineering, Singapore

Abedini S, Schäfer S, Bäcker F, Ludwig C (2012) Geometric figures and potential component families of metal sheets for the use in architecture. In: Advanced Building Skins Conference 2012, Graz

Bäcker F, Ertuğrul M, Groche P (2010) A new process chain for forming individually curved sheet stringers. Int J Mater Form 3(1):837

Chryssolouris G (1996) Flexibility and its measurement. Ann CIRP 45(2):581–587

Diels H, Kranz W (1956) Die Fragmente der Vorsokratiker, 3 vol. 8th edn. Weidmann, Berlin

Erbar M, Schmitt W (2012) Flexibles Spaltprofilieren: Verfahrensentwicklung—implementierung und Prozessregelung. In: 8. Fachtagung Walzprofilieren & 4. Zwischenkolloquium SFB 666, Darmstadt, 14–15 November 2012

Fechner GT (1876) Vorschule der Ästhetik. Redaction of 1925 (posthum), 3rd edn. Breitkopf & Härtel, Leipzig, p 184–202

Grimshaw N, Hurd P (eds) (2011) Grimshaw architecture: the first thirty years. Prestel Publishing, Munich

Groche P, Schmitt W (2010) Verfahrensentwicklung zum flexiblen Spaltprofilieren. In: 3. Zwischenkolloquium SFB 666, Darmstadt, 29–30 September 2010

Groche P, Vucic D (2006) Multi-chambered profiles made from high-strength sheets. Production Engineering, Annals of the WGP 3(1):67–70

Groche P, Vucic D, Jöckel M (2005) Herstellung einteilig verzweigter Blechstrukturen. Wt Werkstattstechnik online 95(10)

Groche P, Ringler J, Vucic D (2007) New forming processes for sheet metal with large plastic deformation. In: Micari F, Geiger M et al (eds) Key engineering, vol 3, Palermo, Italy, p 251–258

Groche P, Taplick C, Özel M, Mahajan P, Stahl S (2016) Benefits of stress superposition in combined bending-linear flow splitting process. International Journal of Material Forming, 2016 (submitted)

Herbert J, Schäfer S (2008) Anwendungspotentiale im Bauwesen für integrale Blechbauweisen höherer Verzweigungsordnung. Der Bauingenieur 83:410–418

Krase W, Fischer V (2002) Opus 41: Nicholas Grimshaw & Partners: Halle 3, Messe Frankfurt am Main, Edition Axel Menges, Stuttgart, 2002

Kummle R, Weber H, Storbeck M, Beiter P, Berner S, Schmitt W, Groche P (2012) Walzprofilieren im Wandel—Bewährtes Verfahren, neue Flexibilität. In: 11. Umformtechnisches Kolloquium Darmstadt, Seeheim, 6–7 March 2012, pp 141–151

Lorenzer T (2010) Mehr Flexibilität zu niedrigeren Kosten. Umfrage zum Produktionszyklus in Fertigungsunternehmen. Eidgenössische Technische Hochschule Zürich, Institut für Werkzeugmaschinen und Fertigung

Mahajan P, Taplick C, Özel M, Groche P (2016) A study on springback of bending linear flow split profiles. International Deep Drawing Research Group (IDDRG) Conference, 2016, Linz, Austria

Neuwirth M, Özel M, Schmitt W, Rullmann F (2014a) Flexibles Spaltprofilieren: Verfahren und numerische Abbildung. In: 9. Fachtagung Walzprofilieren & 5. Zwischenkolloquium SFB 666, Mörfelden-Walldorf, 19–20 November 2014

Neuwirth M, Schmitt W, Groche P (2014b) On the Origin of Specimen: load-adapted integral sheet metal products. In: Procedia Engineering, Nagoya, vol 81, pp 310–315

Neuwirth M, Schmitt W, Groche P (2014c) On the Origin of Specimen: load-adapted integral sheet metal products. In: 11th International Conference on Technology of Plasticity, ICTP 2014, Nagoya, Japan, 19–24 October 2014

Neuwirth M, Özel M, Groche P (2016) Process characteristics in flexible flow splitting, steel research international, Wiley-VCH (submitted)

Özel M, Tijani Y, Rullmann F, Schmitt W, Eufinger J, Groche P, Tobias M (2014) Numerische Abbildung von Spaltprofilierprozessen und Bruchmechanische Beschreibung des Rissfortschrittsverhaltens von spaltprofilierten Strukturen. In: 9. Fachtagung Walzprofilieren & 5. Zwischenkolloquium SFB 666, Mörfelden-Walldorf, 19–20 November 2014

Pottman H et al (2007) Architectural geometry. Bentley Institute Press, Exton

Reising J, Schäfer S (2015) Flächentragwerke aus flexibel spaltprofilierten Blechbauteilen. Bautechnik 92(4):259–263

Schäfer S, Reising J, Abedini S, Bäcker F (2012) Definition of a tool library for the approximation of freeform surfaces. In: Zhang L et al (eds) ICAMMS 2012. International Conference on Advanced Material and Manufacturing Science 2012, Beijing, December 2012, p M1558

Schäfer S, Abedini S, Groche P, Bäcker F, Ludwig C, Abele E, Jalizi B, Müller C, Kaune V (2013) Verbindungstechniken durch die Technologie des SFB 666. Bauingenieur 88(1):8–13

Schmitt W, Rullmann F, Ludwig C, Groche P (2011) Entwicklungsstufen des Spaltprofilierens. In: Merklein M (ed) Tagungsband zum 1. Erlangener Workshop Blechmassivumformung 2011 DFG Transregio 73, Erlangen, 13 October 2011, p 53–76

Schmitt W, Brenneis M, Groche P (2012) On the development of flexible flow splitting. In: Proceedings of the 14th Metal Forming Conference, Wiley-VCH Verlag GmbH & Co., Weinheim, p 55–58

Terkaj W, Tolio T, Valente A (2009) Designing manufacturing flexibility in dynamic production contexts. In: Design of Flexible Production Systems. Springer, Berlin, pp 1–18

Wachsmann K (1959) Wendepunkt im Bauen. Rowohlt, Wiesbaden

Zimmermann A, Abedini S, Schäfer S (2013) Ästhetische Formgestalten in der Architektur, 2nd edn. Meisenbach, Darmstadt

Chapter 8
Finding New Opportunities: Technology Push Approach

C. Wagner, L. Ahmels, S. Gramlich, P. Groche, V. Monnerjahn, C. Müller, and M. Roos

Realizing the benefits of a manufacturing technology is a key challenge that manufacturing engineers and designers face that exceeds conventional aspects of manufacturability and manufacturing compliant solutions. The goal is to comprehensively utilize manufacturing potential through manufacturing-induced properties to find new opportunities for innovative product and process solutions.

Innovation is characterized by the successful realization and introduction of truly new ideas and products to the market. Often, innovations are driven by technologies that allow generation and development of new products or processes (Pahl et al. 2007b). New manufacturing technologies, such as linear flow splitting, can be the basis for fundamental innovation. Utilizing an integrated product and process development approach, manufacturing engineers and designers can systematically realize the benefits of a manufacturing technology by generating innovative product and process ideas.

The characterization of manufacturing technologies by their manufacturing-induced properties is a promising basis for a systematic technology push approach. These properties include material, mechanical, geometrical, and other kinds of properties that are influenced by the technology. Manufacturing-induced properties represent technological potential in a formalized manner. The potential of linear flow splitting in product and process innovation becomes apparent in the form of

C. Wagner (✉) • S. Gramlich • M. Roos
Institute for Product Development and Machine Elements (pmd), Technische Universität Darmstadt, Darmstadt, Germany
e-mail: wagner@pmd.tu-darmstadt.de

L. Ahmels • C. Müller
Physical Metallurgy (PhM), Technische Universität Darmstadt, Darmstadt, Germany

P. Groche • V. Monnerjahn
Institute for Production Engineering and Forming Machines (PtU), Technische Universität Darmstadt, Darmstadt, Germany

© Springer International Publishing AG 2017
P. Groche et al. (eds.), *Manufacturing Integrated Design*,
DOI 10.1007/978-3-319-52377-4_8

the range of possibilities for function integration and process integration combined with different manufacturing technologies in a continuous flow production. By systematically utilizing these technology-specific combinations of properties, functional benefits, new processes, and innovative products and processes can be realized.

8.1   Technology-Pushed Product Innovation

Product innovation is characterized by novel product design and successful introduction into a defined market (Pahl et al. 2007b). To successfully realize technology-pushed product innovation, manufacturing characteristics have to be comprehensively considered during the product design process. Characteristics include possible geometric and material properties that can be realized with the help of the manufacturing technology being considered. By focusing on these manufacturing characteristics, novel ideas for innovative products that systematically utilize technological potentials for design can be deduced. In manufacturing-integrated design, a systematic approach based on manufacturing-induced properties (Chaps. 3 and 4) and their impact on life cycle processes aids the designer to comprehensively realize technological potential, which enables product innovation. The technology-pushed product innovation approach makes the identification of design elements with a specific combination of manufacturing-induced properties necessary. Products consist of a variety of function carriers. The function carriers realize the product functions that are based on implemented working principles. The task of a technology push is to realize a feasible technology-pushed product idea by identifying promising working principles and function carriers, starting with manufacturing-induced design elements (Fig. 8.1). Manufacturing-induced properties have to be systematically matched with function-relevant properties of specific applications. After identifying innovative function carriers, the product idea and potential applications can be concretized by analyzing the feasible value of the manufacturing-induced properties and their impact on the function-relevant properties. The success of a subsequent product innovation process is largely dependent on this initial product idea, making a methodic approach necessary.

Fig. 8.1 Components of a technology-pushed product idea

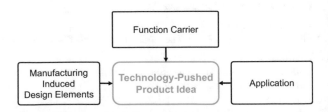

8.1.1 *From Manufacturing-Induced Properties to Product Innovation*

Manufacturing-induced properties of a manufacturing technology are the starting point for technology-pushed product innovation (Fig. 8.2) based on a technology-pushed product idea. Linked to the corresponding manufacturing-induced design elements, they characterize the manufacturing technology. Manufacturing-induced properties provide a way for the designer to assess manufacturing possibilities in the form of geometric, material, and mechanical properties. Manufacturing-induced design elements are determined that can be beneficially utilized to realize function carriers required for the product function for promising product applications due to their manufacturing-induced properties. Starting from a principle technology-pushed product idea, in which suitable function carriers and their manufacturing-induced design elements are defined, the technology-pushed product idea is methodically concretized. In order to realize a cost effective and functional design to achieve product innovation, the designer further focuses on systematically utilizing manufacturing-induced properties. Part of the methodically supported

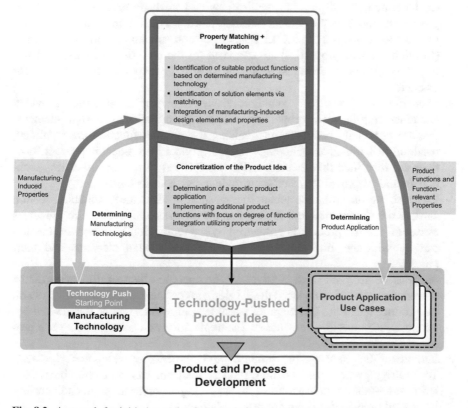

Fig. 8.2 Approach for initiating technology-pushed product innovation

concretization process is the determination of suitable applications from a pool of generally possible ones. Designer experience, market research, and analysis of existing products can all be starting points for the identification of possible applications that fit the determined manufacturing technology. Properties are key elements for concretizing the identified application: the possible range of manufacturing-induced properties and their values help the systematic assessment of the possibilities of the manufacturing technology for product design.

After identifying a manufacturing technology that has the potential for technology-pushed product development, the steps from manufacturing-induced properties to the technology-pushed product idea and subsequent product innovation can be summed up as follows:

1. Determination of a manufacturing technology and manufacturing process chain.
2. (a) Analysis of the manufacturing technology: Based on analysis of the manufacturing technology chosen, its design possibilities are identified and described in a formalized way, using manufacturing-induced design elements and properties.
   (b) Identification of possible product functions that fit the manufacturing-induced design elements.
   (c) Identification of possible applications and analysis of corresponding use processes: Based on the provided manufacturing-induced properties, properties relevant to realizing the product function or corresponding working principles (function-relevant properties) are extracted for the possible applications identified. Possible applications can be identified with the help of market analysis, for example.
3. Matching of identified function-relevant properties to the provided manufacturing-induced properties: Manufacturing-induced design elements that are especially beneficial in realizing the product function are identified, resulting in a principle technology-pushed product idea. Specific product functions are identified that fit the specific possibilities of the manufacturing technology and its provided manufacturing-induced design elements.
4. Concretizing the principle technology-pushed product idea: Suitable function carriers are determined and their influence on the application's use processes assessed. The use process is influenced by the realization of necessary function carriers with the aid of manufacturing-induced design elements and their properties.
5. Analysis of use processes: Analysis of the specific effect of utilization of the chosen manufacturing-induced design elements on the use process, especially on the product function. The range of possible applications can be determined by, for example, experimental studies. The application and corresponding use processes are determined.
6. Initiating the subsequent product and process development process: Technology-pushed product innovation is realized, based on the concretized technology-pushed product idea, by applying a subsequent manufacturing-integrated product development process (Chap. 9).

The following section illustrates this generally applicable approach, using the example of innovative technology-pushed linear guides and linear motion systems initiated by the manufacturing technology linear flow splitting.

8.1.2 Linear Flow Split Linear Guides

Analyzing linear flow splitting reveals its potential for a very cost-efficient manufacturing process due to its possible integration into a continuous flow production. In combination with various technologies in the continuous flow production, complex profiles can be manufactured. High speed cutting (HSC) processes within the continuous flow production can produce manufacturing-induced design elements with integrated drillings or slots (Hirsch et al. 2008); consecutive bending processes enable the realization of closed chambers (Wäldele et al. 2007) (Chap. 3). With the help of these complex manufacturing-induced design elements, various functions that utilize specific properties can be realized in integral construction, leading to products with high quality, functionality, and performance. Linear flow splitting's main potential for technology-pushed innovation lies in its capacity for a high degree of function integration and manufacturing process integration.

This makes multifunctional linear guides a possible application for a technology-pushed product idea based on linear flow splitting. Linear guides are widely used machine elements with many applications where linear motion is necessary, such as transportation, machine tools, and handling (Lommatzsch et al. 2011a). Market analysis shows that there is market potential for products characterized by high technology level and low to medium costs, making continuous flow production with linear flow splitting a promising technological starting point for product innovation.

Linear Flow Split Flanges as Rolling Contact Areas. As in the systematic technology push approach illustrated in Sect. 8.1.1, the first step to identifying a technology-pushed product idea is analysis of the considered manufacturing technology. Typical manufacturing-induced properties provided by linear flow splitting are high hardness and low surface roughness, as well as geometric properties of the linear flow split flanges produced (Chaps. 3 and 4). Rolling contact areas of linear guides require low surface roughness and high hardness to yield satisfactory lifespan and rolling contact quality. Matching the manufacturing-induced properties of linear flow splitting to the specific function-relevant properties of rolling contact areas enables the identification of a product function that can be beneficially realized with the help of a design element that is provided by linear flow splitting: realizing the linear guiding function using linear flow split flanges (Fig. 8.3).

Due to their high hardness, ductility, and surface quality, linear flow split flanges are especially suitable as rolling contact areas (Sect. 4.1.4) (Karin et al. 2010). This fundamental idea of beneficial utilization of property relations can be documented in the form of process-integrated design guidelines (Sects. 4.2 and 5.1.3). Using

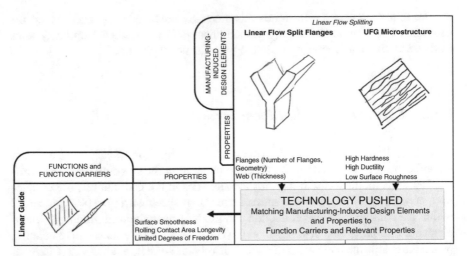

Fig. 8.3 Identification of a technology-pushed principle product idea based on manufacturing-induced properties and function-relevant properties

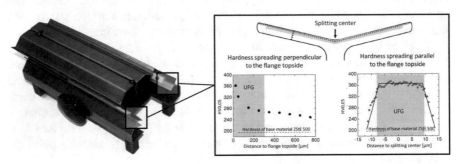

Fig. 8.4 Prototype of a linear guide with linear flow split flanges according to Karin et al. (2010) and Lommatzsch et al. (2011a)

linear guides with linear flow split flanges for the rolling contact area (Fig. 8.4) has various advantages over common linear guides that have to be manufactured in many manufacturing process steps to achieve the desired product quality. Continuous flow production allows a cost-efficient manufacturing process by making subsequent hardening processes unnecessary in many applications due to the existing hardness of the ultrafine-grained microstructure. Huge lightweight potential lies in the possibility of integrating bifurcations into the profiles, realizing steel profiles with high stiffness in an integral construction (Müller et al. 2008). Advantages in manufacturing processes and in later use processes become possible.

Applications requiring high precision, for example, machine tools, necessitate narrow tolerances. Achieving a high stiffness of the linear flow splitting stands and coping with high occurring process forces is very demanding, making the realization of narrow tolerances very difficult. Especially market segments demanding

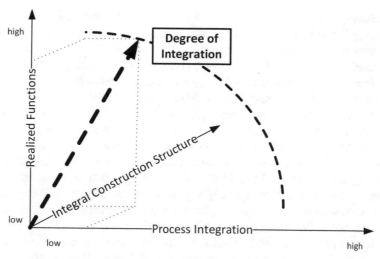

Fig. 8.5 Degree of integration

high functionality and low costs are promising. Linear flow split linear guides combine low surface roughness with high hardness and fatigue strength, leading to long-lasting rolling contact surfaces manufactured at low cost using continuous flow production.

Systematic function integration plays a major part in the development of innovative linear flow split profiles. Multifunctional products in integral construction offer benefits in functionality (and possible applications and markets). In addition, the production in a continuous flow production allows a cost-efficient manufacturing process, as fewer assembly operations and parts are necessary. Functionality and cost-efficiency are key requirements for a product to be successfully introduced to the market. The degree of successfully executed integration can be measured in three dimensions, as shown in Fig. 8.5: the degree of integral construction structure, the number of realized functions, and the degree of process integration.

Linear flow split flanges are especially suitable for use as rolling contact areas due to their specific manufacturing-induced properties and cost-efficient manufacturing, illustrating the benefits of an increased degree of integral construction structure. Starting from this principle technology-pushed product idea, additional functions can be integrated into linear flow split guides with the help of the continuous flow production, increasing the number of realized functions and further increasing the degree of integration. To realize product innovations based on this principle technology-pushed product idea, linear flow split linear guides should provide a wide range of functions that can be used in applications where functionality is most important. In conjunction with additional manufacturing technologies such as bending or welding (Chaps. 3 and 4), the linear flow splitting process chain enables the implementation of additional functions in linear guides to further increase functionality.

Concretizing the Principle Technology-Pushed Product Idea: Integrating Additional Functions. The principle technology-pushed product idea of utilizing linear flow split flanges as rolling contact areas of linear guides has to be further concretized to realize product innovations based on a technology-pushed product idea. Linear flow splitting provides flanges with an ultrafine-grained microstructure that can be beneficially employed in linear guides, realizing the main product function. By integrating the linear guiding function into the bifurcated sheet metal profile in integral construction, linear flow split linear guides offer a high degree of integration due to their high level of integral construction structure. Linear flow split linear guides can also increase the degree of functionality by integrating functions in addition to the linear guiding function. A continuous flow production, like linear flow splitting's process chain, allows the integration of additional functions in linear flow split linear guides without adversely affecting production costs. Not only linear flow splitting's manufacturing-induced design elements and properties can be matched to function-relevant properties and properties relevant to the working principle. Additional design elements and manufacturing-induced properties of different technologies within the continuous flow production can also be considered. Comprehensive analysis of the entire continuous flow production, like the initial analysis of linear flow splitting, reveals possible manufacturing-induced design elements and properties. Systematic utilization of the provided manufacturing-induced properties of these additional technologies enables the realization of bifurcated sheet metal structures with a high degree of integration based on linear flow splitting. With the help of roll forming, welding, or bend splitting (Chap. 3), additional functions can be realized by systematically utilizing the additionally provided manufacturing-induced properties of the aggregated manufacturing process chain. Combined with linear flow splitting, these additional manufacturing technologies allow the realization of closed chambers with thin walls, for example. Adding further functionality to linear flow split linear guides can expand possible applications of the cost-efficient continuous flow production process chain. Additional markets can be reached, based on the linear flow splitting technology.

Mapping manufacturing-induced properties of manufacturing-induced design elements to function-relevant properties of different functions and working principles during the concretization of the principle technology-pushed product idea leads to the identification of potential manufacturing-induced design elements. Specific process-integrated design guidelines help the designer to identify suitable design elements and solutions that can be manufactured in continuous flow production and used to realize the desired product function (Chap. 4). Linking the aggregated manufacturing-induced properties of closed chambers to relevant function-relevant properties of a clamping function, for example, verifies the design element's potential for realizing the additional product function. These patterns can then be documented in a structured way in the form of a matrix, as shown in Fig. 8.6.

Part of the concretization process is analysis of the influence that determination of a specific manufacturing-induced design element has on product function and use processes to determine a specific application. With the help of mathematical

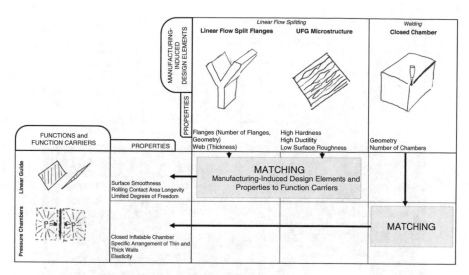

Fig. 8.6 Adding functions utilizing the continuous flow production

equivalent models and experimental validation, the resulting deformation of the closed chamber can be determined (Lommatzsch et al. 2011b). The deformation directly affects the possible braking force of the clamping function and, consequently, the possible product application.

By linking manufacturing-induced properties to properties relevant to the working principle, fundamental (not project specific) guidelines and rules for function integration in a continuous flow production can be derived (Sects. 4.2 and 5.1), allowing the recurring application of fundamental insights in the following projects. By matching the manufacturing-induced properties of a manufacturing-induced design element to the properties of different working elements, design elements capable of realizing more than one working principle and function can be identified. In Fig. 8.6, this can be seen as a matching across multiple rows in the property matrix. With the help of several realized working principles, more functions can be realized, leading to a higher degree of function integration. A clamping function can be realized utilizing thin walls of closed welded chambers, for example. By putting the thin walls under pressure, a thin sheet metal part attached to a sled moving between the closed chambers can be clamped. Matching across multiple columns can also lead to a higher degree of function integration by reducing the number of realized design elements. Identifying different manufacturing-induced design elements that are capable of realizing the same working principle reveals the potential for reducing the number of design elements. Not all possible design elements have to be realized to fulfil the product function.

Figure 8.7 shows the realization of the concretized technology-pushed product idea: a multifunctional linear guide with an integrated clamping function, realized by systematically utilizing the provided manufacturing-induced design elements and manufacturing-induced properties of the continuous flow production to realize

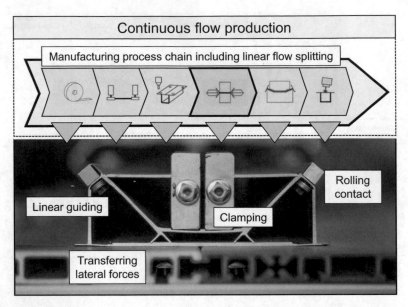

Fig. 8.7 Manufacturing process chain realizing a multifunctional linear guide according to Wagner et al. (2014)

various product functions (Lommatzsch et al. 2011a; Lommatzsch et al. 2011b). In the realization shown in Fig. 8.7, the rolling contact area and the linear guiding function are realized with the help of linear flow split flanges. Integrated bifurcations increase the stiffness of the overall system, transferring lateral forces. Additionally, utilizing these bifurcations, the clamping function is realized with the help of further manufacturing technologies within the continuous flow production. With the help of bending and welding processes, the chambers are formed and sealed, providing a working space for applying pressurized air.

Focusing on the systematic utilization of manufacturing-induced design elements of the entire continuous flow production process chain, for example, integral bifurcations and closed chambers, increases the degree of integration. Integrating auxiliary product functions, like the clamping function, increases the number of integrated functions while maintaining a high degree of process integration and integral construction. The number of possible applications and product quality increase. At the same time, continuous flow production's benefits to cost-efficient manufacturing of large quantities of bifurcated sheet metal profiles at low costs is maintained, making technology-pushed product innovations possible. In addition to applications focused on the linear guiding function, realizing linear flow split linear motion systems with integrated actuator components facilitate the utilization of linear flow split profiles for completely different applications.

Enlarging the Range of Possible Linear Flow Split Product Applications: Innovative Multifunctional Linear Motion Systems. Focusing on manufacturing-induced design elements and properties of a technology supports

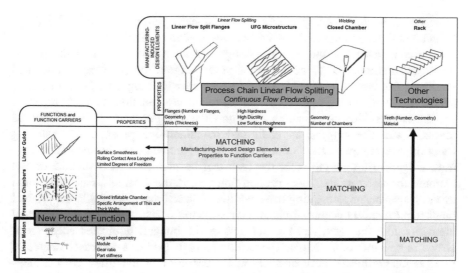

Fig. 8.8 Combining linear flow splitting's process chain with different manufacturing technologies

the realization of products with a high degree of function integration and manufacturing process integration, as they can be realized within the continuous flow production. When manufacturing more complex products for demanding applications, multiple manufacturing technologies, that are not necessarily part of a continuous flow production, are usually required to realize all desired product functions (Wagner et al. 2015). These multiple manufacturing technologies provide the necessary design elements and manufacturing-induced properties, such as a specific geometry or surface finish, necessary to realize the product function in the determined application. In addition to integrating auxiliary functions, the realization of additional main functions can vastly increase the range of possible applications based on linear flow split linear guides. Linear guidance of objects and their motion generation are important in many related applications. The need to provide the product's linear motion requires a new main function, the drive function, in the developed product, which is represented by an additional row in the property matrix (Fig. 8.8). Linear guides with additional sensor and actuator components to realize the drive function are called linear motion systems. These actuator components can be racks of a rack and pinion drive or linear motor components, for example. Linear motion systems have a broader range of possible applications than linear guides. The webs of the linear flow split linear guides can be used to integrate actuator components, manufactured with the help of different manufacturing technologies, into the system. This enables the realization of a linear motion system, based on the linear flow splitting technology.

Actuator components, like the rack of a rack and pinion drive, are usually realized with the help of manufacturing technologies that are not part of the continuous flow production. These additional manufacturing technologies are represented in the property matrix by additional columns with manufacturing-induced design elements and manufacturing-induced properties (Fig. 8.8).

These additional manufacturing design elements, like a rack with specific geometric and material properties, enlarge the range of possible applications and functions that can be realized based on linear flow splitting. Matching the newly acquired manufacturing-induced properties to the new function-relevant properties assists the identification of suitable design elements: For example, a rack can be realized with the help of an additional manufacturing process. By matching function-relevant properties of a linear motion function to the manufacturing-induced properties of the rack, the rack can be identified as a suitable design element for realizing the linear motion function, in the form of a rack and pinion drive. However, implementing these additional functional elements in the product leads to a decreased degree of integral construction structure because of additional elements, like the rack, that are not part of the integral bifurcated sheet metal structure (Wagner et al. 2014). The additional parts lead to a decreased degree of process integration because of additional necessary manufacturing process steps that are not part of the continuous flow production: the manufacturing of the rack and the subsequent joining process that is necessary to integrate the rack into the web of the sheet metal structure. This decrease in the degree of integration can lead to increased manufacturing costs. Comparing different columns (representing the different manufacturing-induced design elements) of the same row (representing a specific product function with function-relevant properties) in the property matrix helps identify alternative solutions for realizing a desired product function. Generally, integrated solutions that comprehensively utilize the benefits of a continuous flow production should be favored. Adding actuator components increases the functionality of linear motion systems with linear flow split linear guides. This makes the integration of actuator components necessary. The cost-efficiency of the continuous flow production process chain can be further improved by implementing an integral joining process into the continuous flow production—especially when occurring mechanisms are utilized, requiring a technology-pushed process innovation.

8.2 Technology-Pushed Process Innovation

In addition to product innovation, process innovation is made possible by systematically utilizing the provided manufacturing-induced properties provided by a specific technology. The product innovation approach is based on matching manufacturing-induced properties to function-relevant properties. Utilizing occurring mechanisms of a technology (or process chain), characterized with the help of manufacturing-induced properties, can lead to a technology-pushed process innovation. A procedure for technology-pushed process innovation, with a manufacturing technology as the starting point, is illustrated in Fig. 8.9, similar to the one described in Sect. 8.1.

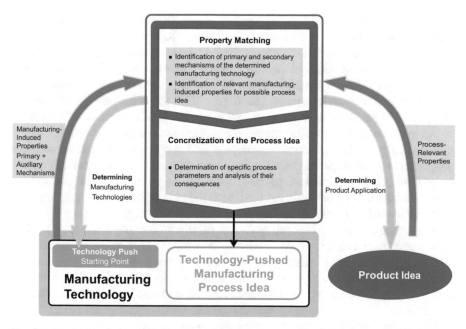

Fig. 8.9 Technology-pushed process innovation based on manufacturing-induced properties

Similar to the technology-pushed product innovation process, a manufacturing technology with potential for technology-pushed innovation is the starting point for the technology-pushed process innovation procedure. After analyzing the chosen manufacturing technology (Step 2a in Sect. 8.1.1), the process innovation procedure requires the identification of processes that fit the identified manufacturing-induced design elements and properties. Manufacturing-induced properties that can be utilized to assess the potential for process innovation are matched to properties of process-relevant elements to acquire a principle process innovation idea. During a concretization phase, process parameters are identified that help adjust the new manufacturing process to specific applications. The resulting technology-pushed manufacturing process idea can then be the starting point for a subsequent manufacturing-integrated product and process development process (Chap. 9).

8.2.1   Process Innovation Driven by Manufacturing-Induced Properties

The manufacturing technology potential for process innovation can be evaluated by analyzing its mechanisms and provided manufacturing-induced properties. Especially beneficial to process innovations is a systematic utilization of already occurring (secondary) mechanisms in manufacturing processes (Wagner et al. 2015). So-called "grind hardening" processes, for example, utilize the dissipated heat of grinding processes to additionally transform the surface structure of the machined element, increasing the hardness of its surface (Kolkwitz et al. 2011). These secondary mechanisms induce properties that have the potential for additional product and process innovation. The process chain of linear flow splitting has shown its potential for product innovation through the systematic focus on the possibility of function integration (Sect. 8.1). Still, there is available and unused technological potential in linear flow splitting in terms of process development.

As in the product innovation procedure, process innovations can be achieved by systematically matching manufacturing-induced properties to function-relevant properties of related applications. During manufacturing processes, not only mechanisms that directly contribute to the manufacturing technology goal occur. Secondary mechanisms like dissipated heat or, in the case of linear flow splitting, high hydrostatic compressive stresses can be observed and used for process innovation. Similar to the distinction between main and auxiliary functions (Pahl et al. 2007a), these mechanisms indirectly contribute to the original goal of the manufacturing technology. To improve existing manufacturing technologies, systematic analysis of already occurring (auxiliary) phenomena has huge potential for process innovation since no additional effort is necessary to realize these manufacturing-induced properties.

The manufacturing-induced properties linked to these mechanisms can be matched to the properties of process-relevant elements of various processes utilized in different applications. Exemplary, Fig. 8.10 shows this matching of manufacturing-induced properties of the continuous flow production's design elements, comprising linear flow splitting and high speed cutting, to the properties of process-relevant elements of joining processes and plasma nitriding processes.

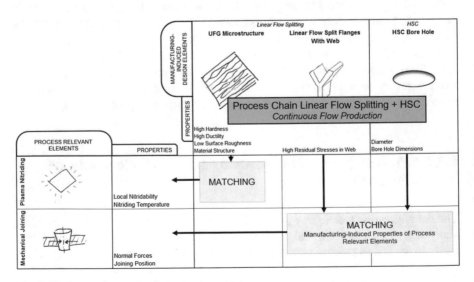

Fig. 8.10 Process innovation by property matching

Fig. 8.11 (a) Hardness of nitrided and untreated flange and web over distance to surface, (b) layer thickness at flange and web for different nitriding temperatures and times (Bödecker 2013)

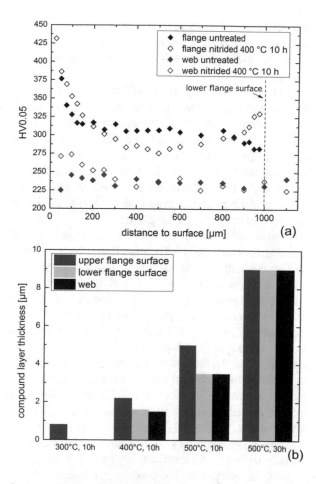

Plasma nitriding is a surface treatment often used to increase the hardness and wear resistance of surfaces. When ferritic steels are subject to a hot nitrogen atmosphere, iron nitrides are formed within the material (Schuster et al. 2012). Depending on the process conditions and the microstructure of the ferritic steel, the nitrides can grow to form a closed compound layer that increases the surface hardness of the steel. Figure 8.11a displays the influence of a plasma nitriding treatment, at 400 °C for 10 h, on the hardness of the flange and the web of an HC 480 LA linear flow split profile. For further information about the treatment, see Bödecker (2013). The treatment causes an increase in hardness of all surfaces and a slight decrease in hardness in the flange volume.

However, the nitridability of the UFG microstructure at the upper flange surface differs from the nitridability of the lower flange surface and web. Figure 8.11b shows that at 300 °C nitriding temperature for 10 h only the upper flange surface of the HC 480 LA linear flow split profile exhibits a closed compound layer. Even at 400 °C and 500 °C, the nitride layer on the upper flange surface is still significantly thicker than on the lower flange surface and the web. However, after 30 h at 500 °C, the differences vanish and the layer thickness is constant throughout the profile. By adjusting the nitriding temperature, it is possible to either partially nitride the profile or create a completely nitrided profile with constant or varying compound layer thickness.

Matching the provided manufacturing-induced properties of linear flow splitting to the process-relevant properties of plasma nitriding shows how the whole manufacturing process can be improved with the help of already occurring mechanisms. This technology-pushed process innovation can then be further concretized and implemented in various applications or act as the starting point for a subsequent product innovation (Sect. 8.3). For example, locally hardened linear flow split profiles in linear guides utilize the increased nitridability of the flange, which allows local hardening of the high-strength parts of the linear flow split profiles, by adjusting the nitriding temperature.

Sources of possible applications for secondary mechanisms in terms of process innovations could be existing manufacturing technologies or insights gained during the concretizations of technology-pushed product ideas. Concretizing a technology-pushed product idea, as described in Sect. 8.1, creates insights into suitable manufacturing-integrated product design and appropriate product applications. Additionally, the developed product can reveal new necessities in manufacturing processes. These can arise due to additional manufacturing-induced properties that could not originally be provided by the chosen manufacturing technology. For example, the development of a multifunctional linear motion system makes the application of additional manufacturing technologies outside the continuous flow production necessary (Sect. 8.1.2). Increased functionality of the product has the cost of decreased degree of integration and consequently increased manufacturing efforts due to additional, necessary assembly steps. A technology-pushed product idea with unsatisfactory realization possibilities sparks the desire for a further development of the manufacturing process through process-integrated joining processes, for example. To develop an integrated joining process, the resulting manufacturing-induced properties of a process chain, which can differ from manufacturing-induced properties provided by an individual manufacturing process of the chain, are matched to the necessary functional properties of a joined functional element in, for example, a pressed connection (Fig. 8.10). Process innovation potential can be revealed: The high hydrostatic stresses that occur during the linear flow splitting process which realizes linear flow split flanges can also be used to realize joining operations (Wagner et al. 2015). This is explained in detail in the following section.

8.2.2 Mechanical Joining by Linear Flow Splitting

Joining by forming is classified by Mori et al. (2013) and Groche and Türk (2011) in mechanical and metallurgical joining. Mechanical joints base on remaining stresses which occur when at least one joining partner is deformed elasto-plastically and the

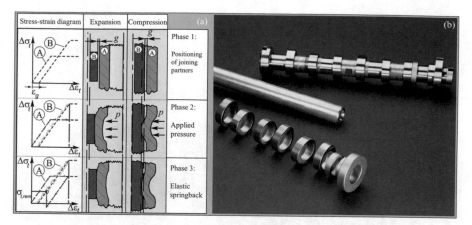

Fig. 8.12 (a) Mechanisms in force locked joints (Marré et al. 2008), (b) manufactured camshaft by hydroforming (Dietz and Grünendick 2004)

other is hampered in recovering after the release of external forces. Main mechanisms of mechanical joints are discussed in Marré et al. (2008), taking into account expansion and compression of tubular joining partners (Fig. 8.12).

Two joining partners A and B are positioned to each other (phase 1) so that an initial gap g results. In the next step, the tube A is deformed plastically by a radial acting pressure p. With the occurring contact between the joining partners, their deformation proceeds until the process is finished (phase 2). Here, tube A is elastoplastically deformed, whereas tube B is purely elastically deformed. After the elimination of the pressure p, the springback of tube B is prevented by the other joining partner and leads to a tangential stress $\sigma_{t,rem}$ (phase 3). Consequently, the main condition for a mechanical joint leads to remaining stresses in the boundary layer of the joining partners.

Various experimental studies are examined regarding the material flow in the process of linear flow splitting, which depends, for example, on process velocity, lubrication, and strip tension (Ludwig et al. 2013). Due to the inhomogeneous material flow in the cross section of the profile, residual stresses result with a characteristic distribution. This characteristic stress distribution can be used for a simultaneous joining of functional elements, e.g., bolts, racks, or RFID transponder. For designing this novel joining process, it is important to know the mechanical joining mechanisms in the process of linear flow splitting in detail. In Groche and Monnerjahn (2017), these main mechanisms are analyzed and presented in the following section.

Stress Distribution in Split Profiles The analysis of numerical results concerning the stress distribution in the web of the split profile during the process (1) and after the process (2) show the correlation between occurring stresses and distance to the splitting area (Groche and Monnerjahn 2017). The node information of these two points in time at an equal section of the profile is shown in Fig. 8.13. During the linear flow splitting process, very high compressive stresses occur near the splitting

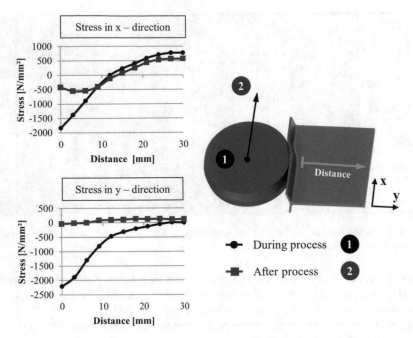

Fig. 8.13 Stress distribution in the web of split profiles (Groche and Monnerjahn 2017)

area, which decrease in direction to the middle of the profile. These compressive stresses in x- and y-direction are part of the hydrostatic compressive stress condition of the linear flow splitting process (Groche et al. 2007).

After a distance of approximately 12 mm to the splitting area, the results also show a change of compressive to tensile stresses in longitudinal direction x. A possible explanation for this phenomenon is the displaced material in longitudinal direction near the splitting zone, whereas the material in the middle of the profile has to be expanded. However, a recovery of stresses occurs after the process, where the stresses in longitudinal direction x maintain maximums of approx. $\pm$ 500 N/mm^2. In contrast, the residual stresses in transverse direction y show no significant values.

The numerical results of the stress distribution in the web of split profiles are verified by measurements with the incremental hole-drilling strain gauge method. Strain gauge rosettes of the type *EA-06-031RE-120* are placed at characteristic areas in the web of the split profiles. A hole with a diameter of 1 mm is drilled into the center of the rosettes and the occurring strains are measured with the aperture *Sint Technology MTS3000* in thickness direction of the sheet metal. The converted stresses in longitudinal direction x and transverse direction y are shown in Fig. 8.14. The measurements confirm the previously presented numerical results. High tensile stresses are measured in the middle of the web (30 mm, marked as diamonds) in longitudinal direction x, whereas high compressive stresses occur near the splitting area (0 mm, marked as triangles). The values for the tensile stresses of approx.

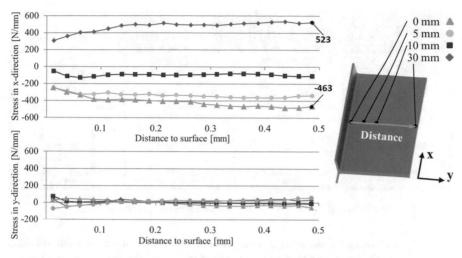

Fig. 8.14 Stress distribution of hole-drilling strain gauge method

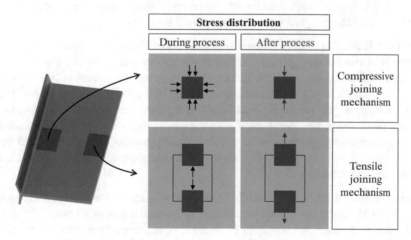

Fig. 8.15 Joining mechanisms in the process of linear flow splitting (Groche and Monnerjahn 2017)

520 N and compressive stresses of approx. −460 N are in good agreement with the numerical values.

Joining Principles. The inhomogeneous stress distribution in the web of the profile offers new possibilities to mechanically join one or more additional functional elements during the linear flow splitting process (Fig. 8.15). In the case of a single joined element, the compressive stresses near the splitting area can be used for joining. During the process, the single element is compressed in longitudinal and transverse direction. Due to the resilience of the sheet material, the element is joined after the process with a force-locking mainly in longitudinal direction. In the case of two coupled elements, the tensile stresses in the middle of the profile can

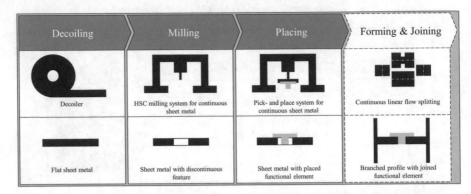

Fig. 8.16 Process chain of linear flow splitting for simultaneous joining

also be used for joining. During the process, the material flows partially into longitudinal direction, which results in an expansion of the middle of the profile and coupled parts. After the resilience of the sheet metal material, high tensile stresses still exist in the middle of the profile, which result in a force locked connection of the coupled parts in longitudinal direction.

Process Chain. A possible process chain for a simultaneous joining of functional elements in the forming process of linear low splitting is shown in Fig. 8.16. In a continuous flat sheet metal, discontinuous features can be milled in-line, in which the functional elements are placed automatically by a pick and place system. In this process step, a gap between sheet metal and functional element still exists. In further processing, the linear flow split profile is formed with the characteristic branched geometry. Simultaneous to this forming process, the gap between functional element and sheet metal is closed. The result is a force locked connection in an integral branched profile.

Verification of the Joining Model. To verify the compressive joining mechanism, cylindrical elements are placed into drilled holes with a gap of approx. 0.05 mm to each other and joined in two stands by linear flow splitting. To determine the joints' strength, the joined cylindrical elements are separated subsequently from the split profile. The separating forces of four different element materials are measured through a piezoelectric sensor and presented in Fig. 8.17. The maximum of the separating force is represented by the elements' material *ZE1100* with approx. 1250 N. The separating force decreases with lower yield strength of the elements' material and it is averaged to approx. 500 N for *DC01*. Joining elements with the same initial sheet metal material *HC800LA* is also possible, which indicates a different joining mechanism in comparison to the conventional joining mechanisms described in Marré et al. (2008). To compare the separating forces with the numerical results, the contact normal stress between cylindrical element and sheet metal is averaged over the circumference. Highest values are also achieved with the elements' material *ZE1100* and decrease with lower yield strengths. However, the percentage reduction of the contact normal stresses is qualitatively conforming to the separating forces.

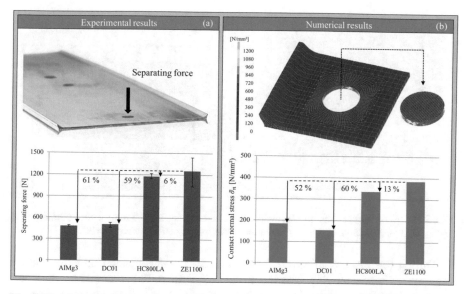

Fig. 8.17 (**a**) Separating forces of experiments, (**b**) contact normal stresses of numerical model (Groche and Monnerjahn 2017)

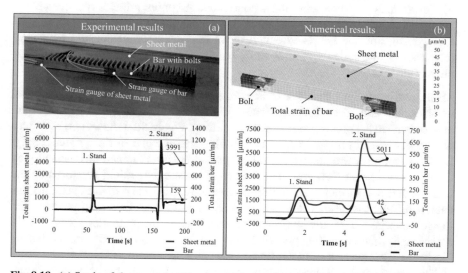

Fig. 8.18 (**a**) Strain of sheet metal and bar in experiment, (**b**) in numerical simulation (Groche and Monnerjahn 2017)

To verify the tensile joining mechanism, a bar with two fixed bolts is placed into two drilled holes with a gap of approx. 0.05 mm. The bar with the two bolts is joined by two splitting stands, while the occurring strain is measured in-line with strain gauges (Fig. 8.18).

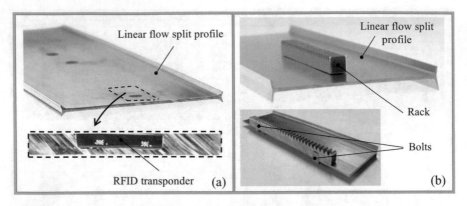

Fig. 8.19 (a) Multifunctional profiles with simultaneous joined RFID transponder (Groche et al. 2015), (b) multifunctional profile with simultaneous joined rack (Monnerjahn 2015)

During the process of linear flow splitting, the sheet metal and the bar are compressed when moving into the tool system. In further processing, the sheet metal and the bar are extended to higher positive values, which are reduced due to the resilience of the materials. With the second stand, the preload of the joint strengthens to higher values. Finally, the bar is mechanically joined and a strain of 159 μm/m is measured after the second stand. A similar effect is observed in the numerical simulations. The sheet metal is also compressed when moving into the tool system, and the bar is extended to 42 μm/m. Experimental and numerical results show qualitatively similar strain curves so that the tensile joining mechanism is also verified.

Multifunctional Split Profiles. In comparison to the mechanisms of conventional mechanical joints, the inhomogeneous stress distribution leads to two different mechanisms in the process of linear flow splitting. Figure 8.19 shows two examples of multifunctional profiles, where the profiles are enriched in their functionality. RFID transponders are placed into milled pocket holes and joined mechanically by the process of linear flow splitting. Important process data might be stored on the transponder, and further manufacturing machines could have direct access to this data at different locations (Groche et al. 2015). In linear flow split profiles, the flange material offers advantages in applications of linear guides (Sect. 8.1.2). Additionally, the drive function of linear motion systems could be provided by mechanically joined racks in the linear flow splitting process.

The simultaneous joining of functional elements in the linear flow splitting process combines joining with a forming operation. In Groche et al. (2016), this approach is defined as *conjoint forming*, which shortens the processing time for higher productivity. On the contrary, this approach extends the continuous manufacturing line of linear flow splitting and roll forming. With an additional assembly line, the functional elements have to be placed on the sheet metal. This offers new possibilities in product design, but also challenges in the synchronization of forming, handling, and joining processes in the continuous flow production line.

8.3 Concurrent Technology-Pushed Product and Process Innovation

While technology-pushed product innovation can initiate a technology-pushed process innovation, the opposite is also possible: Technology-pushed process innovation can be the starting point for product innovation through the utilization of manufacturing-induced properties that are only provided due to process innovation (Fig. 8.20).

The technology-pushed process innovation of joining by linear flow splitting and plasma nitriding (Sect. 8.2) enlarges the range of possible products and applications that can be realized in the linear flow splitting process chain. By integrating mechanical joining into the process chain utilizing occurring mechanisms, new manufacturing-induced design elements and properties become available for product design while the initial functional and economic benefits of the continuous flow production still exist. Thus, completely new designs, applications, and, consequently, technology-pushed product innovations become possible. This results in innovative linear motion systems with integrated RFID chips, for example, and actuator components, like linear motor components, and the rack of a rack and pinion drive (Wagner et al. 2015). The product application is determined (Sect. 8.1) and the manufacturing process concretized accordingly. Product design can be applied to specific applications by systematically influencing necessary assembly

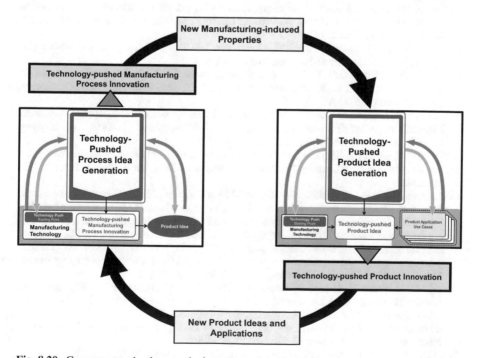

Fig. 8.20 Concurrent technology-pushed process and product innovation

and disassembly forces (by adjusting the fitting point's distance to the flanges) (Wagner et al. 2015). Disassembly forces can be influenced, favoring recycling processes, for example.

Technology-pushed design elements can also be implemented in market-driven development projects. Section 9.2.1 shows the benefits of process-integrated joining of the rack of a rack and pinion drive, leading to cost-efficient manufacturing of the linear motion system of a facade cleaning system, while maintaining the linear flow splitting process chain's specific advantages regarding functionality.

References

Bödecker J (2013) Randschichtmodifikation von integral verzweigten Blechprofilen mit UFG Gradientengefüge. Dissertation, TU Darmstadt

Dietz P, Grünendick T (2004) Umformen verbindet Fügeverfahren für Welle-Nabe-Verbindungen. Institutsmitteilung (29), IMW, TU Clausthal

Groche P, Monnerjahn V (2017) Remote joining by plastic deformation, Journal of Materials Processing and Technology, 2017 (submitted)

Groche P, Türk M (2011) Smart structures assembly through incremental forming. CIRP Ann Manufactur Technol 60(1):21–24

Groche P, Vucic D, Jöckel M (2007) Basics of linear flow splitting. J Mater Process Technol 183 (2–3):249–255

Groche P, Hohmann J, Kessler T, Schreiner J, Krech M, Duschka A, Monnerjahn V, Ćorić M (2015) Industrie 4.0—Forschung am PtU. In: 12. Umformtechnisches Kolloquium Darmstadt, Darmstadt, 10–11 June 2015

Groche P, Wohletz S, Mann A, Krech M, Monnerjahn V (2016) Conjoint forming—technologies for simultaneous forming and joining. Mater Sci Eng 119

Hirsch N, Mauf M, Birkhofer H (2008) Systematic generation of innovative sheet metal profiles using modified TRIZ. In: Roosimölder L (ed) Proceedings of NordDesign 2008 Conference, Tallinn, Estonia, 21–23 August 2008

Karin I, Hößbacher J, Lommatzsch N, Birkhofer H, el Doski C, Hanselka H (2010) Spaltprofilierte Linearführungen auf dem Prüfstand. In: Groche P (ed) Tagungsband 3. Zwischenkolloquium SFB 666, Darmstadt, 29–30 September 2010, p 111–116

Kolkwitz B, Foeckerer T, Heinzel C, Zaeh MF, Brinksmeier E (2011) Experimental and numerical analysis of the surface integrity resulting from outer-diameter grind-hardening. Procedia Eng 19:222–227

Lommatzsch N, Gramlich S, Birkhofer H (2011a) Linear guides of linear flow split components: Development and integration of potential additional functions. In: Proceedings of the 3rd International Conference on Research into Design Engineering, Bangalore, India, p 439–446

Lommatzsch N, Gramlich S, Birkhofer H, Bohn A (2011b) Linear flow-split linear guides: inflating chambers to generate breaking force. In: Culley SJ, Hicks BJ, McAloone TC, Howard TJ, Lindemann U (eds) Proceedings of the 18th International Conference on Engineering Design (ICED 11), Lyngby, Copenhagen, 15–19 August 2011, p 337–346

Ludwig C, Hammen V, Groche P, Kaune V, Müller C (2013) Fertigung qualitätsoptimierter Spaltprofile durch Variation schnell änderbarer Prozessgrößen und deren Einfluss auf die Materialeigenschaften. Mat-wiss u Werkstofftech 44(7):601–611

Marré M, Brosius A, Tekkaya AE (2008) New aspects of joining by compression and expansion of tubular workpieces. Int J Mater Form 1(S1):1295–1298

Monnerjahn V (2015) Fügen per Spaltprofilierung. Umformtechnik: Bleche Rohre Profile (2)

Mori K-i, Bay N, Fratini L, Micari F, Tekkaya AE (2013) Joining by plastic deformation. CIRP Annals—Manufacturing Technology 62(2):673–694

Müller C, Bohn E, Bruder E, Hirsch N, Birkhofer H (2008) Linear flow splitting: profile properties and their implementations in product development. In: Proceedings of the ICTP 2008, 9th International Conference on Technology of Plasticity, Gyeongju, 7–11 September 2008, p 774–779

Pahl G, Beitz W, Feldhusen J, Grote KH (2007a) Engineering design: a systematic approach, 3rd edn. Springer, London

Pahl G, Beitz W, Feldhusen J, Grote KH (2007b) Konstruktionslehre: Grundlagen erfolgreicher Produktentwicklung, 7th edn. Springer, Berlin

Schuster J, Bruder E, Müller C (2012) Plasma nitriding of steels with severely plastic deformed surfaces. J Mater Sci 47(22):7908–7913

Wagner C, Gramlich S, Kloberdanz H (2014) Entwicklung innovativer Produkte durch Verknüpfung von Funktionsintegration und Fertigungsprozessintegration. In: Krause D, Paetzold K, Wartzack S (eds) Proceedings of the 24th Symposium Design for X, Bamberg, 1–2 October 2014, p 361–372

Wagner C, Gramlich S, Monnerjahn V, Groche P, Klobderanz H (2015) Technology pushed process and product innovation: Joining by Linear Flow Splitting. In: Erkoyuncu J (ed) Procedia CIRP, vol 37, CIRPe 2015—Understanding the life cycle implications of manufacturing, p 83–88

Wäldele M, Hisch N, Birkhofer H (2007) Providing properties for the optimization of branched sheet metal products. In: Proceedings of the International Conference on Engineering Design (ICED 07), Paris

Chapter 9
The Result: A New Design Paradigm

M. Roos, S. Abedini, E. Abele, K. Albrecht, R. Anderl, M. Gibbels, S. Gramlich, P. Groche, B. Horn, A. Hoßfeld, S. Köhler, H. Lüthen, I. Mattmann, T. Melz, V. Monnerjahn, C. Müller, M. Neuwirth, J. Niehuesbernd, M. Özel, M. Pfetsch, J. Reising, S. Schäfer, S. Schmidt, E. Turan, S. Ulbrich, C. Wagner, A. Walter, T. Weber Martins, and A. Zimmermann

M. Roos (✉) • S. Gramlich • I. Mattmann • C. Wagner
Institute for Product Development and Machine Elements (pmd), Technische Universität Darmstadt, Darmstadt, Germany
e-mail: roos@pmd.tu-darmstadt.de

S. Abedini • J. Reising • S. Schäfer • A. Zimmermann
Institute of Construction Design and Building Construction (KGBauko), Technische Universität Darmstadt, Darmstadt, Germany

E. Abele • A. Hoßfeld • S. Schmidt • E. Turan
Institute for Production Management, Technology and Machine Tools (PTW), Technische Universität Darmstadt, Darmstadt, Germany

K. Albrecht • R. Anderl • T. Weber Martins
Department of Computer Integrated Design (DiK), Technische Universität Darmstadt, Darmstadt, Germany

M. Gibbels
Research group System Reliability, Adaptive Structures, and Machine Acoustics (SAM), Technische Universität Darmstadt, Darmstadt, Germany

P. Groche • S. Köhler • V. Monnerjahn • M. Neuwirth • M. Özel
Institute for Production Engineering and Forming Machines (PtU), Technische Universität Darmstadt, Darmstadt, Germany

B. Horn • S. Ulbrich • A. Walter
Research Group Nonlinear Optimization (NOpt), Technische Universität Darmstadt, Darmstadt, Germany

H. Lüthen • M. Pfetsch
Research Group Discrete Optimization (DOpt), Technische Universität Darmstadt, Darmstadt, Germany

T. Melz
Research group System Reliability, Adaptive Structures, and Machine Acoustics (SAM), Technische Universität Darmstadt, Darmstadt, Germany

Fraunhofer Institute for Structural Durability and System Reliability LBF (LBF), Fraunhofer-Gesellschaft zur Förderung der angewandten Forschung e.V., Munich, Germany

C. Müller • J. Niehuesbernd
Physical Metallurgy (PhM), Technische Universität Darmstadt, Darmstadt, Germany

© Springer International Publishing AG 2017
P. Groche et al. (eds.), *Manufacturing Integrated Design*,
DOI 10.1007/978-3-319-52377-4_9

301

9.1 Integrated Algorithm-Based Product and Process Development

One of the key challenges faced by engineers is finding, concretizing, and optimizing solutions for a specific technical problem in the context of requirements and constraints (Pahl et al. 2007). Depending on the technical problem's nature, specifically designed products and processes can be its solution with product and processes depending on each other. Although products are usually modeled within the context of their function, consideration of the product's life cycle processes is also essential for design. Processes of the product's life cycle concern realization of the product (e.g., manufacturing processes), processes that are realized with the help of the product itself (e.g., use processes), and processes at the end of the product's life cycle (recycling or disposal). Yet, not just product requirements have to be considered during product development, as requirements regarding product life cycle processes need to be taken into account, too. Provision for manufacturing process requirements plays an important role in realizing the product's manufacturability, quality, costs, and availability (Chap. 3). Further life cycle demands, such as reliability, durability, robustness, and safety, result in additional product and life cycle process requirements. Consequently, the engineer's task of finding optimal product and process solutions to solve a technical problem or to fulfill a customer need is characterized by high complexity, which has to be handled appropriately (Chaps. 5 and 6).

There is also huge potential for innovative design that results from life cycle process opportunities, such as manufacturing technologies which can be the basis for realizing unconventional solutions. This goes beyond applying rules to ensure manufacturability of a specific design. Instead, comprehensive exploitation of all geometric and material design opportunities, in the form of manufacturing-induced properties, offers possibilities, for example, for lightweight design, increased reliability, and functional benefits (Chap. 4). Actual product and process ideas have to be generated with respect to these manufacturing-induced properties to realize market-pulled and technology-pushed product innovation (Chaps. 7 and 8). The *Integrated Algorithm-Based Product and Process Development* approach merges insights from multiple disciplines into a new framework for developing product and process solutions, with a focus on optimality and innovation, by realizing the potential of specific manufacturing technologies, such as linear flow splitting (Sect. 2.5).

9.1.1 The Integrated Approach

The core of the Integrated Algorithm-Based Product and Process Development approach is the strong link between product development and life cycle processes, focusing manufacturing and use processes, in a consistent development procedure.

Special emphasis is on the comprehensive integration of process insights and process requirements into development, by using manufacturing-induced properties. This already takes place during the early development phases, beginning with design task clarification. To handle complexity, solution finding is supported by algorithm-based optimization methods. The basic framework of the approach is shown in Fig. 9.1.

In contrast to conventional product development approaches (Sect. 1.2), the integrated approach consistently focuses on equally determining and designing products and their processes (manufacturing and use) (Fig. 9.1 bold black arrows). This requires the concurrent determination of product properties as well as process parameters. The key is to comprehensively, consistently, early, and continuously integrate the impact of all design decisions on the product and technical processes into the development process (Fig. 9.1 blue box). Based on early anticipation (Fig. 9.1 orange arrows) of manufacturing and use processes (Fig. 9.1 green and violet arrows), manufacturing potential can be realized in product and process design, stakeholder requirements can be fulfilled, and customer value can be increased by integrating anticipated information into the development process. Manufacturing and use process requirements in addition to product requirements lead to an increased complexity of the design task. This complexity can be efficiently handled by applying algorithm-based optimization methods to find solutions. The elements within the framework of the Integrated Algorithm-Based Product and Process Development approach are as follows:

- *Anticipation of technical processes and product behavior* (Chaps. 3 and 4)
  Model-based and experimental process analyses are performed to gain information from manufacturing and use processes for product design purposes. The analyses of manufacturing processes result in the identification of manufacturing-induced properties. The results from use process analyses provide knowledge about the impact of specific manufacturing-induced properties on the product in context of its use processes. Anticipated process information has to be integrated into different steps of the development process, with emphasis on the early development phases.
- *Design task clarification and requirement acquisition* (Chap. 5)
  Product functions, use processes, and product applications are identified and determined. With a focus on integrating product and associated manufacturing and use processes within the development approach, requirements of the product, as well as of manufacturing and use processes are acquired, based on anticipated knowledge on manufacturing and use processes. The acquired product and process requirements are systematically prepared and structured for the subsequent algorithm-based solution finding.
- *Transformation into and formalization of an integrated product and process design task* (Chap. 5)
  Product and (manufacturing and use) process requirements are formally processed as desired properties. Appropriate manufacturing technologies are identified and determined. Manufacturing-initiated solutions, based on

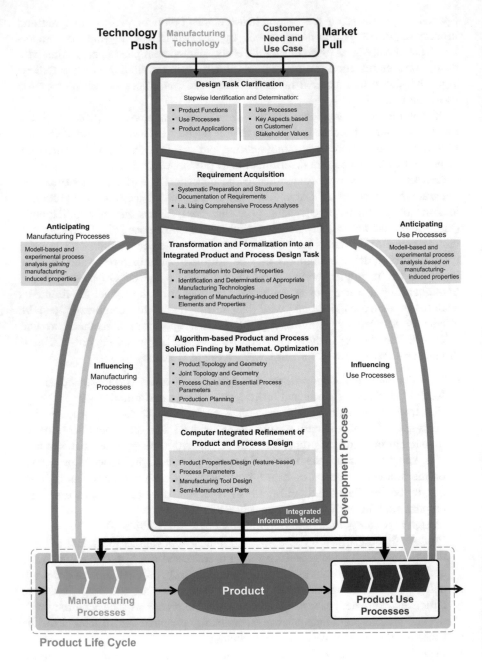

Fig. 9.1 Integrated Algorithm-Based Product and Process Development

anticipated and integrated manufacturing-induced properties, are integrated into the design task to extend the solution space by comprehensively considering geometric and material properties of the chosen manufacturing technology. The prepared information is co-located as a formalized integrated product and process design task, documented as objective functions, design constraints, and design variables for mathematical optimization purposes.

- *Algorithm-based product and process solution finding using mathematical optimization* (Chap. 5)
  Context and purpose-specific algorithm-based mathematical optimization methods are applied to find the optimal solution for problems, presupposing the equal consideration of products and processes. These solutions not only address product topology and geometry, but equally manufacturing processes, e.g., in the form of optimal control parameters.
- *Computer integrated refinement of product and process design* (Chap. 6)
  Optimized product and process solutions are extended using additional product properties and process parameters that originate from anticipated life cycle processes by applying computer integrated models, methods, and tools.
- *Information model* (Chap. 6)
  The information model is used for storing and transferring data between the elements of the integrated approach.
- *Technology push and market pull* (Chaps. 7 and 8)
  The development process can be initiated by a manufacturing technology-pushed or a market-pulled product idea, where both are based on assumptions within a similar procedure, using anticipated process information.

9.1.2   Impacts of the Integrated Approach

Established development approaches provide different ways to consider manufacturing aspects during the design process. Most approaches emphasize embodiment and detail design for ensuring the product's manufacturability (Sect. 1.2). Only approaches that provide an integrated perspective on product development and manufacturing allow the realization of the technology's full potential for product and process design. Approaches such as IPPD integrate product development and manufacturing aspects to a certain extent by representing the product in the context of its life cycle. The novel integrated algorithm-based product and process development approach builds on this interaction between the product and its life cycle processes. It provides a procedure characterized by a concurrent development of product and process solutions. Additionally, the approach includes domain-specific methods and procedures that support the development process. The increased interaction between the product and its life cycle processes enables a multi-domain consideration of the entire technical system, resulting in a new design paradigm. The paradigm comprises new definitions, models, and methods (Chaps. 3–8). Manufacturing no longer serves only geometric and material realization of the product: It significantly determines solution finding in terms of product and process solutions. The paradigm change concerns:

- Increased *links* between different disciplines. Development is no longer just the domain of the product designer but is part of *integrated product and process development*. The basis is a systematic procedure for consistent and integrated process consideration and development.
- *Bidirectional* information flow between development and product life cycle, based on anticipation, determination, and influencing of life cycle processes. The role of manufacturing is not only important in realizing the results of development. There are insights from manufacturing that are required during development to ensure manufacturability and feasibility of processes, and to realize the potential regarding the product function.
- A *product and life cycle process integrated perspective* on development and especially idea generation, resulting in realized product and process innovation.

The paradigm change, associated with the Integrated Algorithm-Based Product and Process Development, is characterized by fundamental changes in the interaction of elements, steps, and procedures within the overall development process, compared to development approaches that mainly focus on the product. These characteristics are caused by the concurrent consideration and integration of life cycle processes (manufacturing and use) during development, which leads to deeper coherence between product and process development. The characteristics are as follows:

- Consistent and integrated product and process consideration, realized by:
    - *Equal determination and concretization of product properties, processes, and process parameters:* Each product cannot be developed and realized without determining manufacturing and use processes. This is even more important for a technology-pushed procedure. Within the three steps design task clarification, requirement acquisition, and subsequent transformation and formalization of the design task, product as well as manufacturing and use process requirements are analyzed to derive and document desired properties. Additional manufacturing technologies are determined during formalization of the design task. Manufacturing-initiated solutions, based on anticipated manufacturing-induced properties, are integrated into the design task. The refinement step concretizes and completes product and process design to ensure feasibility of the product in the context of its processes (manufacturing and use).
    - *Systematic processing and return of product and process information (analysis):* The overall approach comprises anticipation of life cycle process information (manufacturing and use) from various discipline-specific perspectives, such as materials science and structural durability. This information is processed in the form of generally valid statements, such as process integrated design guidelines that can be returned for integration into the

development process. This allows the consideration of process restrictions and manufacturing potential for product and process design.

- *Comprehensive application and integration of manufacturing-induced properties (synthesis):* Insights into manufacturing processes are processed consistently as manufacturing-induced properties. Insights into use processes are processed by analyzing these processes with an emphasis on the potential impact of manufacturing-induced properties. The manufacturing-induced properties are integrated into the development process according to the level of concretization, with a focus on the early phases of solution finding.
- *Continuous and early anticipation and integration of impacts from design decisions on products and processes:* Based on the highly formalized anticipation and integration steps, process information can be applied at almost every stage of the development process, particularly during early phases. Product and process solutions are characterized by life cycle characteristics in the form of manufacturing-initiated solution elements. This allows parallel development and comparison of promising solution alternatives on the basis of different manufacturing technologies, for example, applied within continuous flow production.

- *Ensuring the optimality of product and process solutions:* By formalizing the design task for mathematical optimization algorithms, an automated solution finding procedure is applied to find the best product and process solution. Optimization algorithms are adjusted to the purpose of the product and process design task. Subsequent refinement is also adjusted for consistent formalization to allow consistent concretization of the product and process solution.
- *Efficient generation of solutions and reduction of time-consuming iterations:* Iterations are avoided by early inclusion of anticipated opportunities and restriction of the chosen manufacturing technology in the design task. Formalization of the design task also includes selective data preparation for mathematical optimization. With a focus on the purpose of the optimization, this confinement to relevant data leads to a more efficient optimization procedure. Additional information, stored and provided by the information model, is integrated into design during subsequent refinement steps.
- *Early achievement of high product and process maturity:* High product and process maturity is achieved by comprehensively fulfilling central product and life cycle process requirements. Established development approaches do not ensure comprehensive fulfillment of all relevant life cycle process requirements due to the late determination of applicable manufacturing technologies during development process. The early anticipation and consideration of life cycle process requirements (manufacturing and use) in the development process results in higher product and process maturity during early phases. Formalization of the design task for subsequent optimization ensures comprehensive consideration and fulfillment of all anticipated product and life cycle process requirements.

- *Consistent application of algorithms occurs throughout the entire development process by ensuring compatibility and implementation of efficient methods and tools, with a focus on integrated product and process consideration:* The information model allows smooth and consistent deployment and transfer of information between different perspectives, models, and tools, used by experts from different disciplines within the overall approach. A uniform database, providing information from different disciplines, is a central characteristic of the integrated approach.
- *Systematic development and concretization of manufacturing-initiated ideas, solutions, and process extensions in the context of a combined technology push and market pull procedure:* Market-pulled and technology-pushed product and process ideas are initiated by using anticipated process information. Even though market pull and technology push approaches use different initial information to start the development process, both use similar procedures. The further steps of the development approach support the successful realization of these ideas, in the form of product and process solutions or even process extensions.

To realize the full potential of the Integrated Algorithm-Based Product and Process Development approach, as described by the characteristics mentioned, it can be beneficially applied in combination with innovative manufacturing technologies, such as linear flow splitting in the context of continuous flow production. This application reveals various synergies between the product and its life cycle processes, based on the variety of manufacturing-induced properties provided by continuous flow production.

The result of applying the Integrated Algorithm-Based Product and Process Development approach are optimal product and process solutions that fulfill the acquired product and process requirements. Applying manufacturing-induced properties reveals further potential for product and process design, for example, lightweight design, durability, robustness, safety, and functionality (Sect. 9.2). Use processes are attuned to the product in its function context, which leads to a distinct increase in customer value. The combined consideration of manufacturing and use processes provides advantages, like, for example, improved manufacturability, increased process reliability, shortened planning and processing times, and increased quality. Furthermore, this leads to a shortened time-to-market when combined with algorithm-based development. The solutions developed are characterized by a high degree of innovation, caused by the realization of manufacturing potential and peculiarities. The combination of a technology-pushed and market-pulled procedure contributes to the realization of additional product and process innovation.

The Integrated Algorithm-Based Product and Process Development approach has been applied in real, virtual, and prototypical development projects. The following case studies provide a detailed view of the approach and its impact on development results.

9.2   Case Studies

9.2.1   Facade Cleaning System

The integrated approach shown in Sect. 9.1 is fundamental for the new design paradigm. Not only the product and its function is considered in the development process, but also the product's life cycle processes (manufacturing and use) are taken into account. Focusing on manufacturing, the following section illustrates how restrictions of a technology are considered while at the same time its innovation potential is comprehensively realized for finding product solutions as well as applicable manufacturing process chains. The exemplary development of a *facade cleaning system* goes through all individual steps of the novel development approach to demonstrate the benefits of an integrated perspective on products and processes.

Design Task Clarification. Due to the high risk human cleaning staff is exposed to when cleaning facades, more and more automated cleaning systems are applied. Facade cleaning systems realize cleaning processes especially on high buildings and difficult to access facades. The analysis of existing facade cleaning systems reveals that their functionality highly depends on a few function carriers (Fig. 9.2):

*   *Cleaning elements*: Cleaning elements realize the actual cleaning processes including fluid transport and supply
*   *Guide*: The guide performs a guided motion of the cleaning elements on the facade
*   *Drive*: The drive realizes the movement of the system
*   *Energy supply*: The energy supply provides energy necessary for operating the drive and auxiliary systems

The main challenge in developing facade cleaning systems is to find innovative technical solutions for these function carriers. There are already various existing solutions to carry out the cleaning process by using appropriate cleaning elements,

Fig. 9.2 Idea of a multifunctional facade integrated linear motion system

for example, by the Serbot AG (2016) or iku®windows (2016). Still, there is huge potential for innovative solutions for the remaining function carriers guide, drive, and energy supply, which can be combined in a *facade integrated linear motion system*. The realization of such a system highly depends on the chosen manufacturing technology. Linear flow splitting in the context of a continuous flow production offers huge potential with regard to a high degree of function integration and manufacturing process integration (Chap. 8). Particular benefit can be gained by applying the Integrated Algorithm-Based Product and Process Development to find an optimal solution within the various requirements and by utilizing hidden manufacturing potential.

The application of the technology push approach in Chap. 8 has already shown that the manufacturing technology linear flow splitting is ideally suitable for the production of multifunctional linear systems with a high degree of both function and process integration. Linear guides with high stiffness and an integrated drive function can be used in facade cleaning systems that are seamlessly integrated into building facades. Mechanical joining by linear flow splitting allows a cost-efficient integration of a rack necessary for the realization of a rack and pinion drive (Sect. 8.2) with a large possible travel. Hence, linear flow splitting combined with mechanical joining is the starting point for the technology-pushed development of a multifunctional facade integrated linear motion system.

Requirement Acquisition. Many stakeholders with just as many individual requirements have to be considered in the development of the multifunctional facade integrated linear motion system. The impact of the system on aesthetic aspects of the building facade is one of the key factors of its development. By adjusting the motion system's width to the standard width of vertical rods of a curtain wall (50 mm, e.g., Schüco Germany (2016)), it can be seamlessly integrated into the facade's appearance. Other requirement clusters range from requirements regarding the product function to requirements regarding environmental conditions including low noise, good weathering resistance, easy maintainability, and safety aspects that are particularly important for buildings within the public domain.

In addition to the abovementioned aspects, there are also several product life cycle requirements which play a key role. Mechanical cleaning processes necessitate a specific contact pressure between the cleaning elements and the facade. Manufacturing and assembly processes have to be considered in order to realize the product at low costs, quality, availability, and adaptability. Corresponding requirements concern, for example, the compliance of manufacturing tolerances regarding flange lengths, material thicknesses, or bending angles. Additional requirements arise from the consideration of an easy transport and assembly on site. A selection of clustered requirements with high relevance for the optimization problem is listed in Table 9.1 left.

Transformation and Formalization into an Integrated Product and Process Design. A key challenge of developing the multifunctional facade integrated linear motion system is the comprehensive consideration of all product and process requirements during solution finding. In order to cope with the ensuing complexity, an algorithm-based approach for solution finding is applied. With the help of

Table 9.1 Excerpt from the requirements list of a multifunctional facade integrated linear motion system

Cluster	Product and process requirements	Desired properties (and physical quantities)
Function	Load capacity of 15 kg	Vertical load: 150 N
Safety	Prevent possible human contact to moving parts	Undercut in the linear guide
	Prevent falling down in case of energy shortage	Holding force > weight force
Cleaning/ supply	Provide contact pressure for mechanical cleaning	Lateral force: 50 N on cleaning element
Manufacturing, assembly	High possible batch size, especially for large facades	Profiles with constant cross section
	Linear flow split or linear bend split flanges should not be bifurcated again	Order of bifurcations: ≤ 1
	Low assembly effort on site, favorably integral construction	No joints within the profile's cross sectional structure
	Number of linear bend split flanges should be limited to two per sheet to reduce production effort	Number of linear bend split bifurcations: ≤ 2
Aesthetics	Seamless integration into facade	Width: 50 mm
	Avoid shadowing of underlying windows	Height: <100 mm

mathematic optimization methods, the optimal solution is identified within the solution space that is constrained by the acquired requirements. Requirements in the form of desired properties directly correlate with product properties and can be directly utilized for the mathematical optimization of the linear motion system (Table 9.1 right). The desired properties that are implemented in the mathematic optimization are complemented with physical quantities, such as, for example, external loads, resulting from acquired requirements.

The algorithm-based solution finding for the identified function carriers requires the formalization of the design task in the form of objective functions and constraints to identify optimal solutions within the solution space that is defined by properties (act as design variables) (Chap. 5). The function carrier realizing the guiding function is chosen to demonstrate the formalization and the subsequent optimization and refinement steps. The other function carriers for the drive function and energy supply are indirectly included in the mathematical optimization by considering their respective working surfaces and design spaces in the utilized algorithm.

The facade cleaning system is actuated by a DC-micromotor that satisfies the high requirements regarding the available design space and load capacity. A two-stage gear transmission is used for providing the translational motion of the sled. For safety reasons self-locking is realized with the help of a worm gear as first gear. The second gear transmits the rotational motion to a rack attached to the linear flow split profile. In order to minimize the required design space (and hereby

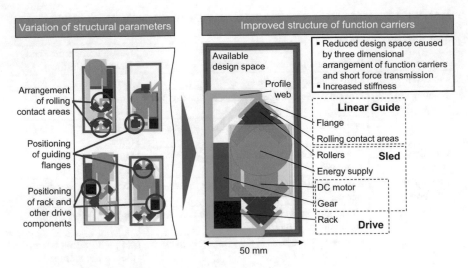

Fig. 9.3 Positioning and arrangement of function carriers

satisfying requirements regarding the product's aesthetic), the arrangement of the mentioned function carriers is determined by systematic variation of their positioning in relation to each other (Fig. 9.3). The resulting design space and working surfaces are considered in the algorithm-based solution finding for the function carrier realizing the linear guiding function.

The solution space of the multifunctional facade integrated linear motion system is mainly defined by the function carrier "linear guide" and its corresponding design variables, such as number and arrangement of chambers, positioning of the rack, profile width, profile height, material, and material thickness. Starting from the defined solution space, constraints and objectives are derived from formalized desired properties originating from product and process requirements (Table 9.1). The basic objective is the maximization of stiffness, i.e., the minimization of the linear guide's compliance (Eq. (5.2)) according to the optimization problem 5.3 (Sect. 5.2).

To grasp the innovation potential of the technology-pushed product idea and successfully exploit it for a product, manufacturing possibilities have to be comprehensively utilized within the development process. By integrating manufacturing information of additional manufacturing processes within the continuous flow production, further functional benefits can be achieved. A systematic property matching reveals promising manufacturing technologies which are complementary to the already chosen manufacturing technology linear flow splitting:

- *High speed cutting*: Realizing discontinuous form elements (holes) within a continuously produced profile structure to prepare mechanical joining of the rack and for joining the profile to the facade
- *Linear bend splitting*: Placing additional bifurcations in the profile to increase stiffness while keeping the thin-walled lightweight design

Problem	Identifier	Design Recommendation	Consequence	Explanation	Solution elements
Guided longitudinal motion	PIDG 9.1.1: Linear flow split flanges as rolling contact area	Use linear flow split flanges with UFG microstructure to realize a rolling contact area of a linear guide.	high longevity and surface quality	UFG microstructure with high hardness and low surface roughness	
Welding	PIDG 9.1.2: Placing welding seams at the end of flanges	Use welding seams positioned at the end of the flanges to assemble structural elements.	minimizing effects of welding on material within the UFG microstructure of the bifurcated profile	Hardness of the flanges decreases towards their end. Thus, welding leads to a less significant reduction in UFG material properties.	
Transfer lateral forces	PIDG 9.1.3: Transferring higher lateral forces	Use linear bend split bifurcations welded to the web of the profile as integral struts.	increased stiffness to transfer lateral forces	The integrated bifurcations have to be welded to the profile's web in order to create closed chambers that increase the overall stiffness.	
Mechanical Joining	PIDG 9.1.4: Mechanical joining of functional elements	Use HSC notches at the middle of the linear flow split web for mechanically joining a functional element with two or more sequential arranged pins.	connection with high separating forces	High residual stresses during the linear flow splitting process can be utilized for joining elements in previously inserted notches or holes.	

Fig. 9.4 Process integrated design guidelines and manufacturing-initiated solution elements for a facade integrated linear motion system

- *Roll forming*: Increasing the geometrical flexibility of the profile structure with respect to a flexible positioning of the rolling contact areas
- *Laser welding*: Closing thin-walled chambers of the bifurcated profile to increase torsional stiffness

The identified and determined manufacturing technologies are beneficially combined with linear flow splitting to realize additional functional benefit for the facade integrated linear motion system as well as to reduce manufacturing effort. With the help of manufacturing-induced properties, the possibilities of the determined manufacturing technologies can be comprehensively utilized by implementing manufacturing-initiated solution elements into the design of the facade integrated linear motion system. Process integrated design guidelines (PIDGs) can be easily implemented in the design process due to their structure based on property relations and concrete and applicable design recommendations (Fig. 9.4 and Sect. 4.2). The integration of manufacturing-induced properties in the form of PIDGs requires a prior systematic anticipation of corresponding information and insights of manufacturing processes and their impact on use processes. Fundamental process analyses that are necessary for the systematic anticipation of linear flow splitting and its surrounding manufacturing technologies within the continuous flow production are discussed in more detail in Chaps. 3 and 4 and Sect. 8.2.2.

The application of PIDG leads to manufacturing-initiated solution elements realizing the working elements of the facade integrated linear motion system (Fig. 9.4). These solution elements are implemented into the algorithm-based solution finding process of the whole system's profile structure and geometry.

For initiating the algorithm-based solution finding, the objective function, design variables, and design constraints are complemented by a design space diagram (Fig. 9.5). This diagram comprises a two-dimensional illustration of the design space with manufacturing-induced design elements as predefined manufacturing-initiated solution elements. The design elements for the facade integrated linear motion system cover flanges for the rolling contact, linear joined rack pin

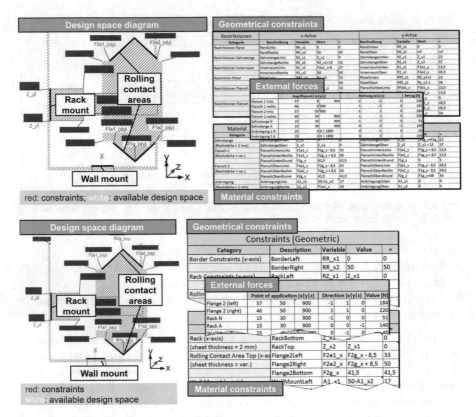

Fig. 9.5 Design space diagram for a facade integrated linear motion system

connections, predefined welding points for closing the profile structure, and linear bend split flanges. The diagram is complemented by geometrical and material constraints (Fig. 9.5). Additional information about external loads according to anticipated use processes is given to find the optimal solution in terms of maximizing the stiffness of the guiding profile. This predefined structure of the facade integrated linear motion system allows for mathematical optimization with specific optimization methods.

Algorithm-Based Solution Finding. Given the design space diagram (Fig. 9.5), the next step is to find the design of the guiding profile such that the compliance (Chap. 5) is as small as possible. A more detailed description of the algorithms described in this paragraph can be found in Göllner et al. (2014). This will be achieved in three steps: topology optimization for the generation of the general structure, the extraction of its solution in the form of a graph, and a geometry optimization (see Fig. 9.6 for the output of all three steps).

The topology is optimized by using a SIMP method (*solid isotropic material with penalization*) (Bendsøe and Sigmund 2004; Andreassen et al. 2001). An example of a solution of this process can be seen in Fig. 9.6a, where the

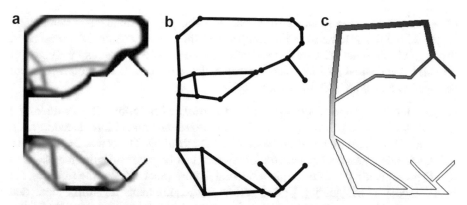

Fig. 9.6 (**a**) Solution of the topology optimization (*grayscale* values represent density), (**b**) extracted graph (unweighted) from the solution of the topology optimization, (**c**) solution to the geometry optimization after a simplification of the graph (*grayscale* values represent the norm of the displacement)

optimization is performed on the 3D object. Therein grayscale values represent the density of the material, meaning the darker a pixel is, the more material should be placed at this position. A white pixel depicts an area that should be free from material.

The topology is described by a weighted graph $G = (V, E)$, where the edge weights c_e are used to model the thickness of edge e in E (compare Fig. 9.6b for an unweighted graph). This graph is created as follows: First the image is converted to a black-and-white picture. Thereby some pixels get rounded to zero if their density is too low resulting in a sparser graph. In this picture a so-called skeleton is calculated, a thin representation in which every segment has a thickness of one pixel without changing the topology. The nodes V are then defined by the intersection and end points of the segments. By enlarging the intersection points and removing them from the skeleton, the remaining objects define the edges E of G. The length of every edge can be calculated as the distance between its end points. After applying different operations that can be found in Göllner et al. (2014), the thickness c_e of edge e is derived. The graph is saved as an XML file with relative thickness in a format which can automatically be read for the creation of a CAD file. In the example, the intermediate steps were corrected manually but generally do not need any user intervention. Figure 9.6b shows the graph for the profile of the facade cleaning system without the weights c_e.

Manufacturing requirements, constraining the bifurcation order and the number of linear bend split bifurcations (Table 9.1) necessitate a simplification of the graph in Fig. 9.6b, in order to be able to produce the profile. To ensure optimality after these simplifications, a geometry optimization (Sect. 5.2) was performed as a last step. The parametrization uses a description based on a triangulation of the profile. The optimization variables are the coordinates of the triangle's vertices and the compliance is again the objective function. The optimization is handled by an algorithm that does not need any manual intervention (Fig. 9.6c), see Chap. 5.

The compliance was reduced by about 44.44%, the maximal displacement by 17.61%, and the volume by 5.1% compared to the starting solution. As prescribed by Fig. 9.5, the functional areas (rack mount, rolling contact areas, and wall mount) are all in the desired position. The algorithm also generates an XML file which can be used to create a CAD file automatically.

Computer Integrated Refinement of the Optimized Solution. The results of the geometry optimization are stored in the formalized data model introduced in Sect. 6.2.2 as XML file format and provided to CAD modeling. The coarse 3D geometry of the facade integrated linear motion system can be automatically generated as CAD model using the algorithmic modeling feature (Sect. 6.2.2). The resulting 3D CAD model of the guiding profile fulfills the requirements regarding loads and constraints. However, it still needs to be adapted and refined to consider further requirements regarding design space, manufacturing process, and restrictions. In collaboration with product design and process planning engineers, the necessary iterations steps were accomplished. Some branches (Fig. 9.7) were removed to reduce the complexity of the facade profile structure. Their influence on the profile stiffness is not sufficiently high to justify the high effort and cost of their manufacturing. Additionally, the design features for linear flow splitting and linear bend splitting (Sect. 6.2.1) as well as bending radii and holes were added.

The result is a fully parametrized 3D CAD model of the guiding profile. Based on this profile, the whole facade integrated linear motion system can be assembled. The resulting CAD models and assemblies provide the framework for effectuating further CAD-Finite-Element-Method (CAD-FEM) and CAD-Multi-Body-Simulation (CAD-MBS) processes. A FEM is necessary to validate the strength and displacement of the final profile geometry since it differs from the mathematical optimization solution as depicted in Fig. 9.6. Figure 9.8 shows the structural displacement of the guiding profile taking the whole assembly and all loads into account. The FEM result shows that the maximum displacement is on the free edges of the profile, in the opposite direction of the facade. Yet, the magnitude of the maximum displacement is small and does not affect the linear guiding function. The CAD-MBS simulates the up and down motion of the sled on the guiding profile under the simplifying assumption that all components are rigid. This scenario anticipates the essential use process of the facade integrated linear motion system. The results of the motion analysis show that components of the assembly (including the linear flow split ones) perform their desired motion and function without any collision with each other.

The final CAD assembly highlights the efficiency, accuracy, and robustness of the automated interface between mathematical optimization and CAD modeling using the developed algorithmic features (Sect. 6.2.2). The developed data model and algorithmic modeling feature is able to deal with and generate 3D CAD models with complex geometries and bifurcations. The challenge is thereby to constantly ensure the validity of the B-Rep data structure of such CAD models (Sect. 6.2.1). Furthermore, the described CAD modeling strategy using parametrization and design features ensures the reusability of the CAD model. It enables an efficient

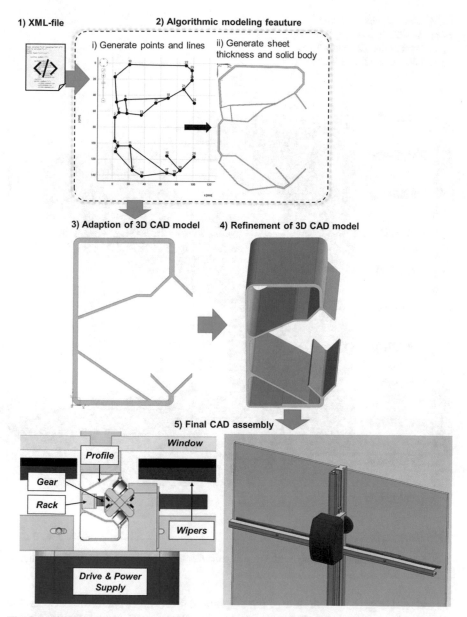

Fig. 9.7 CAD modeling approach for facade integrated linear motion system

change in the design parameters of the model in order to generate new product variants.

During the product creation process of the facade integrated linear motion system, a sequence model based on the integrated information model was

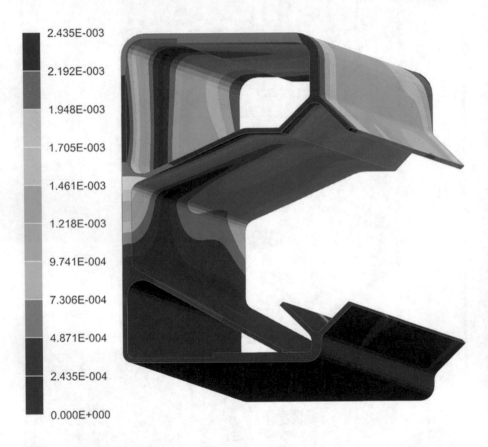

FEM_final : Solution 1 Result
Subcase - Static Loads 1, Static Step 1
Displacement - Nodal, Magnitude
Min : 0.000, Max : 0.213, Units = mm

2.435E-003

2.192E-003

1.948E-003

1.705E-003

1.461E-003

1.218E-003

9.741E-004

7.306E-004

4.871E-004

2.435E-004

0.000E+000

Units = mm

Fig. 9.8 Finite-Element-Method results showing the displacement on the final profile geometry

introduced to give an overview of the communication and cooperation between the
different domains (Albrecht and Anderl 2014). The partial model *ModuleModel*,
especially relations and connections of *MultifunctionalModule* (Sect. 6.1.1), can be
used to describe the facade integrated linear motion system and its structure at the
information level. This information is used by design and manufacturing engineers
for further processing.

**Continuous and Integral Manufacturing Process of the Multifunctional
Facade Integrated Linear Motion System.** The linear motion system of the

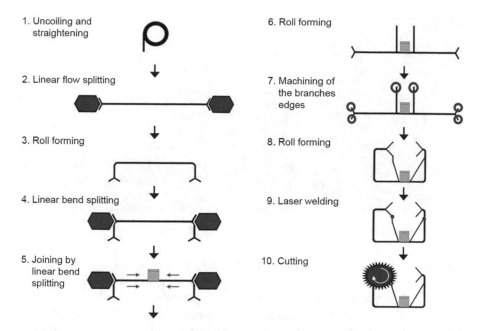

1. Uncoiling and straightening

2. Linear flow splitting

3. Roll forming

4. Linear bend splitting

5. Joining by linear bend splitting

6. Roll forming

7. Machining of the branches edges

8. Roll forming

9. Laser welding

10. Cutting

Fig. 9.9 Process chain for the guiding profile of the facade integrated linear motion system

facade cleaning system can be produced on a continuous flow production line such as shown in Sect. 3.2. Figure 9.9 illustrates the production sequence of the linear guiding profile. First, the sheet metal is uncoiled and deformed to a branched profile geometry by the process of linear flow splitting (Sect. 3.1). The branched profile is bent through roll forming (Sect. 3.2.1) to a U-profile geometry, followed by the linear bend splitting process. After a few stages, discontinuous milling operations are performed by the HSC sheet milling machine tool (Sect. 3.2.2) to prepare the following joining processes. A gear rack is inserted into the milled grooves by a pick-and-place robot system, which operates within the continuous flow production line. By using the specific stress distribution in the web of the profile during the linear flow or bend splitting process (Sect. 8.2.2), the gear rack is mechanically joined. Before the sheet metal profile is bent into its final geometry, the band edges and flanges are machined to ensure a linear shape and to prepare the band edges for the laser welding process. The used milling tools were developed with regard to the cutting forces, the machined surface quality, a small burr, and a long tool life (Sect. 3.2.2). Through various roll forming stages, the final profile geometry is formed. By a laser welding system (Sect. 3.2.3), the profile geometry is finally closed and cut to length by a flying saw unit. The final geometry of the realized guiding profile is shown in Fig. 9.10.

Benefits of Applying the Manufacturing Integrated Algorithm-Based Product and Process Development. Initiated by a technology-pushed product idea, the consistent application of the Manufacturing Integrated Algorithm-Based Product and Process Development leads to the realization of a highly integrated facade

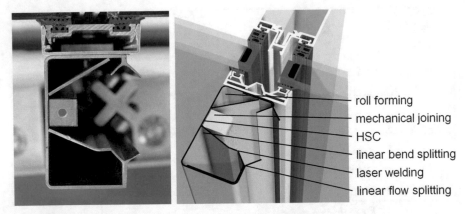

roll forming

mechanical joining

HSC

linear bend splitting

laser welding

linear flow splitting

Fig. 9.10 Prototype of the guiding profile (*left*: realized; *right*: virtual)

integrated linear motion system with high stiffness and small dimensions. The linear motion system's properties as well as corresponding processes have been determined and concretized. Early integration of manufacturing information in the form of process integrated design guidelines supports the comprehensive realization of manufacturing potential. The implementation of integral rolling contact areas with UFG microstructure, mechanically joined function carriers, and the continuously manufactured integral construction illustrate a comprehensive utilization of the manufacturing-induced properties provided by the continuous flow production.

Two elements are essential for the manufacturing-integrated development: knowledge of the designed products and the anticipation and implementation of specific manufacturing technological knowledge of the continuous flow production. By applying optimization algorithms, an optimal topology and geometry of the guiding profile with maximized stiffness is realized. The information model with specific tools and methods allows the inclusion of details like bending radii and surface roughness into the solution in the subsequent refinement step. This leads to an efficient optimization process while guaranteeing completeness of the solution that has been successfully manufactured in continuous flow production.

9.2.2 Integrated Fastenings for Hangers

Considering manufacturing potential for the design of fastening elements is expedient, to meet assembly and disassembly requirements. The complexity of this design task, in which manufacturing and product use processes have to be considered, can be handled efficiently by applying algorithmic optimization methods in the context of the Integrated Algorithm-Based Product and Process Development approach. Starting from the technology-pushed product idea of *profile integrated*

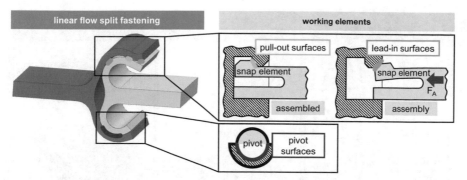

Fig. 9.11 CAD model of a linear flow split fastening and working elements with highlighted working surfaces using linear flow split flanges according to Mattmann et al. (2014)

linear flow split fastenings, the integrated approach is applied, with a focus on anticipation and integration of manufacturing insights in the early phases of algorithm-based product design. A profile prototype is realized and used for gaining manufacturing insights. These insights are further processed and integrated into the design of the linear flow split fastening. An integrated linear flow split fastening prototype demonstrates the comprehensive realization of manufacturing potential.

Technology-Pushed Product Idea—Integrated Fastenings. Manufacturing-induced properties are a starting point when identifying promising applications for linear flow split products. Systematic property matching (Chaps. 4, 5, and 8) reveals the applicability of linear flow split profiles for realizing profile integrated detachable fastenings. Assembly, use, and disassembly processes necessitate various working surfaces (process relevant elements). The linear flow split flanges in combination with the high geometric flexibility of roll forming are excellently suitable for realizing multiple working surfaces within an integrated profile structure (Chap. 5 and Fig. 9.11). The fastening's working surfaces need special material properties, such as high hardness, which is beneficial when realizing a line contact between the fastening partners during assembly. Using flow split flanges to realize the working surfaces stands to reason, considering that the increase in hardness at the flange surface (Sect. 4.1.1) is one of the key manufacturing-induced properties. An overview of the geometric variety that arises from linear flow split fastenings is shown in Fig. 7.29.

Figure 9.11 shows a CAD model of a profile integrated detachable fastening. The working surfaces (blue: lead-in surfaces; yellow: pull-out surfaces; violet: pivot surfaces) play a key role in ensuring smooth assembly and disassembly processes of the fastening partners (Mattmann et al. 2014). A pivot allows rotating assembly while the resulting force F_A leads to bending in the snap element. The snap element holds the assembled parts in position. Application of the linear flow split flanges (manufacturing-induced design elements) is important to ensure product function and realization of the working elements and working surfaces for assembly and disassembly processes. At the same time, the manufacturing restrictions for combined linear flow splitting and roll forming are taken into account.

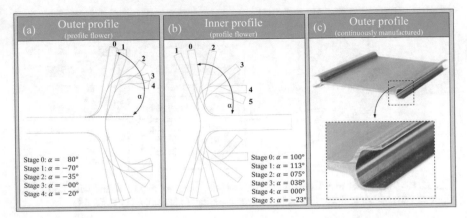

Fig. 9.12 (**a**) Flower pattern of the outer profile, (**b**) flower pattern of the inner profile, (**c**) continuously manufactured profile prototype

The flower patterns for the inner and outer profile geometries have to be designed accurately to manufacture the profile prototype (Fig. 9.11) in a continuous roll forming process. The main challenge of designing a roll forming process is the strain that occurs at the band edge (Sect. 3.2.1). The roll forming process is analyzed in a numerical model, with four stages of the outer profile and five stages of the inner profile (Figs. 9.12a, b). The linear flow split profile from stage 10 (Sect. 3.1.1) is used as the initial geometry (stage 0 of the roll forming process) for both profiles. For the designed flower patterns, the maximum total strain at the band edge in longitudinal direction is less than 0.01. The prototype of the continuously manufactured outer profile, with ten linear flow splitting stands and four roll forming stands, is shown in Fig. 9.12c.

Processing of Insights for the Realization of Linear Flow Split Fastenings. The analysis of the prototype in the context of its manufacturing processes reveals several insights:

- *Local material hardening through flow splitting*: These zones can be used for highly stressed surfaces, for example, when high surface pressure occurs.
- *Increased strength through flow splitting:* These zones can be used for highly stressed regions, for example, when bending stresses occur.
- *Small bending radii*: The flanges need to be heat treated locally prior to the forming process to realize the desired shape without component failure. Appropriated laser heat treatment parameters are described in Sect. 4.1.2.
- *Maximal flange length*: The maximal flange length determines the required number of linear flow splitting stands.

The processed manufacturing insights are prepared as process integrated design guidelines, comprising the manufacturing-induced design elements and properties, as in Chaps. 4 and 5 (Fig. 9.13). This is the basis for integrating manufacturing-induced properties and solution elements into the design of profile integrated fastenings.

Problem	Identifier	Design Recommendation	Consequence	Explanation	Solution elements
Detachable fastenings	PIDG 9.2.1: Linear flow split flanges as snap elements	Use the linear flow split flanges or linear bend split flanges with increased strength as snap element of detachable fastenings.	realizing increased elastic disassembly displacements	The linear flow split flanges have an increased strength which allows higher displacement at the same Young's Modulus.	
Detachable fastenings	PIDG 9.2.2: Linear flow split flanges as lead-in/pull-out surfaces	Use the upper surface of linear flow split flanges or linear bend split flanges with increased hardness as lead-in or pull-out surfaces of detachable fastenings.	realizing line contact	The upper surface of linear flow split flanges have an increased hardness.	
Locking degrees of freedom	PIDG 9.2.3: Roll formed linear flow split flanges to lock degrees of freedom	Use roll formed linear flow split flanges or roll formed linear bend split flanges with specific bending radii (> 1 mm) for providing multiple working surfaces for detachable fastenings.	lock additional degrees of freedom	Bending radii at heat treated linear flow split flanges have to be > 1 mm to avoid cracks at the upper surface of the linear flow split flanges.	

Fig. 9.13 Process integrated design guidelines and solution elements for linear flow split fastenings (Roos et al. 2016)

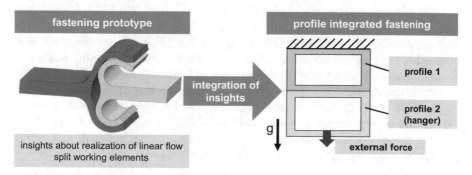

Fig. 9.14 Use case of a profile integrated linear flow split fastening with integration of insights

Application of Linear Flow Split Fastenings—Hangers with Integrated Snap-Fit. Additional benefit to a ceiling mounted hanger system is realized by integrating a detachable fastening system, for which the processed insights into linear flow split fastenings are comprehensively considered (Fig. 9.14).

Essential requirements for developing the integrated fastening are derived from analyses of use, assembly, disassembly, and manufacturing processes. Analyzing the use process reveals that providing a high stiffness is essential for a fastening system in order to bear high external loads and fulfill safety demands. Nonetheless, the stiffness cannot be limitlessly increased due to neighboring systems which restrict the maximum dimensions of the fastening. Analyses of manufacturing processes show that a specific sheet thickness ratio between sheet and web has to be considered. Generally, the design of fastening systems is dominated by assembly aspects. Maximum assembly forces depend on the joining principle, in this case manually joining the fastening partners. A selection of requirements clustered with respect to life cycle processes is shown in Table 9.2. These requirements are transformed and formalized into constraints and objectives when applying engineering design optimization methods that can handle the design task complexity (Sect. 5.1).

Table 9.2 Selection of essential requirements of an integrated linear flow split fastening

Cluster	Description	Value	Explanation
Manufacturing	Sheet thickness ratio	1:2 (flange:web)	Based on process parameters
	Flange bending radius	≥ 1 mm	Roll forming of the flanges
Use	Applicable external force	≥ 1000 N	Equal to 100 kg load
	Stiffness of the profiles	maximize	Referring to use case
	Fastening width	≤ 64 mm	Referring to use case
	Fastening height	≤ 50 mm	Referring to use case
	Fastening length	20 mm	Multiple fastening elements along the continuously manufactured profile 1
Assembly	Assembly forces	≤ 500 N	Manual assembly without tools

The objective is set as minimizing the compliance (Sect. 5.2) which is equivalent to maximizing the structural stiffness. Minimizing assembly and disassembly forces are also appropriate objectives for optimizing a fastening system. An implementation of these objectives necessitates FE simulations of assembly and disassembly processes which leads to increased computational cost. By implementing assembly and disassembly forces in the form of design constraints (restricting the angle of the working surfaces and the maximum web length), the optimization can be performed very efficiently without additional FE simulations (Roos et al. 2016). Further design constraints are derived from manufacturing processes (e.g., sheet thickness) and use processes (e.g., maximum dimensions).

Integration of Manufacturing-Induced Properties into Algorithm-Based Solution Finding. The design space diagram (Fig. 9.15), a graphical representation of the formalized design task, contains information about the constrained design space. A topological draft of the linear flow split fastening (Fig. 9.15 grey elements in design space diagram) comprises integrated solution elements for the fastening partners characterized by manufacturing-induced properties (Fig. 9.13).

The formalized design task is the starting point for shape optimization, with a special focus on modeling and optimizing the contact problem between the fastening elements (Fig. 9.15). The shape optimization process is completely implemented in MATLAB®.

Optimized and Refined Product Design. The optimized product design representation of the integrated fastening is stored using the XML file format, based on the data model introduced in Sect. 6.2.2 (Fig. 9.16). Based on the XML file, a 3D CAD model of the optimized product design is generated, using the algorithmic modeling feature. The resulting CAD model represents the results of the mathematical optimization accurately and provides the coarse shape of the final product design. It stills needs refinement to comprehensively consider the full range of manufacturing-induced design elements, such as bending radii, which are added in the refinement process for efficiency reasons. The developed interface is based on the continuous data and information exchange between mathematical

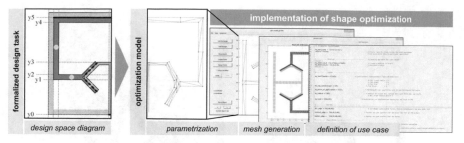

Fig. 9.15 Formalized design task, in the form of a design space diagram and transfer for optimization for a snap-fit fastening

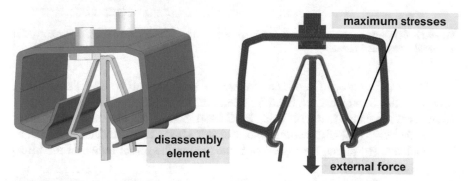

Fig. 9.16 Optimized and refined fastening loaded by an external force

optimization and CAD modeling. Modeling iterations can be avoided since the initial 3D geometry is given, which means that the time required to model 3D geometries can be significantly reduced. 3D CAD design and modeling of optimized product designs can be facilitated.

The optimized and refined snap-fit fastening (maximum dimensions of 36 mm width, 45 mm depth, and 20 mm length of Profile 2) is characterized by significantly increased stiffness (about 99.14% reduced compliance compared to the starting geometry) at almost the same mass as the starting geometry. Thus, the lightweight potential of linear flow split profiles in combination with optimized design is exploited. The snap-fit fastening (material: HC 480 LA) can bear external forces of up to 1200 N (starting geometry: 500 N); assembly forces are reduced to 300 N (starting geometry: 700 N).

Further manufacturing potential can be realized during refinement by applying additional geometry related manufacturing-induced properties. The linear flow split flanges are used to add disassembly elements to lower the required lateral disassembly forces.

Realization of the Integrated Snap-Fit. Some design areas of the fastening have to be considered as a priority to realize continuous flow production. In area *A* in Fig. 9.17a, the bending of the lower flange creating the hook is located directly at

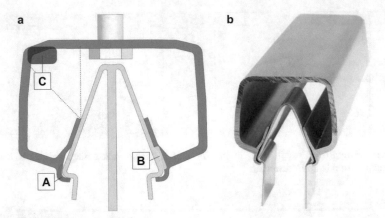

Fig. 9.17 (**a**) Design areas of the snap-fit with regard to manufacturing processes, (**b**) prototype of the profile integrated fastening

the bifurcation zone and represents a narrow bending radius. The flange has a straight outlet to support the bending operation for producing the desired radius. Afterwards the flange is cut off using HSC (Sect. 3.2.2). The clearance in area B is located directly at the bifurcation zone, which is characterized by an initial radius. The implemented radius for clearance is adapted to the initial radius of the bifurcation zone.

During the profile bending process, the bifurcations and fastening zones of both arms create an undercut in area C. Performing a proper support from the inside is a big challenge in a continuous roll forming process. For an accurately formed cross section, this support is necessary, since the bending of these areas (C) directly influences the gap of the fastening elements. For that reason, a tool that provides supporting contact from the inside is used in the final bending step. A prototype of the profile integrated fastening is shown in Fig. 9.17b.

9.2.3   Nonlinear Skywalk Beam

With the introduction of the flexible flow splitting technology, a new range of products with nonlinear geometry can be produced (Chap. 7). However, additional parameters and constraints (e.g., additional degrees of freedom) need to be considered for product and process development, especially with regard to mathematical optimization, geometric modeling, and simulation. The inclusion of additional parameters and constraints increases the complexity of the development process. To demonstrate the beneficial use of the flexible flow splitting technology, the Integrated Algorithm-Based Product and Process Development approach is applied to the development of a nonlinear bifurcated *beam construction* for a *skywalk*. A special focus is put on the refinement steps of the optimized product and process solution.

Design Task Clarification and Formalization. The connection between two buildings can be carried out through spatially closed transitions, called "skywalks." The vertical loads of skywalks are transferred to the exterior walls of the two connected buildings. However, listed buildings are typically not designed to take up additional loads of this magnitude. Therefore, it may be necessary to transfer the loads exclusively to one of the two buildings. A design concept for a beam structure with four beams connecting a listed building to a new building is shown in Fig. 9.18a. To keep the mechanical load on the new building as low as possible, the weight of such a skywalk has to be minimized. At the same time, a sufficient stiffness is required to keep the displacement at the free end at a minimum. Beam geometries with a variable profile height are ideally suitable for this task, as they provide an option to achieve a maximum stiffness with minimum weight. A method

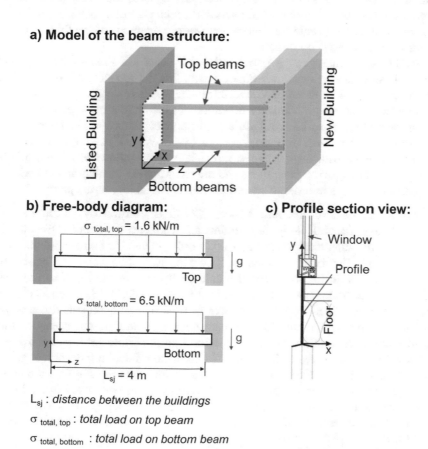

a) Model of the beam structure:

Top beams

Listed Building

New Building

y

x

z

Bottom beams

b) Free-body diagram:

$\sigma_{total,\,top}$ = 1.6 kN/m

g

Top

$\sigma_{total,\,bottom}$ = 6.5 kN/m

g

y

z

Bottom

L_{sj} = 4 m

L_{sj} : *distance between the buildings*

$\sigma_{total,\,top}$: *total load on top beam*

$\sigma_{total,\,bottom}$: *total load on bottom beam*

c) Profile section view:

y

Window

Profile

Floor

x

Fig. 9.18 (**a**) Model of the beam structure, (**b**) free-body diagram of the beam construction, (**c**) profile view for the bottom side

Table 9.3 Selection of relevant requirements

Cluster	Description	Value
Design	Shape of the joining surface between beam and window	Straight
	Beam length	4000 mm (according to distance between buildings)
Boundary conditions (manufacturing)	Angle of inclination	max. 15°
	Arc radius r	$r \geq 350$ mm
	Local profile height h	100 mm $\leq h \leq$ 1000 mm

to realize such a nonlinear profile from sheet metal is the flexible flow splitting technology (Sect. 7.1.2).

Figure 9.18b shows the free-body diagram for the beam construction. It also illustrates the load scenario which is a necessary input for the subsequent mathematical optimization. Figure 9.18c depicts the section view of a double Y-profile for the bottom side. The profile realizes the load-carrier function and the fixation of the windows in an integral construction.

Following the Integrated Algorithm-Based Product and Process Development (Fig. 9.1) approach, relevant information regarding the product and its life cycle processes needs to be determined and integrated in subsequent steps. All defined requirements comprising boundary conditions of the chosen manufacturing technology flexible flow splitting regarding the nonlinear bifurcated beam parts are documented in the form of a clustered list (Table 9.3). These requirements are further processed with respect to the subsequent optimization steps (Sect. 5.1.2). The geometric properties have to be determined without violating process related boundary conditions in order to ensure the manufacturability of the parts.

Algorithm-Based Product and Process Solution Finding by Mathematical Optimization. Once all design constraints and variables are defined, the mathematical optimization methods are applied in order to find the optimal solution for the design task. The starting geometries for the top and bottom beams are generated in accordance with the documented requirement list (Fig. 9.19a). The physical behavior of the beams under the given load scenario (Fig. 9.18) is modeled as a system of PDEs (Sect. 5.2). This system is used as a subproblem for a shape optimization problem to find the optimal design. In the present scenario, this could either mean to minimize the weight for a given stiffness or to maximize the stiffness for a given weight. The latter will be demonstrated in the following. For the parametrization of the beam, cubic splines with four control points are used. These points serve as optimization variables for the nonlinear curves of the lower surface of the bottom beam and the upper surface of the top beam. To solve the optimization problem, the *ARCex* algorithm (Sect. 5.2.1) with three mesh levels is applied. The compliance serves as the objective function and has been reduced for the optimal design by 91.48% for the beam at the bottom of the skywalk and 87.06% for the beam at the top. Figure 9.19b depicts the optimized geometries of

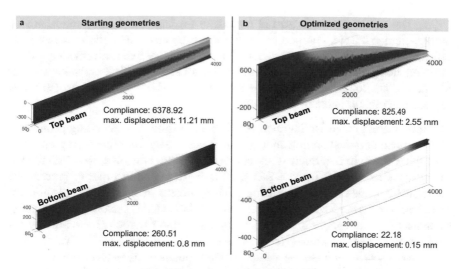

Fig. 9.19 (a) Starting geometry, (b) results of the shape optimization. *Red color* corresponds maximal displacement while *blue color* represents minimal displacement

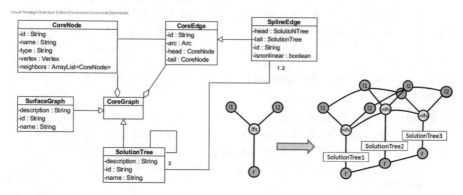

Fig. 9.20 Information model for representation of nonlinearity (Albrecht et al. 2014a; Albrecht et al. 2014b)

the top and bottom beams with a nonlinear shape in longitudinal direction represented by cubic splines.

Integrated Information Model. The integrated information model (Sect. 6.1.1) enables the continuous data exchange between mathematical optimization and CAD modeling for nonlinear bifurcated parts. To integrate the representation of nonlinear bifurcations, the core model is adapted (Fig. 9.20). Therefore, an additional class *SplineEdge* was integrated into the core model of the information model (Albrecht et al. 2014a). *SplineEdge* is a specialization of *CoreEdge*. It has the same relationships as *CoreEdge* and integrates the additional information about nonlinear representation. To define *SplineEdges* B-Splines are used (Hoschek and Lasser

1992). Due to the available partial models of the integrated information model, manufacturing-induced properties used for the process simulation of nonlinear bifurcations are provided (Sect. 6.1.1). Based on the adapted information model, the data model represented in an XML-schema (Sect. 6.2.2) is extended (Weber Martins et al. 2015a). Element tags are added as specified above to store and represent nonlinear geometries as B-Splines.

Computer Integrated Refinement of Product Design. The resulting geometries from the mathematical optimization are characterized by a nonlinear progression of the profile's web in longitudinal direction (Fig. 9.19b). However, the CAD modeling features (Sects. 6.2.1 and 6.2.2) need to be extended in order to process the additional parameters for visualizing such nonlinear geometries. Regarding the algorithm to read the XML file and generate the part in CAD systems, programming methods to create splines are added according to the parameters used during the mathematical optimization (Weber Martins et al. 2015a). In this case, cubic B-Splines ($p = 3$) and all necessary control points (x,y,z-coordinates) are set up. Hence, the CAD model (Fig. 9.21) is automatically generated based on the results of the optimization (Fig. 9.19b).

The direct modeling features (Sect. 6.2.2) are extended to enable the positioning of UDFs for linear flow or linear bend splitting operations on nonplanar faces. A suitable solution is to use normal vectors as positioning reference. Tangential vectors are used to extract the nonlinear trajection guide for the sweep operation to generate the 3D CAD model. Corresponding parameters and methods are implemented into the design feature for CAD systems (Weber Martins et al. 2015b). The CAD modeling procedure for the skywalk beams consists of three steps (Fig. 9.21):

1. Import of optimization data and generation of the optimized geometry: The algorithmic modeling feature is applied to generate the CAD model based on the imported results of the mathematic optimization. Within this step, the nonlinear bifurcated parts for the top and bottom of the skywalk's use case are generated.
2. Adaption and refinement: The resulting CAD models are adapted to integrate additional manufacturing-induced properties (e.g., material specifications and radii) (Chap. 4) and manufacturing process parameters. The CAD models are refined by applying design features for nonlinear flow splitting and adding bend operations to it.
3. Final CAD assembly: The CAD model parts of the skywalk structure are put together into an assembly model.

Computer Integrated Refinement of Process Design. To avoid expensive experimental test series at the prototype stage or over dimensioning of the tool system, a FEM-based process simulation can be used to predict the required forces and number of forming steps to generate the optimized and refined geometry. For this purpose, an FE model is generated to design the tool set needed to produce the desired beam geometries (Fig. 9.22). To validate the predictive capabilities in terms

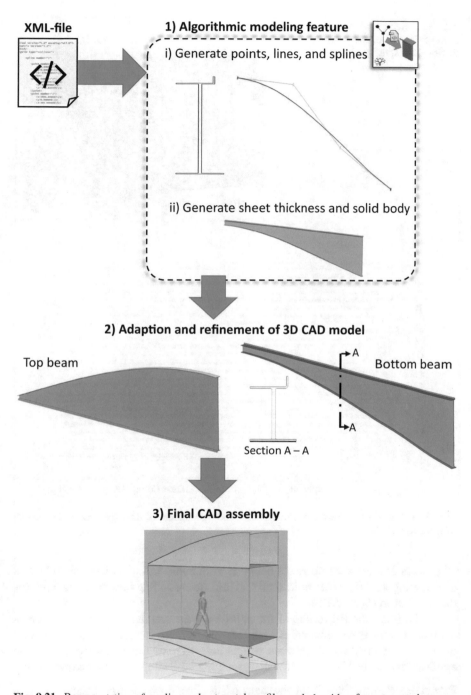

Fig. 9.21 Representation of nonlinear sheet metal profiles and algorithm for automated generation of CAD models

Fig. 9.22 Geometry used
in the FE model of the
flexible flow splitting
process

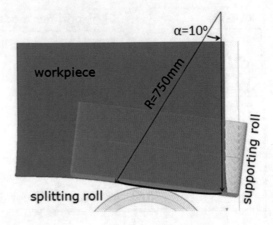

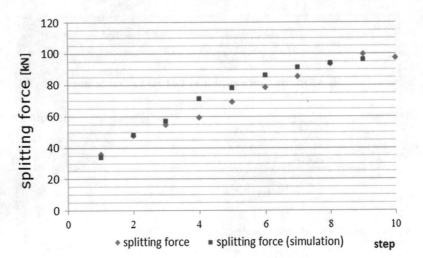

Fig. 9.23 Comparison between experiments and simulation of the splitting force results
(Neuwirth et al. 2014)

of process forces, a comparison between numerical and experimental values is
drawn using a DD11 grade mild steel and the corresponding flow curve as input for
the simulation (Sect. 7.1.2).

In Fig. 9.23, the FE results of the splitting force and the corresponding forces
resulting from experiments are shown for nine forming steps. It can be stated that
there is a very good match between simulations and experiments, which shows the
applicability of the FE models for designing the tool setup in further experiments.

Prototype Production. With the help of flexible flow splitting (Sect. 7.1.2), a
scaled-down sheet metal prototype of the bottom beam has been manufactured
(Fig. 9.24). In accordance with the geometry of the bottom beam, the prototype is

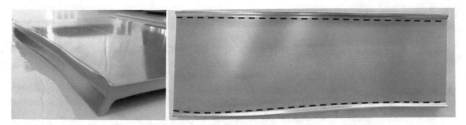

Fig. 9.24 Prototype of bottom beam with variable height profile

characterized by a parallel orientation of the bifurcated band edges at the point of fixity to the wall (Fig. 9.21). The sheet metal part, possessing a thickness of 6 mm, was formed within 10 steps with an incremental splitting depth of 1 mm and work piece velocity of 1 m/min. The available tooling setup used to manufacture the prototype allows the realization of a local profile height between 350 mm (min.) and 550 mm (max.).

Benefits of Applying the Manufacturing Integrated Algorithm-Based Product and Process Development. The application example of a beam construction for a skywalk demonstrates that the Integrated Algorithm-Based Product and Process Development approach is eligible for nonlinear sheet metal technology such as flexible flow splitting. To enable the applicability of this development approach, optimization methods, information model, and the modeling features for CAD have been extended and adapted to ensure the accurate representation of nonlinear bifurcated parts. A FEM-based process simulation was applied to predict the necessary process forces as well as the forming steps of the developed sheet metal parts. A scaled-down prototype of the nonlinear bifurcated beam was produced using a tooling setup under real conditions.

Applying the Integrated Algorithm-Based Product and Process Development approach for developing nonlinear sheet metal parts underlines the efficient solution generation and avoidance of time-consuming iterations. In particular, the mathematic optimization and CAD modeling steps blend seamlessly into each other since their data exchange is coordinated. CAD models are generated within a few seconds based on the optimized geometry. Therefore, the Integrated Algorithm-Based Product and Process Development approach provides a framework for engineers, enabling an efficient development of complex products, especially in the context of their manufacturing processes. The flexible flow splitting technology enables the realization of beam geometries with a variable height profile for the skywalk. Hereby, an almost homogeneous stress distribution among the beams can be achieved.

References

Albrecht K, Anderl R (2014) Erweiterung des Informationsmodells zur Integration von Informationen aus der Produktentstehung. In: Groche P (ed) Tagungsband 5. Zwischenkolloquium SFB 666, Mörfelden-Walldorf, 19–20 November 2014

Albrecht K, Weber Martins T, Anderl R (2014a) An information model representing nonlinear flow splitting of bifurcated sheet metal. In: Proceedings of the ACEAT—Annual Conference on Engineering and Technology, Osaka, Japan, p 313–321

Albrecht K, Weber Martins T, Anderl R (2014b) Nichtlineares Spaltprofilieren von Blechprofilen im rechnergestützten Produktentwicklungsprozess. In: Brökel KF, Grote KH, Rieg F, Stelzer R (eds) 12. Gemeinsames Kolloquium Konstruktionstechnik (KT 2014), Methoden in der Produktentwicklung: Kopplung von Strategie und Werkzeugen im Produktentwicklungsprozess, Bayreuth, 16–17 October 2014, p 203–212

Andreassen E, Clausen A, Schevenels M, Lazarov B, Sigmund O (2001) Efficient topology optimization in MATLAB using 88 lines of code. Struct Multidiscip Optim 43(1):1–16

Bendsøe M, Sigmund O (2004) Topology optimization: theory, methods, and applications, 2nd edn. Springer, Berlin

Schüco Germany (2016) https://www.schueco.com/web2/com. Accessed 08 Sept 2016

Göllner T, Lüthen H, Pfetsch ME, Ulbrich S (2014) Profiloptimierung im Rahmen eines durchgängigen Produktentstehungsprozesses. In: Groche P (ed) Tagungsband: 5. Zwischenkolloquium SFB 666, Mörfelden-Walldorf, 19–20 November 2014, p 15–24

Hoschek J, Lasser D (1992) Grundlagen der geometrischen Datenverarbeitung, 2. neubearb. und erw. Auflage. Teubner, Stuttgart

Iku®windows (2016) Willkommen bei iku®window. http://www.iku-windows.com/. Accessed 08 Sept 2016

Mattmann I, Roos M, Gramlich S (2014) Transformation und Integration von Marktanforderungen und fertigungstechnologischen Erkenntnissen in die Produktentwicklung. In: Groche P (ed) Tagungsband 5. Zwischenkolloquium SFB 666, Mörfelden-Walldorf, 19–20 November 2014, p 5–14

Neuwirth M, Özel M, Schmitt W, Rullmann F (2014) Flexibles Spaltprofilieren: Verfahren und numerische Abbildung. In: 9. Fachtagung Walzprofilieren & 5. Zwischenkolloquium SFB 666, Mörfelden-Walldorf, 19–20 November 2014

Pahl G, Beitz W, Feldhusen J, Grote KH (2007) Engineering design: a systematic approach, 3rd edn. Springer, London

Roos M, Horn B, Gramlich S, Ulbrich S, Kloberdanz H (2016) Manufacturing integrated algorithm-based product design: case study of a snap-fit fastening. In: Wang L, Kjellberg R (eds) Procedia CIRP, vol 50, 26th CIRP Design Conference, p 123–128

Serbot (2016) Fassadenreinigung. http://serbot.ch/de/fassadenreinigung. Accessed 08 Sept 2016

Weber Martins T, Albrecht K, Anderl R (2015a) Automated Import of XML-Files Containing Optimized Geometric Data to 3D–CAD-Models of Non-Linear Integral Bifurcated Sheet Metal Parts. In: Proceedings of the ASME 2015—International Design Engineering Technical Conferences and Computers and Information in Engineering Conference, Boston, Massachusetts, 2–5 August 2015

Weber Martins T, Albrecht K, Anderl R (2015b) Entwicklung von Modellierungsfeatures für nichtlineares Spaltprofilieren und Spaltbiegen in CAD-Systemen. In: Tagungsband der 12. Magdeburger Maschinenbau-Tage, Magdeburg

Index

© Springer International Publishing AG 2017
P. Groche et al. (eds.), *Manufacturing Integrated Design*,
DOI 10.1007/978-3-319-52377-4